Study Guide

Belk | Borden

Human Biology

Catherine Podeszwa

Benjamin Cummings
San Francisco Boston New York
Cape Town Hong Kong London Madrid Mexico City
Montreal Munich Paris Singapore Sydney Tokyo Toronto

Acquisitions Editor: Gary Carlson
Editorial Assistant: Kaci Smith
Managing Editor: Deborah Cogan
Production Supervisor: Mary O'Connell
Manufacturing Buyer: Michael Penne
Marketing Manager: Gordon Lee
Supplement Cover Designer: 17th Street Studios
Main Text Cover Designer: Studio A
Cover Photograph: Getty Images/Jimmy Chin
Design and Composition: Progressive Publishing Alternatives

ISBN-10: 0-13-148132-0
ISBN-13: 978-0-13-148132-9

Benjamin Cummings
is an imprint of

1 2 3 4 5 6 7 8 9 10 — EB — 12 11 10 09 08
www.pearsonhighered.com

TABLE OF CONTENTS

CHAPTER 1

THE SCIENTIFIC METHOD: PROVEN EFFECTIVE

Learning Goals

1. Describe the characteristics of a scientific hypothesis and give examples of ideas that cannot be tested scientifically.
2. Compare and contrast the terms *scientific hypothesis* and *scientific theory*.
3. Distinguish between inductive and deductive reasoning.
4. Describe why the truth of a hypothesis can't be proven conclusively via deductive reasoning.
5. Describe the features of a controlled experiment, and explain how controlled experiments eliminate alternative hypotheses for the results.
6. List the techniques scientists can use to minimize bias in experimental design.
7. Define *correlation* and explain the benefits and limitations of using this technique to test hypotheses.
8. Describe the kinds of information that statistical tests can provide.
9. List the factors that can influence whether or not an experiment will return statistically significant results.
10. Summarize the techniques you can use to evaluate scientific information from secondary sources.

Chapter Outline

I. An Introduction to the Scientific Process
  - A. The Nature of Hypotheses
    1. A hypothesis is a proposed explanation for one or more observations.
    2. Hypotheses must be testable by scientific methods.
    3. Hypotheses must be falsifiable, which means able to be proven false.
  - B. Scientific Theories
    1. A scientific theory is based on well-supported hypotheses.
      - a) Hypotheses must be related to a single concept.
      - b) Hypotheses must be supported by independent lines of research.
    2. Germ theory is a scientific theory.
      - a) Germ theory examines the link between microorganisms and disease.
      - b) It arose from work by scientists such as Pasteur and Koch.
        - (1) Pasteur observed the link between bacteria and sour milk.
        - (2) Koch observed the link between anthrax bacteria and fatal symptoms in mice.
      - c) Germ theory is supported by the observation that antibiotics can treat diseases caused by bacteria.
      - d) Germ theory was used to verify the link between *H. pylori* and stomach ulcers.
  - C. The Theory of Evolution and the Theory of Natural Selection
    1. The theory of evolution states that all modern organisms are descended from a common ancestor.
      - a) Evidence for this theory is found in the similarities among living things.
      - b) Evidence for this theory is found in the pattern of shared traits among groups of organisms.
    2. The theory of natural selection deals with traits in populations.
      - a) There are variations between individual organisms.
      - b) Some variations help individuals to survive and reproduce.
      - c) Variations that increase survival and reproduction will become more common in a population.
      - d) Less advantageous variations will become less common in a population.
    3. Both theories have abundant observational and experimental support.
  - D. The Logic of Hypothesis Testing
    1. Inductive reasoning: using a series of observations to construct a hypothesis.
      - a) Example: Observations about the health benefits of chocolate help you to form a hypothesis that chocolate reduces the risk of cardiovascular disease.
      - b) Inductively deduced hypotheses may make sense but turn out to be false.

2. Deductive reasoning: using "if/then" statements to decide how to test a hypothesis.
   a) Example: If consuming dark chocolate reduces cardiovascular disease, then people who consume dark chocolate will experience fewer heart attacks than people who do not.
   b) You reject the hypothesis if you do not get the predicted result.
   c) You cannot automatically accept the hypothesis if you do not reject it.
      (1) There may be alternative explanations.
      (2) Example: People who eat dark chocolate may have less stress, and less stress causes them to have fewer heart attacks.

II. Hypothesis Testing
  A. The Experimental Method
    1. Experiments are actions or observations that test hypotheses.
       a) An independent variable is manipulated during an experiment.
       b) The effect on a dependent variable is measured.
       c) Information collected during an experiment is called data.
    2. Some hypotheses cannot be tested experimentally.
       a) Example: Hypotheses about events that happened in the past.
       b) These hypotheses can often be tested through observation of the natural world.
  B. Controlled Experiments
    1. Use groups that are not exposed to experimental treatment.
    2. Control for independent variables to eliminate alternative hypotheses.
       a) Use random assignment of individuals to put together control and experimental groups.
       b) Use placebos or give both groups the same treatment except for the variable being studied.
  C. Minimizing Bias in Experimental Design
    1. Blind Experiment
       a) Subjects do not know if they belong to control or experimental group.
       b) Subjects will not behave differently based on treatment group.
    2. Double-Blind Experiment
       a) Neither subjects nor researchers know who belongs to control or experimental groups.
       b) Subjects will not behave differently based on treatment group.
       c) Researchers will not treat subjects differently based on treatment group.
  D. Using Correlation to Test Hypotheses
    1. Experiments on Model Organisms
       a) Use mammals—organisms that are closely related to humans.
       b) Allow experimentation without putting humans at risk.
       c) Bring up ethical issues.
          (1) Animal welfare is a concern.
          (2) Functional differences between other animals and humans mean that experimental results may not be applicable to humans.
    2. Looking for Relationships Between Factors
       a) Data can be used to find a correlation, or relationship between two variables.
       b) The correlation may arise from observational data, not data from a controlled experiment.
       c) A correlation does not prove a cause-and-effect relationship between variables.
       d) Correlations are often used in epidemiological studies—studies on the distribution and causes of disease.

III. Understanding Statistics
  A. What Statistical Tests Can Tell Us
    1. A small subgroup, or a sample, of a population is chosen for an experiment.
    2. The sample is chosen to reflect characteristics of the general population.
    3. Statistical tests can tell scientists how likely it is that the sample represents the population.
  B. Statistical Significance: A Definition
    1. Chance can interfere with accurate statistical results.
       a) A sampling error is a difference between the sample and the population as a whole.
       b) Statistics allow scientists to determine the probability that their results are due to a sampling error.
    2. Error can be calculated.
       a) Standard error is a statistical measure of variability in a sample.
       b) A confidence interval uses standard error to give the highest and lowest likely average values for a population.

C. Factors Influencing Statistical Significance
 1. Sample size can influence conclusions.
  a) Small sample size
   (1) Can lead researchers to conclude there is an effect where there isn't one.
   (2) Can cause results of a study to be questioned.
  b) Large sample size
   (1) Makes it more likely that the research effect seen is a true effect.
   (2) Makes it more likely that the researcher will find statistical significance between control and experimental groups when there is a true difference.
D. What Statistical Tests Cannot Tell Us
 1. Statistical tests calculate sampling error, not observer error.
  a) A poorly designed experiment may produce misleading results.
 2. A statistically significant result may not be meaningful or important.
  a) A large sample size may reveal very small effects.
  b) Significance to human health can be misinterpreted when study results are reported.

IV. Evaluating Scientific Information
A. Information from anecdotes
 1. Anecdotal evidence is based on personal experience.
 2. Anecdotes are not as reliable as results from well-designed experiments.
B. Science in the News
 1. Consider the source of media reports.
 2. Carefully evaluate claims made by advertisers.
 3. Evaluate web sites carefully.
  a) Use web sites from reputable medical establishments.
  b) Evaluate whether the web site is using scientific information to sell a product.
  c) Ensure that the information is up to date.
  d) Be on the lookout for unsubstantiated claims.
C. Understanding Science from Secondary Sources
 1. Determine if the source presents experimentally tested results or untested hypotheses.
 2. Look for clues that the source is being appropriately cautious or overemphasizing results.
 3. Be skeptical about studies that are controversial.

Practice Questions

Matching

1. hypothesis
2. bias
3. inductive reasoning
4. deductive reasoning
5. independent variable
6. dependent variable
7. scientific theory
8. experiment
9. double-blind
10. falsifiable
11. prediction
12. scientific method
13. placebo
14. testable
15. data

a. the result that is expected when a hypothesis is tested
b. information collected during hypothesis testing
c. able to be tested using scientific observations
d. a set of actions designed to test a specific hypothesis
e. a set of related hypotheses that are supported by independent lines of research
f. special knowledge or treatment that may influence the results of an experiment
g. a factor that is manipulated by the investigator during an experiment
h. a proposed explanation for one or more observations
i. a process that allows observations to be tested and verified
j. combining a series of observations to discern a general principle
k. an ineffective treatment that is given to the control group in some investigations
l. a set of observations that could be proved to be false
m. using a general principle to predict an expected observation
n. a factor that may respond when other factors are changed during an experiment
o. an experiment in which neither research subjects or investigators are aware of the hypothesis

Fill-in-the-Blank

16. A ____________ uses experimental groups and control groups to test a ____________.
17. When testing a hypothesis, scientists use a ____________ of a population.
18. A small ____________ has a high probability of containing the true ____________ average.
19. A company that markets a weight-loss drug by showing two or three people who successfully lost weight on the drug is using ____________ in its marketing.
20. Rats and dogs used in human biology experiments are considered to be ____________.
21. In an investigation that tests whether cold medicines successfully treat common cold symptoms, the type of cold medicine is the ____________ variable and each patient's response to the cold medicine treatment is the ____________ variable.
22. When you increase the sample size in an ____________, you generally decrease the amount of ____________.
23. After ____________ that are collected during a scientific investigation are processed, the investigator writes an article and submits it to a scientific ____________. Editors send the article out to be ____________ before the article is published.
24. Two variables may show a ____________, and yet there may be no ____________ relationship between the two variables.
25. In a controlled experiment, subjects in the ____________ group should have the same general overall health as subjects in each ____________ group. Subjects should be ____________ to groups in order to minimize differences between the groups.

Labeling

Use the terms below to label Figure 1.1.

very likely
likely
somewhat likely
unlikely
very unlikely

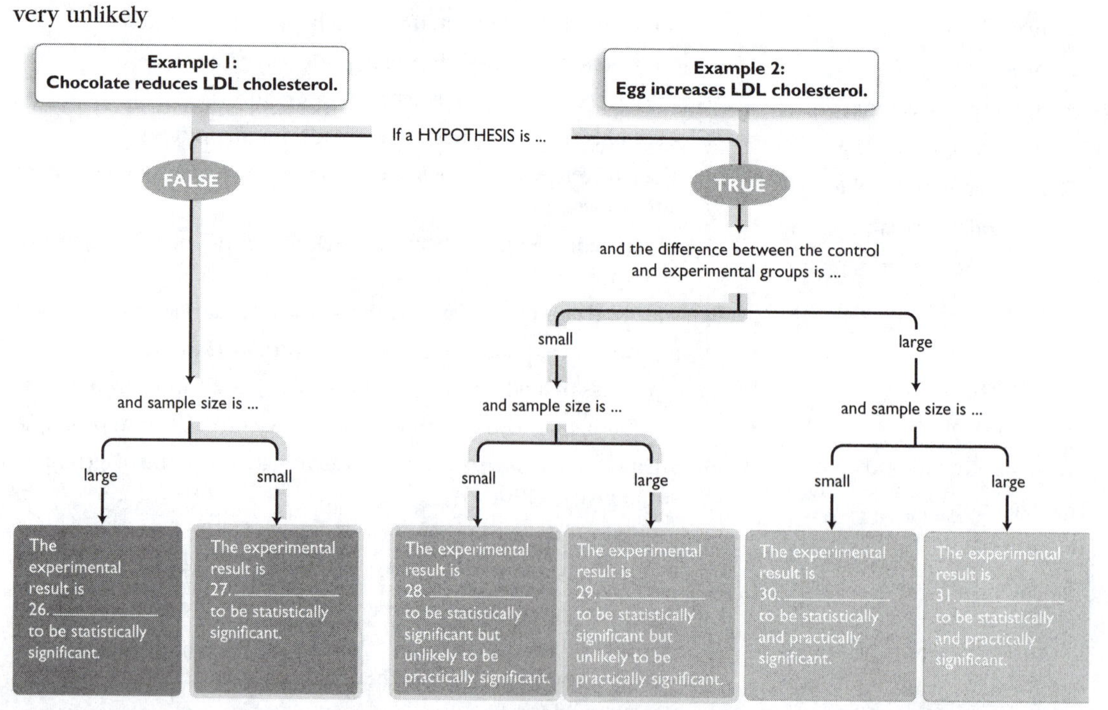

Figure 1.1

Use the terms below to label Figure 1.2.

hypothesis
observation
logic
question

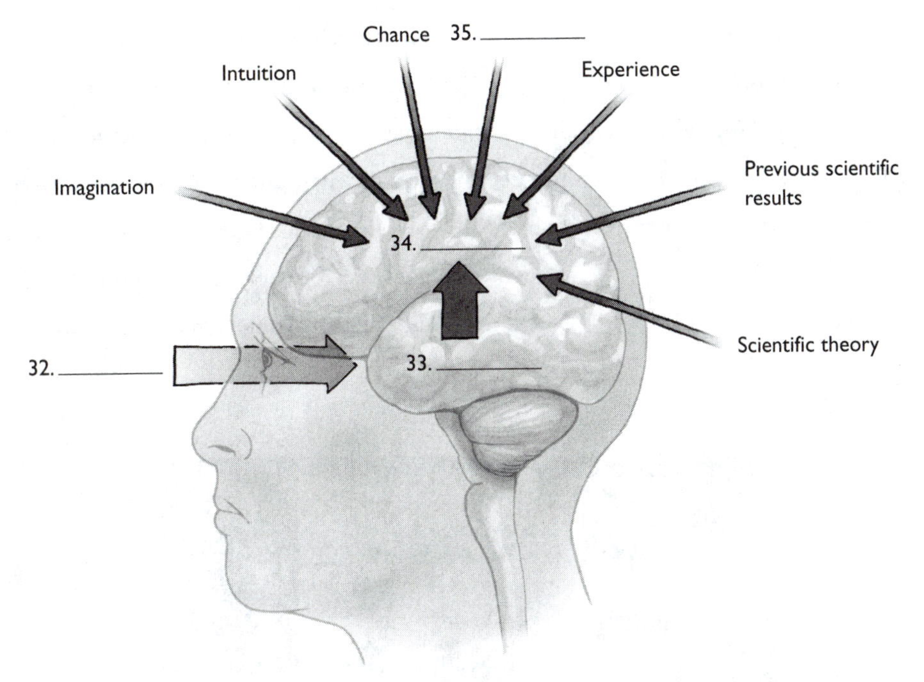

Figure 1.2

Roots to Remember

Use your knowledge of the root words presented in this chapter to circle the term that correctly completes each sentence.

36. Hypothyroidism is a condition in which the thyroid gland produces (too much/too little) thyroid hormone.
37. A scientist who uses generally accepted information to construct a hypothesis is using (deductive/inductive) reasoning.
38. Microbiology is the (study of/experimentation with) microscopic (objects/organisms).

Paragraph Completion

Use the following terms to complete the paragraphs. Terms used more than once are listed multiple times.

contains
controlled experiment
correlation
data
does not contain
double-blind
hypothesis
randomly assigned
reject
sample size
statistically significant
statistics

A variety of studies indicate that oatmeal may help to prevent some cardiovascular diseases. A scientist decides to test the (39) ____________ that the consumption of oatmeal may lower LDL ("bad") cholesterol. The scientist sets up a (40) ____________ in which human subjects are (41) ____________ to control and experimental groups. Subjects in the control group are given cereal that (42) ____________ oatmeal. Subjects in the experimental group are given cereal that (43) ____________ oatmeal. Each subject is given a box of their assigned cereal and told to eat 1 cup of cereal with 1/2 cup of milk each morning for a month. The investigation is (44) ____________, so neither the subjects nor the investigators know who is getting which cereal.

The investigator collects (45) ____________ on cholesterol levels each week for each subject. After a month, the investigator uses (46) ____________ to determine if there is any (47) ____________ between a change in cholesterol levels and consumption of oatmeal. The investigator finds that both groups had an overall drop in cholesterol levels. The experimental group experienced a larger overall drop in cholesterol than did the control group. However, the difference between the groups was not (48) ____________. The (49) ____________ in the experiment was relatively small. Therefore, the investigator could not necessarily (50) ____________ the hypothesis that oatmeal may lower LDL cholesterol.

Crossword Puzzle

Across

51. A 95% confidence interval is very likely to contain both sample and population ____________.
52. A scientist uses observations to ____________ a hypothesis.
53. Sample ____________ can affect the statistical significance of any difference between control and experimental groups.
54. Personal testimonial or personal experience.
55. Standard ____________ is a statistical measure of the amount of variability in a sample.
56. An experiment can use model organisms to eliminate the experimental ____________ to human health.

Down

57. When one variable ____________ with another, the two variables have a relationship.
58. Collected during an experiment.
59. Independent and dependent.
60. A newspaper or internet source.
61. Colleagues who review scientific papers.

Word Choice

Circle the word or phrase that correctly completes each sentence.

62. Scientific information in newspapers is (more reliable/less reliable) than scientific information in peer-reviewed journals.
63. In a confidence interval, the highest likely value for the population average is equal to the sample average (plus/minus) the standard error.
64. A good sample is (representative of/different from) the general population.
65. When variables are correlated, one variable (has an influence on/has a relationship with) the other variable.
66. A double-blind experiment specifically controls for (experimental conditions/observer bias).

Critical Thinking

67. Prions are normal proteins found in the brain that, when misfolded, can cause spongiform encephalopathy. One form of spongiform encephalopathy is also known as mad cow disease. A prion is not a microorganism, and yet it causes disease. Does the discovery of prions as disease agents disprove germ theory?
68. Some advertisements for skin cream claim that the cream erases fine lines and deep wrinkles. These products are marketed online with before shots showing extreme wrinkles and after shots showing smooth, wrinkle-free skin. Why should you be skeptical of the claims made by manufacturers of these products?
69. A researcher sets up an experiment to test a new cholesterol drug. She recruits a large sample group of people and randomly assigns each person to a treatment or placebo group. She assigns a number to each person. People in the placebo group are given numbers that start with 1,000. People in the treatment group are given numbers that start with 2,000. People in both groups are given the same instructions for taking the drugs, and are not told which treatment they have been assigned to. What is a possible flaw in this experiment's design, and how could it be fixed?

Practice Test

1. A human health study tests the effects of a protein-rich diet on blood sugar levels of subjects who have diabetes. Which of the following is the dependent variable in this study?
   A. type of diabetes
   B. blood sugar level
   C. age of the subjects
   D. amount of protein given

2. Which of the following is minimized in a well-controlled experiment?
   A. differences between subjects
   B. alternative hypotheses
   C. observer bias
   D. all of the above

3. A scientist is studying the effects of a new drug on blood pressure. Which of the following factors is *least* important for the scientist to consider as subjects for the study are being chosen?
   A. weight of the potential subjects
   B. age of the potential subjects
   C. activity level of the potential subjects
   D. interests of the potential subjects

4. Which of the following is an example of deductive reasoning?
   A. A magazine article investigates the observation that people currently in their sixties are healthier than people in their sixties 20 years ago.

B. A scientist predicts that if yoga reduces stress, then people who practice yoga will experience fewer cardiovascular problems.
C. A doctor uses information from a variety of studies on digestive system health to create a new diet.
D. A student researches information on the negative health effects of dairy products to hypothesize that a vegan (all plant) diet improves general health.

5. Which model organism would be *least* effective in a human health study?
A. mice
B. chimpanzees
C. pigeons
D. rats

6. Which of the following is *not* true of a blind experiment?
A. Researchers do not know which subjects are part of the control and experimental groups.
B. Placebos are used to control for bias.
C. Research subjects do not know if they are part of the control or experimental group.
D. Subjects are randomly assigned to control and treatment groups.

7. A teacher observes that kids who eat fruits and vegetables for lunch are rarely absent from school. Which of the following is a reasonable conclusion for the teacher to make?
A. Fruits and vegetables can prevent illness.
B. Kids who eat healthy foods do better in school.
C. Consumption of healthy foods may be correlated with good health.
D. The vitamins in fruits and vegetables help kids to stay healthy.

8. A television news show reports on the apparent connection between heart disease and poor sleeping habits. Which of the following questions would be *least* useful to ask when evaluating this report?
A. Do poor sleeping habits cause heart disease or are these factors merely correlated?
B. How many years of poor sleep does it take before you experience heart disease?
C. Is the report based on multiple peer-reviewed investigations?
D. Is there an alternative hypothesis that might give a better explanation of the link between sleep and heart disease?

9. What is one way to decrease sampling error?
A. increase sample size
B. calculate a confidence interval
C. decrease sample size
D. reduce bias

10. What is a scientific theory?
A. a statement based on deductive reasoning
B. an untested hypothesis
C. a model based on well-supported hypotheses
D. an induction based on limited facts

11. Which of the following would contain the most reliable scientific information?
A. a peer-reviewed scientific journal
B. a web site from a trusted medical source
C. a magazine ad from a pharmaceutical company
D. a television news report by a doctor

12. Which of the following is a testable hypothesis?
A. Ghosts exist in this world.
B. The paintings of Matisse are very beautiful.
C. Meditation can help you to obtain inner peace.
D. Daily exercise can help you to avoid obesity.

13. A scientist observes the following: (1) Children exposed to secondhand smoke tend to have more colds than children who are not exposed; and (2) smoking causes respiratory problems in adults. The scientist decides to test a hypothesis that secondhand smoke can cause respiratory problems in nonsmoking adults who were exposed as children. To construct this hypothesis, the scientist used
    A. deductive reasoning.
    B. inductive reasoning.
    C. controlled experiments.
    D. tested correlations.

14. A sample
    A. should reflect the characteristics of the larger population.
    B. should be small enough to be easily tested.
    C. should include people in all states of health.
    D. should be used only when necessary.

15. A local election contains a referendum that will determine whether a new health clinic will be built. A poll indicates that 53% of survey respondents favor the construction of the new health clinic; 45% of survey respondents do not favor the clinic's construction. The poll has a standard error of 4%. Which of the following can you conclude from the poll?
    A. The referendum will pass.
    B. The referendum will fail.
    C. The vote is statistically too close to call.
    D. Other factors will prevent passage of the referendum.

Answer Key

Matching

Answers: **1.** h; **2.** f; **3.** j; **4.** m; **5.** g; **6.** n; **7.** e; **8.** d; **9.** o; **10.** l; **11.** a; **12.** i; **13.** k; **14.** c; **15.** b

Fill-in-the-Blank

16. controlled experiment; hypothesis; **17.** sample; **18.** confidence interval; population; **19.** anecdotal evidence; **20.** model organisms; **21.** independent; dependent; **22.** experiment; sampling error; **23.** data; journal; peer reviewed; **24.** correlation; cause-and-effect; **25.** control; experimental; randomly assigned

Labeling

26. very unlikely; **27.** unlikely; **28.** somewhat likely; **29.** likely; **30.** likely; **31.** very likely; **32.** observation; **33.** question; **34.** hypothesis; **35.** logic

Roots to Remember

36. too little; **37.** deductive; **38.** study of; organisms

Paragraph Completion

39. hypothesis; **40.** controlled experiment; **41.** randomly assigned; **42.** does not contain; **43.** contains; **44.** double-blind; **45.** data; **46.** statistics; **47.** correlation; **48.** statistically significant; **49.** sample size; **50.** reject

Crossword Puzzle

Across

51. averages; **52.** create; **53.** size; **54.** anecdotal; **55.** error; **56.** risk

Down

57. correlates; **58.** data; **59.** variables; **60.** secondary; **61.** peers

Word Choice

62. less reliable; **63.** plus; **64.** representative of; **65.** has a relationship with; **66.** observer bias

Critical Thinking

67. It does not disprove the theory, because the theory supports the idea that all diseases are not necessarily caused by microorganisms. It does modify the theory to include abnormal proteins as potential disease agents.

68. These claims are part of a paid advertising campaign, so the advertiser may be exaggerating the product's effectiveness. Skin creams are not drugs, so they are not regulated by the FDA and their effectiveness has not necessarily been tested with controlled studies. Before-and-after shots can be exaggerated with computer enhancements, so these pictures may not accurately show the effects of the cream.

69. The assigned numbers identify the treatment and placebo groups, so the researchers who are analyzing the data would know which results are associated with which treatment. They should make this a double-blind study by randomizing the numbers assigned, so they are not associated with a specific group.

Practice Test

1. B; **2.** D; **3.** D; **4.** B; **5.** C; **6.** A; **7.** C; **8.** B; **9.** A; **10.** C; **11.** A; **12.** D; **13.** B; **14.** A; **15.** C

CHAPTER 2

THE CHEMISTRY OF LIFE: DRINK TO YOUR HEALTH?

Learning Goals

1. Describe the components of an atom and how they interact.
2. State several properties of water that make it a good solvent.
3. Compare and contrast covalent, ionic, and hydrogen bonds.
4. Describe the pH scale and what makes a substance an acid or a base.
5. Describe how buffer systems in the human body can help maintain homeostasis.
6. List the building-block molecules of a carbohydrate, a protein, and a fat.
7. Discuss the manner in which proteins fold.
8. Describe the process of hydrogenation.
9. Make a rough sketch of the structure of a DNA molecule, and label each of the following: a nucleotide, a sugar, a nitrogenous base.
10. Compare and contrast the classes of micronutrients: vitamins, minerals, and antioxidants.

Chapter Outline

I. Water: Essential to Life
   A. Human Health
      1. A person can survive only for a few days without water.
      2. Dehydration can lead to impaired physical and mental abilities.
      3. Water must be replenished daily by drinking and eating.
   B. The Building Blocks of Water
      1. Water is composed of the elements hydrogen and oxygen.
      2. Elements are composed of atoms.
         a) Atoms are the smallest units of matter.
         b) Atoms are composed of subatomic particles.
            (1) Protons
               (a) Have a positive charge.
               (b) Are found in the atom's nucleus.
            (2) Neutrons
               (a) Have no charge.
               (b) Are found in the atom's nucleus.
            (3) Electrons
               (a) Have a negative charge.
               (b) Are found outside the nucleus in an electron cloud.
         c) Atoms can be neutral or have a charge.
            (1) A neutral atom has an equal number of protons and electrons.
            (2) An ion has a charge because its number of protons does not equal its number of electrons.
         d) Atoms have mass.
            (1) The mass number is equal to the number of protons and neutrons in the atom's nucleus.
            (2) Isotopes are different versions of a chemical element.
               (a) Isotopes contain the same number of protons but different numbers of neutrons.
               (b) Carbon-14 is a commonly known isotope of carbon.

(c) Radioactive isotopes are unstable and break down over time.
(i) As radioactive isotopes break down, they release energy.
(ii) Some radioactive isotopes are used to diagnose or treat disease.

C. The Structure of Water
1. Water is a compound made up of molecules.
a) A molecule consists of two or more atoms joined by chemical bonds.
b) A molecule can be composed of the same or of different atoms.
c) Each water molecule contains two hydrogen atoms and one oxygen atom that are chemically joined.
d) Water is a compound because its molecules contain more than one type of atom.
2. Electrons within atoms occupy different energy levels.
a) Energy increases as the electrons get farther away from the nucleus.
(1) Electrons in the first energy level, or first electron shell, are closest to the nucleus and have the lowest energy.
(2) Electrons in the successive energy level or shell have a little more energy.
b) Each energy level or shell holds a specific maximum number of electrons.
(1) The first shell holds two electrons.
(2) The second and third shells each hold a maximum of eight electrons.
c) Electrons fill lower energy shells before filling higher energy shells.
d) The outermost shell of an atom is called the valence shell.
3. Atoms with the same number of electrons in their valence shell exhibit similar chemical behaviors.
a) Atoms with a full valence shell will not normally form chemical bonds with other atoms.
b) Atoms whose valence shells are not full can form chemical bonds.
4. The oxygen and hydrogen atoms in a water molecule form a covalent bond.
a) A covalent bond forms when atoms share electrons to complete their valence shells.
b) Atoms can share one, two, or three valence electrons with each other.
(1) A short line indicating a shared pair of electrons represents a single covalent bond.
(2) A double line represents a double covalent bond.
(3) A triple horizontal line represents a triple covalent bond.
5. Ionic bonds form between charged atoms that are attracted to each other.
a) Atoms with one, two, or three electrons in their valence shell tend to lose electrons and become positively charged.
b) Atoms with six or seven electrons in their valence shell tend to gain electrons and become negatively charged.
c) In an ionic bond between two atoms, the negatively charged atom loses electrons, and the positively charged atom gains electrons.
d) More than two atoms can be involved in an ionic bond.
e) Water can cause ions in molecules to dissociate.

D. Water is a Good Solvent
1. Water can dissolve a wide variety of substances.
a) Water is called a solvent when it dissolves a substance.
b) Substances dissolved by water are called solutes.
c) A solution contains solute particles in solvent.
2. Water has polar molecules.
a) On a polar molecule, different regions have different charges.
(1) Oxygen has a partial negative charge in a water molecule.
(2) Hydrogen has a partial positive charge in a water molecule.
b) Hydrophilic substances contain charged atoms and can be easily dissolved by water.
c) Hydrophobic substances do not contain charged atoms and do not mix easily with water.
3. Water can carry many dissolved substances around the body.
a) The liquid portion of blood is made primarily of water.
b) Dehydration can interfere with the body's transport of substances.

E. Water Facilitates Chemical Reactions
1. Reactants come into contact with each other when dissolved in water.
2. Water itself can react with solutes (e.g., formation of carbonic acid).

F. Water is Cohesive
1. Hydrogen bonds are formed between water molecules.
a) In a hydrogen bond, a partially positive hydrogen atom is attracted to a partially negative oxygen atom.

b) Hydrogen bonds can occur between molecules (intermolecular) or within a molecule (intramolecular).
c) Hydrogen bonds are weak and easily broken.
d) Hydrogen bonds make water cohesive or "sticky."
(1) Cohesion helps water to move in a continuous column through plant stems.
(2) Cohesion helps the water in blood to move through the circulatory system.
2. Hydrogen bonds allow water to absorb heat without increasing in temperature.
a) Only after hydrogen bonds are broken does water temperature increase.
b) Water can absorb a high amount of heat energy before it becomes hot.
c) This property of water helps to stabilize body temperature.
(1) Water can stabilize temperatures within cells where chemical reactions are constantly producing heat.
(2) Water can remove heat from the body through the evaporation of sweat.

G. Bottled or Tap?
1. Bottled and tap water are held to the same EPA water quality standards.
2. 40% of bottled waters are filtered tap water.
a) Filtering methods can remove fluoride.
b) Some dentists are concerned that replacement of tap water with bottled water may increase tooth decay in children.
3. To manufacture bottles used for water requires 1.5 million barrels of oil per year.
4. Around 86% of all water bottles go to landfills each year.

II. Acids, Bases, and Salts
A. pH: Measuring the Activity of Ions
1. Hydrogen tends to dissociate from the water molecule.
a) This dissociation forms H^+ and OH^- ions.
b) These ions can react with other charged particles.
c) H^+ and OH^- ions are constantly dissociating and re-associating in pure water.
2. The pH scale is a measure of the concentration of H^+ ions in a solution.
a) The scale ranges from 0 to 14.
b) Low numbers represent acids; high numbers represent bases.
(1) An acid donates H^+ ions to the solution.
(2) A base decreases H^+ ions by releasing OH^- ions to bond with them.
(3) Neutral water (pH 7) has an equal concentration of H^+ and OH^- ions.
3. The maintenance of blood pH is important to human homeostasis.
a) Buffers in blood keep pH within a narrow range.
(1) Carbonic acid in blood dissociates into bicarbonate to lower pH.
(2) Bicarbonate reacts with hydrogen ions to form carbonic acid and raise pH.

B. Electrolytes
1. Electrolytes are ions such as sodium or potassium.
2. Electrolytes form when a substance dissociates into ions in solution.
3. Electrolytes perform important functions, such as conducting electrical impulses along nerves and through muscle.
4. Electrolytes lost through sweat must be replaced.
a) Table salt adds sodium.
b) Fruits and vegetables add potassium.
c) Sports drinks contain electrolytes and may be better than plain water during intense exercise.
5. Electrolytes are added to water as salts.
a) Salts are any neutral compound composed of positive and negative ions.
b) Electrically neutral table salt consists of Na^+ and Cl^- ions.

III. Structure and Function of Macromolecules
A. Carbon and Macromolecules
1. Organic molecules are molecules that contain carbon.
2. Carbon is a flexible element.
a) Carbon has only four electrons in its valence shell.
b) Carbon can bond with up to four other elements.
c) Carbon-containing molecules can be chains, rings, branched chains, or many other shapes.

3. The organic chemicals found in living organisms are called macromolecules.
   a) Macromolecules include carbohydrates, lipids, proteins, and nucleic acids.
   b) Macromolecules are composed of subunits, called monomers, that are joined together to produce polymers.
      (1) Dehydration synthesis is the chemical reaction that produces polymers.
      (2) Hydrolysis is the reverse reaction that breaks polymers into component monomers.

B. Carbohydrates
1. Carbohydrates are commonly called sugars.
   a) Glucose is a simple sugar—or monosaccharide—with a single ring-shaped structure.
   b) Lactose or sucrose is a disaccharide with two rings.
   c) Cellulose is a polysaccharide.
      (1) A polysaccharide is a polymer.
      (2) Starch in potatoes and glycogen stored in muscles and the liver are polymers of glucose.
      (3) Cell walls and insect exoskeletons are composed of polysaccharides.
2. Complex carbohydrates are composed of many branching chains of sugar monomers.
   a) Complex carbohydrates are often stored for later use.
   b) They are digested more slowly than simple sugars because they have more chemical bonds.
3. Dietary fiber is composed mainly of cellulose.
   a) Humans cannot break down dietary fiber into monosaccharides.
   b) Some fiber is broken down by bacteria in the digestive tract.
   c) Fiber helps to move potentially harmful substances through the large intestine.
   d) Fiber helps protect blood vessels.
4. A healthful diet is rich in complex carbohydrates.
   a) Fruits, vegetables, and grains provide complex carbohydrates and fiber.
   b) Sports drinks supply carbohydrates that may be unnecessary.

C. Proteins
1. Proteins are molecules composed of amino acid subunits.
   a) There are 20 commonly occurring amino acids.
      (1) Your body can synthesize most of these amino acids.
      (2) Essential amino acids must be acquired from food.
   b) Amino acids have a distinctive structure.
      (1) Each contains nitrogen as part of an amino group on one end.
      (2) Each contains a carboxyl group on the other end.
      (3) Each amino acid has its own distinctive side groups.
2. Long polymers of amino acids are called polypeptides.
   a) Peptide bonds are covalent bonds that join adjacent amino acids.
   b) Polypeptides have distinctive structures.
      (1) The protein's primary structure is its precise sequence of amino acids.
         (a) These amino acids are held together by peptide bonds.
         (b) Different polypeptides have different linear orders of amino acids.
      (2) The protein's secondary structure is the way that it folds.
         (a) Alpha helices are helical structures.
         (b) Beta-pleated sheets are accordion-like.
         (c) Hydrogen bonds form between the amino group of one amino acid and the carboxyl group of another to create the folding.
      (3) The protein's tertiary structure is its globular, three-dimensional shape.
         (a) Tertiary structure is due to the interaction between side groups.
         (b) Hydrogen, ionic, and covalent bonds are involved in the development of tertiary structure.
         (c) Each polypeptide has a unique tertiary structure.
      (4) A protein's quaternary structure is created by the interaction of the protein's polypeptide chains.
         (a) Only proteins with more than one polypeptide chain have a quaternary structure.
         (b) A quaternary structure is a three-dimensional structure.
3. The structure of a protein is influenced by its environment.
   a) Temperature and pH can affect interactions among amino acid side chains.

b) Heating can break peptide bonds.
   (1) A heated protein is said to have been denatured.
   (2) Heating an egg white denatures the protein and changes the egg white from a clear gel to a hard white substance.

4. Some foods are rich in protein.
   a) Beef, poultry, fish, eggs, and dairy products are protein-rich animal products.
   b) Beans and nuts are protein-rich plant products.
5. Digestion breaks down proteins.
   a) Proteins from food are broken down into component amino acids.
   b) Protein enzymes from supplements are also broken down.
6. Energy drinks provide amino acids.
   a) Taurine is advertised as an energy booster.
      (1) Taurine can be synthesized by the body.
      (2) There is no scientific evidence that taurine supplements boost energy.
   b) Proline and tryptophan are other commonly added amino acids.
   c) Supplements are unnecessary for people who are eating a well-balanced diet.
   d) Amino acid supplements may increase muscle mass in bodybuilders.

D. Lipids
1. Lipids are partially or entirely hydrophobic.
2. There are three types of lipids: fats, phospholipids, and steroids.
   a) Fats consist of three hydrocarbon chains attached to a glycerol molecule.
      (1) The hydrocarbon chains are called fatty acids.
         (a) Saturated fats are fatty acids that are bound to as many hydrogen atoms as possible.
            (i) Saturated fats lie flat and can pack together tightly.
            (ii) These fats tend to be solid at room temperature.
         (b) Unsaturated fats are fatty acids that have carbon-to-carbon double bonds.
            (i) The double bonds in unsaturated fats make these fats kink so they cannot lie flat.
            (ii) These fats tend to be liquid at room temperature.
            (iii) When a fat contains many unsaturated carbons, it is called polyunsaturated.
            (iv) Unsaturated fats have fewer negative health effects than saturated fats.
            (v) Unsaturated fats can be made saturated through hydrogenation.
               (a) Hydrogenation produces trans fats.
               (b) Trans fats can increase health risks.
      (2) The body can burn fatty acids to release energy.
      (3) The body can synthesize all fatty acids except omega-3 and omega-6 acids.
         (a) Omega-3 acids have heart-health benefits.
         (b) These fatty acids can be obtained from fish, walnuts, flaxseed, canola oil, or soybean oil.
   b) Phospholipids have two fatty acid chains and a phosphate group attached to a glycerol molecule.
      (1) The phosphate group creates a hydrophilic head for the molecule.
      (2) When combined with water, phospholipids assemble into spheres with the hydrophilic head facing out.
      (3) Phospholipids are important for cells.
   c) Steroids are composed of four carbon-containing rings with variable side groups.
      (1) Cholesterol is a steroid that helps cells maintain fluid membranes.
      (2) Cholesterol is carried through the body by lipoproteins.
         (a) Low-density lipoproteins (LDLs) are low in protein and high in cholesterol.
            (i) LDLs carry cholesterol that has been consumed in food or made by the liver.
            (ii) LDLs distribute cholesterol to cells.
         (b) High-density lipoproteins (HDL) are high in protein and low in cholesterol.
            (i) HDLs scavenge excess cholesterol from the body and return it to the liver.
            (ii) The liver uses cholesterol to make bile, which is released into the small intestine.
         (c) A low LDL:HDL ratio is healthy.
         (d) Saturated fats can increase the LDL:HDL ratio.

E. Nucleic Acids
1. Nucleic acids are molecules that are concentrated in the nucleus of a cell.

2. Deoxyribonucleic acid (DNA) is a nucleic acid that stores genetic information.
   a) DNA is a double helix.
      (1) Its strands are made of monomers called nucleotides.
         (a) Nucleotides are joined along their length by covalent bonds.
         (b) Each nucleotide has a sugar, a phosphate, and a nitrogenous base.
            (i) The sugar is deoxyribose.
            (ii) The bases may be adenine (A), guanine (G), thymine (T), or cytosine (C).
               (a) A and G are purines composed of two rings.
               (b) T and C are pyrimidines composed of a single ring.
      (2) Each strand has a sugar-phosphate backbone.
         (a) Sugars and phosphates alternate along the length of the helix.
         (b) Phosphate groups give the helix a negative charge.
      (3) Strands are antiparallel.
      (4) Nitrogenous bases from the two strands form hydrogen bonds.
         (a) Adenine (A) always bonds with thymine (T).
         (b) Guanine (G) always bonds with cytosine (C).
3. Ribonucleic acid (RNA) is found in cells.
   a) RNA is single stranded.
      (1) Its backbone contains the sugar called ribose.
      (2) It has uracil (U) as a base instead of thymine (T).

IV. Micronutrients
   A. Micronutrients are essential to good health.
      1. Micronutrients are needed in very small amounts.
      2. They are not destroyed or burned for energy.
      3. Vitamins and minerals are micronutrients.
   B. Vitamins
      1. Vitamins are organic substances that generally cannot be synthesized by the body.
      2. Vitamins function as coenzymes that speed up chemical reactions in the body.
      3. Vitamin deficiencies can affect body function.
         a) Vitamin C deficiency can affect iron absorption.
         b) Vitamin D deficiency can disrupt bone growth.
      4. Some vitamins are synthesized within the body.
         a) Human cells use sunlight to synthesize vitamin D.
         b) Bacteria in the large intestines synthesize vitamin K.
      5. Vitamins are either hydrophilic or hydrophobic.
         a) Hydrophilic—or water-soluble—vitamins are not stored.
         b) Hydrophobic vitamins are stored in body fat.
            (1) These vitamins include A, D, E, and K.
            (2) It is possible to have an excess of these vitamins.
   C. Minerals
      1. Minerals are inorganic substances that are essential for cell function.
      2. Minerals include calcium, chlorine, magnesium, phosphorus, potassium, sodium, and sulfur.
      3. Minerals are important for fluid balance, muscle contraction, nerve impulse conduction, and bone strength.
   D. Antioxidants
      1. Antioxidants are vitamins or minerals that may play a role in preventing disease.
      2. Antioxidants protect cells against reactive substances called free radicals.
         a) Free radicals have an incomplete valence shell and remove electrons from molecules in the body.
         b) Antioxidants bind to free radicals.
      3. Antioxidants are found in fruits, vegetables, nuts, and grains.
   E. What Are You Paying For?
      1. Sports drinks may contain supplemental micronutrients.
      2. A healthful diet is a better source of micronutrients.
      3. Sports drinks may be high in calories.
      4. The claims made by sports drinks have not been evaluated by the U.S. Food and Drug Administration.

Practice Questions

Matching

1. electronegative	a. a substance with pH > 7
2. lipid	b. an ion such as sodium
3. electron shell	c. bond between amino acids
4. covalent bond	d. organic micronutrients
5. vitamin	e. glucose or starch
6. fiber	f. when water molecules stick together
7. electrolyte	g. energy level in an atom where electrons are located
8. peptide bond	h. a hydrophobic molecule composed of hydrocarbons
9. antiparallel	i. carbohydrate that cannot be digested by the human body
10. base	j. a substance containing carbon and hydrogen
11. cholesterol	k. a steroid in the cell membrane
12. cohesion	l. element that is attractive to electrons
13. neutron	m. the orientation of two DNA strands on the double helix
14. organic molecule	n. bond that forms when two atoms share electrons
15. carbohydrate	o. a subatomic particle in the nucleus

Fill-in-the-Blank

16. ____________ is the main component of dietary fiber.

17. The mass number of an atom equals the sum of its ____________ and ____________.

18. ____________ bonds hold the hydrogen and oxygen atoms together in a water molecule. ____________ bonds help water molecules to stick together.

19. A(n) ____________ molecule is double stranded, whereas a(n) ____________ is single stranded.

20. Omega-6 is a(n) ____________ fatty acid that cannot be synthesized by the body.

21. The oxygen atom in a water molecule has a ____________ charge, and each hydrogen atom in the molecule has a ____________ charge.

22. The process of ____________ produces trans fats that are common in fast foods.

23. When a protein is heated, it can become ____________.

24. Complex ____________ are found in many fruits and vegetables. These macromolecules take longer to digest than simple ____________.

25. ____________ are certain vitamins and minerals that help to prevent disease by binding to ____________.

Labeling

Use the terms below to label Figure 2.1.

fatty acid tails
glycerol
hydrophilic
hydrophobic
phospholipid

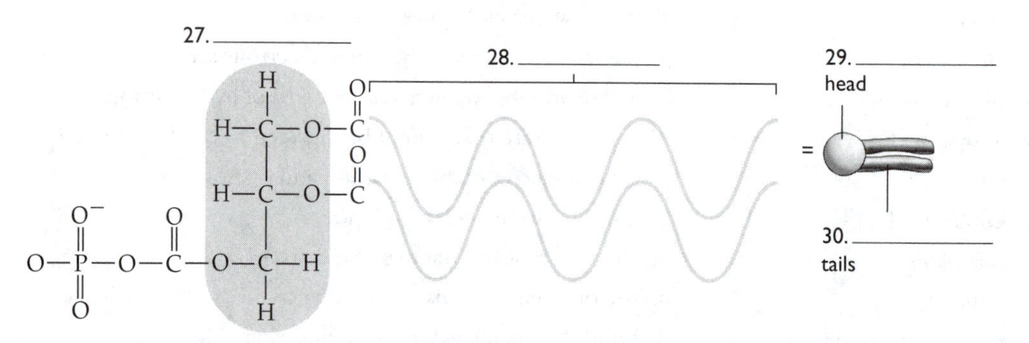

Figure 2.1

Use the terms below to label Figure 2.2.

covalent bonds
hydrogen bonds
nucleotide
nitrogenous base
sugar-phosphate backbone

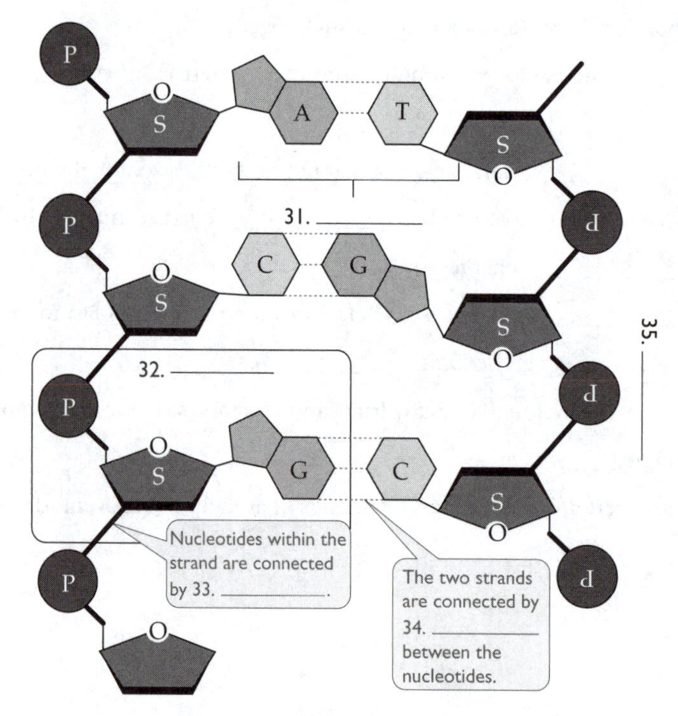

Figure 2.2

Roots to Remember

Use your knowledge of the root words presented in this chapter to answer the following questions.

36. What does the word *polymer* mean?
37. What process uses water to break polymers apart into their component monomers?
38. What root is the opposite of *micro-*?

Word Choice

Circle the word or phrase that correctly completes each sentence.

39. A calcium (Ca^{2+}) atom combines with two chloride (Cl^{-}) atoms to form (ionic/covalent) bonds.
40. Proteins are formed from (monomer/polymer) subunits called amino acids.
41. Acidic lemon juice has a (lower/higher) pH than does milk.
42. Lipid molecules are either partially or completely (hydrophilic/hydrophobic).
43. Minerals are (inorganic/organic) micronutrients that are essential for cell function.
44. (HDL/LDL) lipoproteins scavenge excess cholesterol from the body and return it to the liver.
45. A fat that has carbon double bonds is called (saturated/unsaturated).
46. The base (guanine/thymine) always pairs with the base cytosine in DNA.

Table Completion

Name each macromolecule described in the table.

Macromolecule	Description
47.	a molecule with ribose as its backbone
48.	a constituent of the cell membrane that has both hydrophilic and hydrophobic components
49.	a disaccharide found in milk
50.	an essential fatty acid found in fish
51.	a polymer found in rice and potatoes
52.	a monosaccharide that is found in sucrose and lactose
53.	a molecule composed of amino acids

Paragraph Completion

Use the following terms to complete the paragraphs. Terms used more than once are listed multiple times.

amino
amino acid
amino acids
carboxyl
folded
peptide
polypeptides
primary structure
secondary structure
side

Proteins are important macromolecules that fulfill many functions in the body. Proteins are formed by monomer subunits called (54)____________. Each subunit has a(n) (55)____________ group attached to one end. This group contains nitrogen. On the other end is a(n) (56)____________ group. A(n) (57)____________ group that is unique to that (58)____________ is found in the middle of the subunit. Proteins are often called (59)____________, due to their (60)____________ bonds. Different proteins have different sequences of amino acids. Each protein's unique sequence is called its (61)____________. The protein's (62)____________ relates to the way that the protein is (63)____________. Some proteins have accordion-like pleats. Others may have a helical structure.

Critical Thinking

64. A young girl has very low iron levels in her blood. What might be the symptoms of low iron and why?
65. How would water be different if it didn't form hydrogen bonds between its molecules? How would this change affect organisms?

Practice Test

1. The human body is able to synthesize
   A. essential amino acids.
   B. trans fats.
   C. minerals.
   D. cholesterol.

2. Which meal listed below is likely to provide the most antioxidants?
   A. vegetable stir fry with cashews and brown rice
   B. tuna salad on whole-wheat toast
   C. a hamburger with pickles and lettuce
   D. chicken soup with a side of whole-grain bread

3. Which protein structure depends on the interaction of side groups?
   A. primary structure
   B. secondary structure
   C. tertiary structure
   D. quaternary structure

4. Dietary fiber
   A. reduces the amount of time food spends in the large intestine.
   B. helps to remove cholesterol from the body.
   C. provides essential nutrients to the body.
   D. is easily digested in the stomach and small intestine.

5. All isotopes of the same element
   A. contain different numbers of protons.
   B. contain different numbers of neutrons.
   C. contain different numbers of electrons.
   D. are unstable and break down over time.

6. Which of the following would be the healthiest oil to use when cooking?
   A. unsalted butter
   B. hydrogenated soybean oil
   C. liquid corn oil
   D. low-fat margarine

7. Electrons in an atom have the highest energy level in the
   A. first shell.
   B. second shell.

C. third shell.
D. valence shell.

8. Which of the following is *not* true about hydrogen bonds?
   A. They facilitate cohesion in water.
   B. They are strong and hard to break.
   C. They can occur between molecules or within molecules.
   D. They allow water to absorb energy without getting hot.

9. How could you increase the pH of a solution?
   A. Add an acid to the solution.
   B. Add a base to the solution.
   C. Add water to the solution.
   D. Remove OH^- ions from the solution.

10. A patient has an LDL level of 120. What is a doctor likely to say to this patient?
    A. "Your cholesterol level is okay. Keep eating the same diet but start exercising."
    B. "You are doing a good job at maintaining healthy cholesterol levels."
    C. "Your cholesterol is a little high. Try cutting down on saturated fats."
    D. "Your cholesterol is dangerously high. You should start eating only plant-based foods."

11. Which nitrogenous base is found in RNA but not DNA?
    A. uracil
    B. thymine
    C. adenine
    D. cytosine

12. What could you ingest to make sure you have enough vitamin K?
    A. ample fruits and vegetables
    B. a multivitamin
    C. yogurt with live bacterial cultures
    D. fortified milk

13. The chemical reaction that builds polymers from monomers is called
    A. hydrolysis.
    B. dehydration synthesis.
    C. ionic bonding.
    D. hydrogenation.

14. The building blocks of carbohydrates are called
    A. polymers.
    B. monosaccharides.
    C. carboxyl groups.
    D. hydrocarbons.

15. A new fitness water claims to provide calcium and vitamins A and C. Why might you question this claim?
    A. Vitamin A is hydrophobic, so it would not mix into water.
    B. Calcium is a mineral that cannot be dissolved in water.
    C. Vitamin C is unstable when it is in a liquid.
    D. All of the above.

Answer Key

Matching

1. l; **2.** h; **3.** g; **4.** n; **5.** d; **6.** i; **7.** b; **8.** c; **9.** m; **10.** a; **11.** k; **12.** f; **13.** o; **14.** j; **15.** e

Fill-in-the-Blank

16. Cellulose; **17.** protons; neutrons; **18.** Covalent; Hydrogen; **19.** DNA; RNA; **20.** essential; **21.** partial negative; partial positive; **22.** hydrogenation; **23.** denatured; **24.** carbohydrates; sugars; **25.** Antioxidants; free radicals

Labeling

26. phospholipid; **27.** glycerol; **28.** fatty acid tails; **29.** hydrophilic; **30.** hydrophobic; **31.** nitrogenous base; **32.** nucleotide; **33.** covalent bonds; **34.** hydrogen bonds; **35.** sugar-phosphate backbone

Roots to Remember

36. many subunits; **37.** hydrolysis; **38.** macro-

Word Choice

39. ionic; **40.** monomer; **41.** lower; **42.** hydrophobic; **43.** inorganic; **44.** HDL; **45.** unsaturated; **46.** guanine

Table Completion

47. RNA; **48.** phospholipid; **49.** lactose; **50.** omega-3; **51.** starch; **52.** glucose; **53.** protein

Paragraph Completion

54. amino acids; **55.** amino; **56.** carboxyl; **57.** side; **58.** amino acid; **59.** polypeptides; **60.** peptide; **61.** primary structure; **62.** secondary structure; **63.** folded

Critical Thinking

64. Lack of energy, tiredness, and breathlessness, due to the lowered capacity of the blood to carry oxygen.

65. Sample Answer: Its molecules would not stick together, so it wouldn't form puddles or stick together within cells or tissues. It wouldn't be able to move substances effectively within cells or tissues. It would not retain its ability to absorb large amounts of energy without changing temperature. This would make it ineffective for thermoregulation.

Practice Test

1. D; **2.** A; **3.** C; **4.** A; **5.** B; **6.** C; **7.** D; **8.** B; **9.** B; **10.** C; **11.** A; **12.** C; **13.** B; **14.** B; **15.** A

CHAPTER 3

CELL STRUCTURE AND METABOLISM: DIET

Learning Goals

1. Name the macromolecules that provide the main dietary sources of energy.
2. Explain the difference between a calorie and a Calorie.
3. Describe how macromolecules store energy.
4. Describe how ATP energizes cellular reactions.
5. Describe the structure and function of the nucleus, ribosomes, endoplasmic reticulum, mitochondrion, and Golgi apparatus.
6. Explain how enzymes speed up the rate of metabolic reactions.
7. Describe the differences between active and passive transport.
8. Describe the kinds of substances that can pass freely through the membrane. Describe the types that require help to pass through membranes.
9. Explain how cellular respiration and breathing respiration are related.
10. Write the chemical equation for cellular respiration.

Chapter Outline

I. Food and Energy
 A. A Source of Energy
  1. Food provides molecules that can be broken down.
  2. Energy is stored in the chemical bonds of foods.
   a) Energy is released when bonds are broken.
   b) Released energy is converted into energy the body can use.
  3. Energy is stored within the body.
   a) Glycogen stores energy in the muscles and liver.
   b) Additional energy is stored as fat.
    (1) Fat is stored in adipose tissue.
    (2) Adipose tissue is found in the abdomen, buttocks, and hips.
 B. ATP Is the Cell's Energy Currency
  1. Adenosine triphosphate (ATP) is the chemical that powers most cell activities.
   a) ATP is a nucleotide that contains adenine, ribose, and three phosphates.
   b) Releasing a phosphate group from ATP liberates energy.
   c) ATP is converted to ADP (adenosine diphosphate) when it loses a phosphate.
   d) The liberated phosphate group is added to another molecule, in a process called phosphorylation.
  2. ATP powers mechanical, transport, and chemical work.

II. Cell Structure and Function
 A. Cells are the fundamental units of life.
 B. Most cells are in the 1–100 micrometer (one-millionth of a meter) range.
  1. Cells must be viewed with a microscope.
   a) Light microscopes use lenses and light to produce a magnified image.
   b) Electron microscopes use magnetic lenses and electrons to project an image onto film or a computer screen.
    (1) Transmission electron microscopes produce 2-D images.
    (2) Scanning electron microscopes produce 3-D images.

2. Cells are small due to the ratio of surface area to volume.
   a) The greater the surface area, the higher the rate of exchange of nutrients and wastes.
   b) As a cell gets larger, it needs to exchange more substances, but its surface area does not increase proportionately to allow this.

C. Cell Structures
1. Organelles are subcellular structures that perform specific functions.
   a) A watery matrix called cytosol surrounds the organelles.
      (1) Cytosol contains salts and enzymes necessary for cellular reactions.
      (2) Cytosol and the organelles it surrounds are called cytoplasm.
   b) The nucleus houses DNA and serves as the cell's control center.
      (1) The nucleus is surrounded by a nuclear envelope containing nuclear pores that allow substances to move in and out of the nucleus.
      (2) The nucleus contains
         (a) chromatin composed of DNA and proteins,
         (b) a nucleolus that produces ribosomes, and
         (c) fluid called nucleoplasm.
   c) Mitochondria are energy-harvesting organelles.
      (1) Mitochondria contain an inner and outer membrane.
      (2) The inner membrane contains proteins involved in producing ATP.
   d) Lysosomes are organelles that release energy from nutrients.
      (1) Lysosomes are membrane-enclosed sacs containing enzymes.
      (2) Enzymes in the lysosome break down proteins, carbohydrates, and fats.
      (3) Lysosomes engulf and break down nutrients and nonfunctioning organelles.
   e) Ribosomes are sites where amino acids join together to produce proteins.
      (1) Ribosomes are built in the nucleus and sent out to the cytosol.
      (2) They may float free or be tethered to the endoplasmic reticulum (ER).
   f) ER is a network of membranes.
      (1) Rough ER has ribosomes and synthesizes proteins.
      (2) Smooth ER detoxifies substances or synthesizes lipids.
   g) The Golgi apparatus is a stack of membranous sacs.
      (1) Vesicles containing proteins from the ER fuse with the Golgi apparatus.
      (2) The Golgi apparatus modifies, sorts, and exports the proteins.
   h) Centrioles are structures involved in microtubule formation and the movement of chromosomes during cell division.
   i) The cytoskeleton is a network of filaments and tubules in the cytoplasm.
      (1) Cytoskeleton elements are made of protein.
      (2) Cytoskeleton protein fibers help a cell to maintain structure and move.

III. Membrane Structure and Function

A. Membrane Structure
1. The plasma membrane is the outermost boundary of a cell.
   a) This membrane isolates the cell from its environment.
   b) It controls the exchange of substances between the cell and the outside environment.
2. Membranes are composed of a phospholipid bilayer.
   a) Phospholipid tails of this bilayer interact with themselves and exclude water.
   b) Phospholipid heads maximize exposure to water inside and outside of the membrane.
   c) Proteins are imbedded in the phospholipid layer.
      (1) Transport proteins move substances from one side of the membrane to the other.
      (2) Other proteins serve as receptors that communicate outside conditions to the cell.
      (3) Still other proteins anchor the cytoskeleton to the plasma membrane.
   d) Cholesterol in the bilayer maintains the fluidity of the membrane.
   e) Glycoproteins serve as cell markers in the membrane.

B. Transporting Substances Across Membranes
1. Membranes are selectively permeable.
2. Some substances move across the membrane without help.
3. Some substances require transport proteins and energy input for movement across the membrane.

C. Passive Transport: Diffusion, Facilitated Diffusion, and Osmosis
1. Passive transport does not require energy input.
2. Some substances move across the cell membrane through diffusion.
   a) Diffusion moves substances down their concentration gradient.

b) Only small, hydrophobic molecules can cross the membrane through diffusion.
c) Fats, carbon dioxide, and oxygen can diffuse across the cell membrane.
3. Some substances move across the cell membrane through facilitated diffusion.
a) Facilitated diffusion moves substances down their concentration gradient.
b) Membrane proteins facilitate diffusion.
4. Osmosis is the movement of water across the cell membrane.
a) Osmosis is a form of diffusion.
b) Different solution concentrations affect osmosis in body cells.
(1) In an isotonic solution, osmosis does not occur because there is an equal concentration of solute on both sides of the cell membrane.
(2) In a hypertonic solution, the solution contains a higher concentration of solute than the cell does.
(a) Water leaves the cell by osmosis.
(b) The cell shrivels.
(3) In a hypotonic solution, the solution contains a lower concentration of solute than the cell does.
(a) Water enters the cell.
(b) The cell may burst.

D. Active Transport: Pumping Substances Across the Membrane
1. Active transport uses proteins powered by ATP to move substances against a concentration gradient.
a) Proteins involved in active transport are often called pumps.
b) One type of pump moves sodium ions out of the cell and potassium ions in.

E. Exocytosis and Endocytosis: Movement of Large Molecules Across the Membrane
1. Large molecules cannot be transported across the cell membrane by passive or active transport.
2. Large molecules are moved by membrane-bound vesicles.
a) Exocytosis occurs when a membrane-bound vesicle fuses with the plasma membrane and secretes its contents into the cell exterior.
b) Endocytosis occurs when the cell membrane buds inward to bring a substance into a cell.

IV. Metabolism—Chemical Reactions in the Body

A. Enzymes
1. Metabolism describes all the chemical reactions in the body.
2. Enzymes are proteins that regulate metabolism.
3. Enzymes speed up, or catalyze, the rate of reactions.
a) Enzymes help synthesize polymers.
b) Enzymes help liberate energy stored in chemical bonds.
c) Enzymes decrease the heat required to start a reaction.
4. Different enzymes catalyze different reactions, due to specificity.
a) The specificity of an enzyme is the result of its shape.
b) Each enzyme has a unique sequence of amino acids that creates its shape.
5. Enzymes are generally named for the reaction they catalyze.

B. Activation Energy
1. A metabolic reaction cannot start until it has enough energy to surpass the activation energy barrier.
2. Enzymes decrease the activation energy barrier.
a) Enzymes can bind to a substrate and place stress on chemical bonds.
(1) The region where the substrate binds is called the enzyme's active site.
(2) When the substrate binds it changes the enzyme's shape, causing an induced fit.
(3) The change in shape stresses chemical bonds on the substrate, making them easier to break.

C. Cellular Respiration
1. Cellular respiration converts the energy stored in the chemical bonds of food into energy that can be used by the cell.
2. Cellular respiration is also called aerobic respiration, because some steps require oxygen.
3. Carbon dioxide is a waste product of cellular respiration.

D. A General Overview of Cellular Respiration
1. Glucose is broken down during cellular respiration.
2. The energy released during the conversion of glucose to carbon dioxide and water is used to synthesize ATP.
3. Many of the chemical reactions of cellular respiration occur in the mitochondria.

E. Glycolysis, the Citric Acid Cycle, and Electron Transport
  1. Glycolysis
    a) Glycolysis occurs in the cytosol.
    b) It does not require oxygen and produces a small amount of ATP.
    c) During glycolysis,
      (1) The 6-carbon glucose molecule is broken into two 3-carbon pyruvic acid molecules;
      (2) NAD^+ (an electron carrier) picks up two hydrogen atoms and releases one proton to become NADH;
      (3) NADH transports electrons to the electron transport chain; and
      (4) pyruvic acid is decarboxylated and a 2-carbon fragment is transported to a mitochondrion.
  2. Citric Acid Cycle
    a) This cycle is a series of reactions that are catalyzed by eight different enzymes.
    b) The citric acid cycle occurs in the matrix of each mitochondrion.
    c) The citric acid cycle
      (1) breaks down the 2-carbon fragment produced by glycolysis,
      (2) harvests electrons, and
      (3) releases carbon dioxide.
  3. The Electron Transport Chain
    a) The electron transport chain generates ATP.
      (1) Electrons move through a chain of proteins in the inner mitochondrial membrane.
      (2) Electrons are pulled to the bottom of the chain by oxygen.
      (3) As each protein moves an electron, it also moves a hydrogen ion across the inner mitochondrial membrane.
      (4) Hydrogen ions move back across the membrane through a protein channel called ATP synthase.
      (5) ATP synthase uses the energy generated by the moving hydrogen ions to synthesize ATP from ADP and a phosphate.
    b) At the bottom of the electron transport chain, the electrons combine with the oxygen and two hydrogen atoms to produce water.
F. Metabolism of Proteins and Fats
  1. Excess proteins are broken down into amino acids, which are reused.
    a) Excess proteins are broken down when fats or carbohydrates are unavailable.
    b) Energy is produced from an amino acid by
      (1) removing the amino group;
      (2) converting the amino group to urea, which is excreted in urine; and
      (3) feeding the remaining elements in the amino acid into the citric acid cycle.
  2. Excess fats are broken down into glycerol and fatty acids by enzymes.
    a) Fatty acids are broken into smaller units that undergo cellular respiration.
    b) Fats are only broken down after carbohydrate supplies are depleted.
G. Fermentation
  1. Anaerobic respiration can be used by cells to generate energy in the absence of oxygen.
    a) Glycolysis provides small amounts of ATP.
    b) Fermentation regenerates NAD^+.
      (1) Fermentation does not produce ATP.
      (2) Fermentation produces lactic acid, which can build up and cause muscle soreness.
        (a) Lactic acid is produced when NADH donates electrons to the pyruvic acid produced by glycolysis.
        (b) Lactic acid is transported to the liver where it is broken down when oxygen is again available.
H. Calories and Metabolic Rate
  1. There is a difference between calories and Calories.
    a) A calorie is the amount of energy required to raise the temperature of 1 gram of water by 1 degree Celsius.
    b) A Calorie is 1,000 calories of energy.
  2. When Calorie intake is greater than demand, excess Calories can be stored as body fat.
  3. Metabolic rate is a measure of the body's energy use.
    a) Basal metabolic rate is resting energy use in an alert person.
    b) Metabolic rate changes with activity level.
    c) Exercise habits, body weight, age, genetics, diet, and sex can affect metabolic rate.

V. Health and Body Weight
   A. Underweight
      1. Anorexia is self-starvation.
         a) Low body fat can cause amenorrhea, or cessation of menstruation.
         b) Anorexia can reduce bone density and increase the risk for osteoporosis.
         c) Anorexia can weaken heart muscle.
      2. Bulimia is binge eating followed by purging.
         a) Bulimics can have the same health problems as anorexics.
         b) Forced vomiting can rupture the stomach and damage teeth.
   B. Obesity
      1. Obesity is the accumulation of excess body fat.
         a) Obesity can cause high blood pressure, heart attack, stroke, joint problems, and diabetes.
         b) People who store fat in their upper bodies are more at risk for illness than those who store fat in their lower bodies.
         c) The body mass index (BMI) can be used to assess healthy body weight.

Practice Questions

Matching

1. activation energy
2. bulimia
3. passive transport
4. Golgi apparatus
5. glycolysis
6. ribosome
7. metabolism
8. catalyze
9. hypertonic
10. cytoplasm
11. substrate
12. citric acid cycle
13. isotonic
14. facilitated diffusion
15. centriole

a. amount of energy needed to start a chemical reaction
b. all of the chemical reactions in the body
c. organelle attached to rough ER
d. a solution that contains a higher concentration of solutes than an adjoining cell does
e. a solution that contains the same concentration of solutes as an adjoining cell
f. movement of molecules across the cell membrane with the help of proteins
g. organelle that modifies and sorts proteins
h. substance acted on by a specific enzyme
i. the cytosol and organelles of a cell
j. a disease characterized by binging and purging
k. to speed up a reaction
l. cell structure involved in the movement of chromosomes during cell division
m. movement of molecules across the cell membrane that does not require energy
n. process occurring in the cytosol that produces two ATP molecules
o. pathway that releases carbon dioxide

Fill-in-the-Blank

16. The ____________ of a cell is a network of ____________ filaments and tubules that provide support for the cell.
17. ____________ is a type of ____________ transport that involves water.
18. The ____________ breaks down a 2-carbon molecule and produces ____________ that transfers electrons to the electron transport chain.
19. The process of ____________ is used to regenerate NAD^+ when oxygen is not available.
20. Ribosomes are either floating in the ____________ or attached to ____________.

21. The body mass index (BMI) uses ____________ and ____________ to determine a value that estimates the amount of ____________ that a person has.

22. Glycolysis occurs in the ____________ of the cell, and the citric acid cycle occurs in the ____________ of the mitochondria.

23. Resting energy use of a sedentary person is called the ____________.

24. One ____________ of energy equals 1,000 ____________ of energy.

25. During fermentation, electrons from NADH are added to ____________ acid to produce ____________ acid.

26. Body fat that is stored in the ____________ body can cause more health problems than body fat that is stored in the ____________ body.

27. The products of cellular respiration are carbon dioxide and ____________.

28. Very small, ____________ molecules are able to diffuse through the plasma membrane.

Labeling

Use the terms below to label Figure 3.1.

active site
enzyme
stressed bond
substrate

31. ________: sucrose
32. ________
Glucose
Fructose
30. ________
29. ________
sucrase

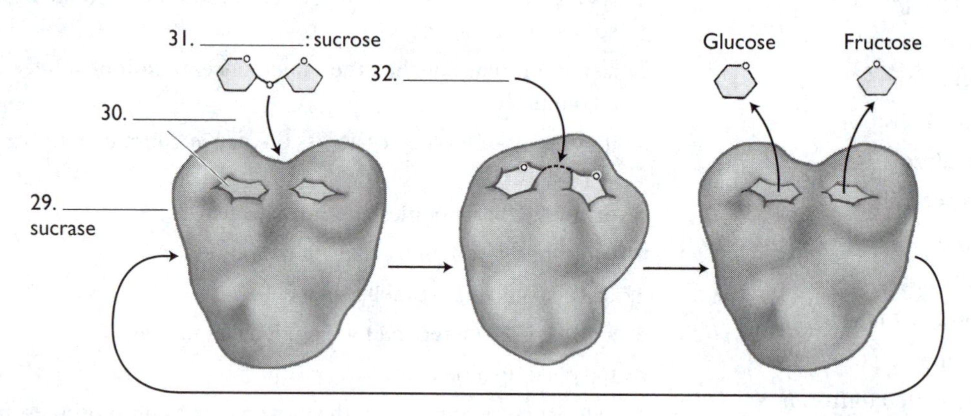

Figure 3.1

Use the terms below to label Figure 3.2.

higher
hypertonic
hypotonic
isotonic
lower
same

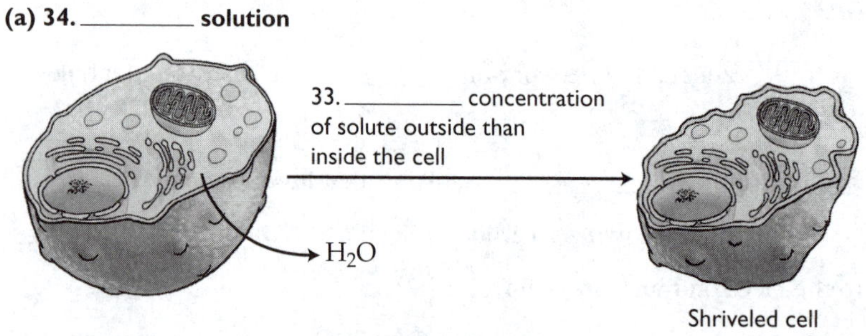

(a) 34. ________ solution
33. ________ concentration of solute outside than inside the cell
H_2O
Shriveled cell

Figure 3.2 ***(Continued)***

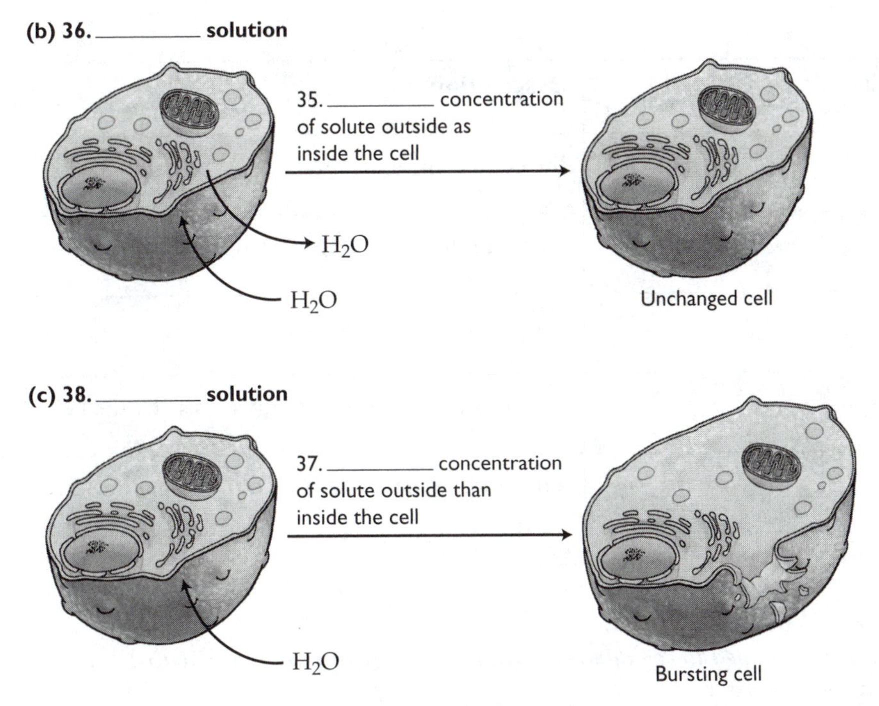

Figure 3.2 ***(Continued)***

Roots to Remember

Use your knowledge of the root words presented in this chapter to answer the following questions.

39. Which process delivers large molecules to the inside of a cell?
40. Which enzyme helps to break down lactose?
41. What is the definition of the word *cytoplasm* using the meaning of its roots?

Word Choice

Circle the word or phrase that correctly completes each sentence.

42. Substances diffuse (down/up) a concentration gradient.
43. Water will move out of a cell when it is immersed in a (hypertonic/hypotonic) solution.
44. Energy is liberated for use by the cell when a phosphate is released from (ADP/ATP).
45. (Anorexia/Obesity) can put a person at a higher risk for osteoporosis.
46. Enzymes are (lipids/proteins).
47. (Lysosomes/Ribosomes) in the cell engulf and release energy from nutrients.
48. Cells move large molecules to their exterior through the process of (exocytosis/endocytosis).
49. Ribosomes are produced in the (chromatin/nucleolus) within the nucleus of a cell.

Table Completion

Name each organelle described in the table.

Organelle	Description
50.	modifies and sorts proteins from the ER
51.	a membrane-enclosed sac of enzymes

continued

continued

Organelle	Description
52.	the cell's control center
53.	where proteins are constructed from amino acids
54.	detoxifies harmful substances
55.	reactions here produce energy for use by the cell
56.	a network of membranes with ribosomes attached

Sequencing

Place these events related to cellular respiration in the correct sequence, with 1 being the first event and 8 being the last event.

57. ____________ The citric acid cycle breaks down a 2-carbon carbohydrate fragment.

58. ____________ Hydrogen flows through ATP synthase protein.

59. ____________ Pyruvic acid is decarboxylated into a 2-carbon carbohydrate fragment.

60. ____________ Electrons move through the electron transport chain.

61. ____________ ATP synthase protein converts ADP and P to ATP.

62. ____________ Glucose is broken down into pyruvic acid.

63. ____________ NADH carries electrons from the citric acid cycle to the electron transport chain.

64. ____________ Proteins pump hydrogen ions into the intermembrane space of the mitochondrion.

Critical Thinking

65. Two abnormal daughter cells that have no cytoskeleton are produced through a mutation in the parent cell's DNA. These cells are part of the connective tissue in the integumentary system. What is likely to happen to these cells and why?
66. Specialized organelles in plant cells can produce glucose for food, through the process of photosynthesis. Why do plant cells also need to perform cellular respiration?
67. An athlete builds muscle and bone mass through exercise. How might he or she be misclassified by the BMI chart?

Practice Test

1. All of the following are found in the plasma membrane *except*
   A. cholesterol.
   B. phospholipids.
   C. cytosol.
   D. proteins.

2. Which nitrogenous base is found in ATP?
   A. adenine
   B. cytosine
   C. guanine
   D. thymine

3. Which organelle is most like a garbage truck?
   A. nucleus
   B. mitochondrion
   C. lysosome
   D. rough ER

4. During a lab experiment, you place a piece of Elodea plant into a test tube that contains a hypotonic solution. What will happen to the cells in the Elodea plant?
   A. The cells will shrivel.
   B. The cells will not change shape.
   C. The cells will burst.
   D. You cannot predict what will happen to the cells.

5. What type of transport moves potassium ions up a concentration gradient across the cell membrane?
   A. diffusion
   B. active transport
   C. osmosis
   D. facilitated diffusion

6. Which equation below represents the breakdown of glucose through cellular respiration?
   A. $6CO_2 + 6H_2O \rightarrow C_6H_{12}O_6 + 6O_2$
   B. $C_6H_{12}O_6 + 6H_2O \rightarrow 6CO_2 + 6O_2$
   C. $C_6H_{12}O_6 + 6CO_2 \rightarrow 6O_2 + 6H_2O$
   D. $C_6H_{12}O_6 + 6O_2 \rightarrow 6CO_2 + 6H_2O$

7. In human males, fat is most often stored in the adipose tissue in the
   A. buttocks.
   B. abdomen.
   C. hips.
   D. chest.

8. What must be available to convert lactic acid to pyruvic acid?
   A. electrons
   B. oxygen
   C. hydrogen
   D. NADH

9. Where is ATP synthase located in the mitochondrion?
   A. in the inner membrane
   B. in the nucleus
   C. in the matrix
   D. in the intermembrane space

10. When all of the macromolecules listed below are available, which does the body generally break down for energy first?
    A. carbohydrates
    B. proteins
    C. fats
    D. nucleic acids

11. Transport vesicles carrying proteins bud off from the
    A. smooth ER and Golgi apparatus.
    B. lysosomes and ribosomes.
    C. centrioles and ribosomes.
    D. Golgi apparatus and rough ER.

12. The size of a body cell is between
    A. 10^{-2} m and 10^{-1} m.
    B. 10^{-3} m and 10^{-2} m.

C. 10^{-4} m and 10^{-3} m.
D. 10^{-5} m and 10^{-4} m.

13. The addition of a phosphate to another molecule is called
   A. glycolysis.
   B. catalysis.
   C. phosphorylation.
   D. endocytosis.

14. What would happen if a lactose substrate encountered a sucrase enzyme?
   A. The sucrase enzyme would split the lactose substrate into its component sugars.
   B. The substrate would not fit into the active sites on the sucrase enzyme.
   C. The substrate would bind with the enzyme and inhibit the enzyme's function.
   D. The lactose substrate would bend and break the sucrase enzyme.

15. When NAD^+ is converted to NADH, it releases
   A. one proton.
   B. one electron.
   C. two protons.
   D. two electrons.

Answer Key

Matching

1. a; **2.** j; **3.** m; **4.** g; **5.** n; **6.** c; **7.** b; **8.** k; **9.** d; **10.** i; **11.** h; **12.** o; **13.** e; **14.** f; **15.** l

Fill-in-the-Blank

16. cytoskeleton; protein; **17.** Osmosis; passive; **18.** citric acid cycle; NADH; **19.** fermentation; **20.** cytosol; rough ER; **21.** height; weight; body fat; **22.** cytosol; matrix; **23.** basal metabolic rate; **24.** Calorie; calories; **25.** pyruvic; lactic; **26.** upper; lower; **27.** water; **28.** hydrophobic

Labeling

29. enzyme; **30.** active site; **31.** substrate; **32.** stressed bond; **33.** higher; **34.** hypertonic; **35.** same; **36.** isotonic; **37.** lower; **38.** hypotonic

Roots to Remember

39. endocytosis; **40.** lactase; **41.** cell fluid

Word Choice

42. down; **43.** hypertonic; **44.** ATP; **45.** Anorexia; **46.** proteins; **47.** lysosomes; **48.** exocytosis; **49.** nucleolus

Table Completion

50. Golgi apparatus; **51.** lysosome; **52.** nucleus; **53.** ribosome; **54.** smooth ER; **55.** mitochondria; **56.** rough ER

Sequencing

57. 3; **58.** 7; **59.** 2; **60.** 5; **61.** 8; **62.** 1; **63.** 4; **64.** 6

Critical Thinking

65. The cells are likely to be compressed by other cells, because they have no cytoskeleton to help them to maintain their shape.
66. They need to perform cellular respiration to release energy from the glucose to fuel cell processes.
67. Muscle and bone would add to the athlete's weight, so the athlete may be misclassified as overweight even though he or she is really at a healthy weight.

Practice Test

1. C; **2.** A; **3.** C; **4.** C; **5.** B; **6.** D; **7.** B; **8.** B; **9.** A; **10.** A; **11.** D; **12.** D; **13.** C; **14.** B; **15.** A

CHAPTER 4

GENES—TRANSCRIPTION, TRANSLATION, MUTATION, AND CLONING: GENETICALLY MODIFIED FOODS

Learning Goals

1. Describe the structure and function of a gene.
2. Describe the synthesis of a protein from transcription through translation.
3. Describe which organelles and structures are involved in transcription and translation.
4. Compare and contrast mRNA, tRNA, and rRNA.
5. Describe the structure and function of transfer RNA (tRNA).
6. Describe how mutations affect gene expression.
7. Explain how two different genes would differ from each other.
8. Describe what makes any two proteins different from each other.
9. Explain how gene expression is regulated.
10. Describe the process of cloning a gene using bacterial cells.

Chapter Outline

I. What Is a Gene?
    A. Genes are segments of DNA.
    B. Genes provide the instructions for making polypeptides.
        1. Polypeptides form functional proteins.
        2. Proteins give cells their genetically determined characteristics.
    C. Genes are located on chromosomes.
        1. Chromosomes carry thousands of genes.
        2. Chromosomes are composed of DNA wrapped around proteins.
        3. An organism's full set of chromosomes is called its genome.

II. Protein Synthesis and Gene Expression
    A. From Gene to Protein
        1. Protein synthesis involves using the instructions from a gene to build a polypeptide.
        2. Protein synthesis involves DNA, proteins, and RNA.
            a) DNA is a double-stranded polymer of nucleotides.
                (1) The sugar in DNA is deoxyribose.
                (2) Nitrogenous base pairs include A-T and C-G.
            b) Proteins are large molecules composed of amino acids.
            c) RNA is a single-stranded polymer of nucleotides.
                (1) The sugar in RNA is ribose.
                (2) Nitrogenous base pairs include A-U and C-G.
        3. Transcription involves the production of an RNA copy of the required gene.
        4. Translation involves decoding the RNA sequence into a string of amino acids that forms a protein.
    B. Transcription—Copying the Gene
        1. Transcription occurs in the nucleus.
        2. The RNA polymerase enzyme binds to a promoter, which is a nucleotide sequence at the beginning of each gene.

3. Once it is attached, RNA polymerase unzips the double helix.
4. Hydrogen bonding occurs between the DNA gene and RNA nucleotides.
5. RNA polymerase performs dehydration synthesis to create covalent bonds between the RNA nucleotides.
6. A single-stranded messenger RNA (mRNA) is formed.
   a) The mRNA carries the message from the gene.
   b) The mRNA consists of introns that do not code for proteins and exons that carry instructions for building proteins.
   c) Introns are spliced out before mRNA leaves the nucleus.

C. Translation—Using the Message to Synthesize a Protein
1. Overview
   a) Translation occurs in the cytoplasm.
   b) Translation involves mRNA, amino acids, ribosomes, and tRNA.
2. Ribosomes
   a) Ribosomes are composed of ribosomal RNA (rRNA) wrapped around many different proteins.
   b) Each ribosome is composed of a small and large unit.
   c) The mRNA threads between the ribosome units.
   d) The ribosome is able to bind to transfer RNA (tRNA) molecules that carry amino acids.
3. Transfer RNA
   a) Each tRNA molecule carries one specific amino acid.
   b) Each tRNA molecule interacts with mRNA to place amino acids in the correct location on a growing polypeptide.
      (1) Sequences of three nucleotides on mRNA, called codons, encode a particular amino acid.
      (2) Sequences of three nucleotides on tRNA, called anticodons, bind to complementary mRNA codons.
      (3) A covalent bond forms when a tRNA anticodon binds to an mRNA codon.
      (4) The ribosome adds the amino acid carried by the tRNA to a growing chain of amino acids.
      (5) The chain of amino acids eventually constitutes a finished protein.
   c) Protein synthesis ends when a stop codon that does not code for an amino acid moves through the ribosome.
   d) After the stop codon appears, the protein is released and folds up on itself.
4. Genetic Code
   a) Shows which codons code for which amino acids.
   b) Has 61 codons that code for amino acids and 3 stop codons.
   c) Each specific amino acid can be coded for by more than one codon.
   d) There is no case where a given codon can code for more than one specific amino acid.
   e) The genetic code is universal, which means that the same codons specify the same amino acids in all organisms.

D. Mutations
1. Can affect the order of amino acids that encode a protein.
2. Can result in a nonfunctional protein or a protein that is different from the one coded by a normal gene.
3. A point mutation is a change to or deletion of a single nucleotide.
   a) A missense mutation is a point mutation that changes one codon to another one that codes for a different amino acid, resulting in a change to the protein that is produced.
   b) A nonsense mutation is a point mutation that changes a codon that codes for an amino acid to a stop codon, resulting in production of an abbreviated protein.
   c) A neutral mutation is a point mutation that changes one codon to another one that codes for the same amino acid, resulting in no change to the protein produced.
   d) A frameshift mutation is a mutation resulting from the insertion or deletion of a nucleotide that causes nucleotides to be regrouped into different codons.

E. Regulating Gene Expression
1. Overview
   a) Different cell types transcribe and translate different genes.
   b) Gene expression is regulated in response to the cell's needs.
2. Regulation of Transcription
   a) This is the most common form of gene expression regulation.
   b) Proteins called activators help RNA polymerase to bind to a promoter on a gene, facilitating gene expression.

c) The presence of a substance can sometimes increase transcription of a gene that is involved in the breakdown of that substance (e.g., alcohol).

3. Regulation of Chromosome Condensation
   a) Gene expression can be regulated by regulating condensing and decondensing of chromosomes.
   b) Lampbrush chromosomes in a developing egg facilitate rapid synthesis of mRNA so that large amounts of protein can be produced to supply egg development.
4. Regulation of mRNA Degradation
   a) Enzymes called nucleases float in the cytoplasm and break down RNA molecules.
   b) mRNA molecules with long tails survive longer and translate more times.
5. Regulation of Translation
   a) Binding of mRNA to the ribosome can be slowed down or speeded up.
   b) The speed of mRNA movement through the ribosome can also be controlled.
   c) A polyribosome can produce several proteins from a single mRNA.
6. Regulation of Protein Degradation
   a) Enzymes in the cell can degrade proteins.
   b) The activity of these enzymes can be regulated.

III. Producing Recombinant Proteins
  A. Recombinant Bovine Growth Hormone (*r*BGH)
    1. An organism undergoes recombination when it receives a new gene or when its arrangement of genes is changed.
    2. The recombinant bovine growth hormone (*rBGH)* gene is produced when the *BGH* gene from a cow is placed in a bacterial cell.
    3. The *rBGH* gene produces large amounts of the recombinant bovine growth hormone (*r*BGH).
    4. This growth hormone can be injected into cows to stimulate milk production.
  B. Cloning a Gene Using Bacteria
    1. Step 1: Remove the gene from the cow chromosome.
       a) Cow DNA is exposed to restriction enzymes, which cut gene sequences.
          (1) Restriction enzymes cut only palindromes.
          (2) Restriction enzymes leave unpaired bases called sticky ends.
          (3) Unpaired bases can form bonds with complementary bases with which they come into contact.
       b) One DNA fragment contains the *BGH* gene.
    2. Step 2: Insert the *BGH* gene into the bacterial plasmid.
       a) The *BGH* gene is inserted into a bacterial plasmid using restriction enzymes.
       b) The sticky ends from the gene and plasmid pair with each other.
       c) The *BGH* gene is now a recombinant gene, or *rBGH*.
    3. Step 3: Insert the recombinant plasmid into a bacterial cell.
       a) Bacteria are treated so that their cell membranes become porous.
       b) When bacteria are placed in a suspension of plasmids, the plasmids move into the cytoplasm of the cells.
       c) Plasmids are replicated inside the bacterial cell.
       d) The bacteria replicate, producing thousands of copies of the *rBGH* gene.
       e) Each bacterium is considered transgenic, or a genetically modified organism (GMO), because it contains a gene from another organism.
  C. FDA Regulations
    1. The FDA is charged with ensuring the safety of foods.
    2. Any food or food additive that is not generally recognized as safe (GRAS) must obtain FDA approval.
    3. In 1993, the FDA approved milk from *r*BGH-treated cows as safe for human consumption.
       a) The FDA found milk from treated and untreated cows to be indistinguishable.
       b) The FDA does not require labeling of milk from *r*BGH-treated cows.
    4. The use of *r*BGH has been banned in Europe and Canada, due in part to animal-welfare concerns.

IV. Genetically Modified (GM) Crops
  A. Potential Benefits of GM Crops
    1. Can be engineered to ripen more slowly and have a longer shelf life.
    2. Can be engineered to produce a higher yield.

a) Crops may be engineered to have genes that produce resistance to pesticides and herbicides.
b) Crops can be engineered to have genes that produce resistance to pests.
3. Can be engineered to have a higher nutritional value.

B. Prevalence of GM Crops
1. GM crops are commonly used in U.S. products.
a) Over 80% of the soybean crop is modified for herbicide resistance.
b) Close to 40% of the U.S. corn crop is modified to produce its own pesticide against certain caterpillars.
c) Over 70% of the U.S. cotton crop is modified to produce its own pesticide against caterpillars.
2. Products using non-GM crops are often labeled.

C. How Are GM Foods Evaluated for Safety?
1. The FDA considers a GM food to be equivalent to the food from which it was derived.
2. A GM food is tested only if there is a concern about allergens or toxicity.
a) Scientists test the protein produced by the modified organism.
b) The food is not approved if the protein is toxic or causes an allergic reaction.

Practice Questions

Matching

1. genome
2. polyribosome
3. recombination
4. anticodon
5. stop codon
6. chromosome
7. activators
8. restriction enzyme
9. codon
10. cloning
11. plasmid
12. mRNA
13. gene
14. tRNA
15. point mutation

a. a cluster that can produce several proteins from one mRNA
b. a piece of circular DNA
c. a molecule that carries instructions from a gene for building a protein
d. a package that carries thousands of genes
e. a sequence of three nucleotides at the base of a tRNA molecule
f. a segment of DNA that carries information about a trait
g. an organism's full set of chromosomes
h. a sequence of three nucleotides that code for a specific amino acid
i. a sequence of three nucleotides that do not code for an amino acid
j. making many exact copies of a gene
k. a molecule that cuts DNA at specific sequences
l. a molecule that carries amino acids
m. the change or deletion of a single nucleotide
n. modification of the arrangement of genes in an organism
o. proteins that help RNA polymerase bind to a promoter

Fill-in-the-Blank

16. ____________ is the enzyme that facilitates transcription.

17. ____________ are composed of DNA and are packaged in ____________.

18. A ____________ organism, or genetically ____________ organism is produced when genes from one organism are transferred into another.

19. A ____________ mutation has no effect on protein synthesis.

20. The anticodon for AGC is ____________.

21. A growing protein is released when a ____________ moves through the ____________.

22. When a point ____________ occurs and a nucleotide is inserted, nucleotides that follow are regrouped into different ____________. This is called a ____________ mutation.

23. An ____________ molecule copies the nucleotide sequence from a gene.

24. In the genetic code, a sequence of ____________ nucleotides in mRNA (called a codon) codes for each amino acid.

25. ____________ RNA carries one amino acid and interacts with ____________ RNA.

26. A chromosome must ____________ before RNA ____________ can get access to a gene.

27. A restriction enzyme ____________ DNA and leaves unpaired bases called ____________.

28. The recombinant bovine growth hormone (*r*BGH) is produced by genetically engineered ____________.

Labeling

Use the terms below to label Figure 4.1.

mRNA
promoter
RNA nucleotides
RNA polymerase

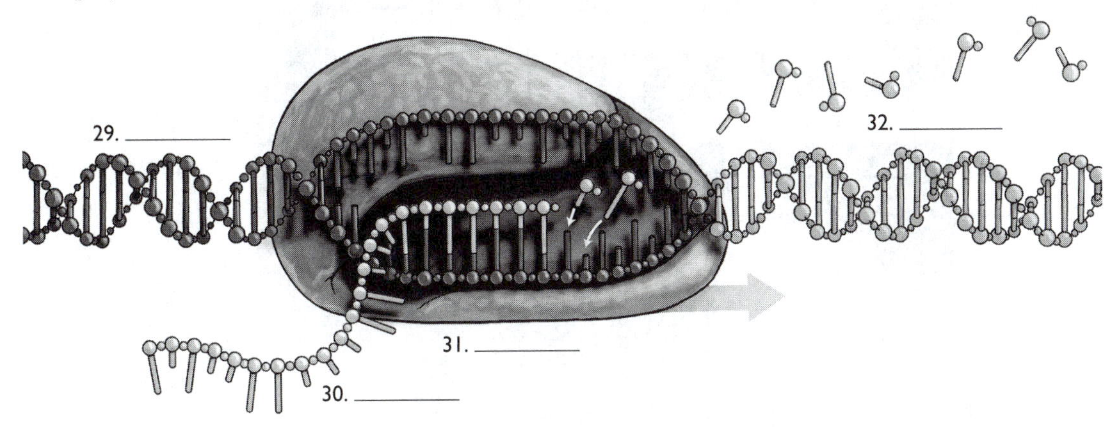

Figure 4.1

Use the terms below to label Figure 4.2.

amino acid
anticodon
codon
mRNA
tRNA

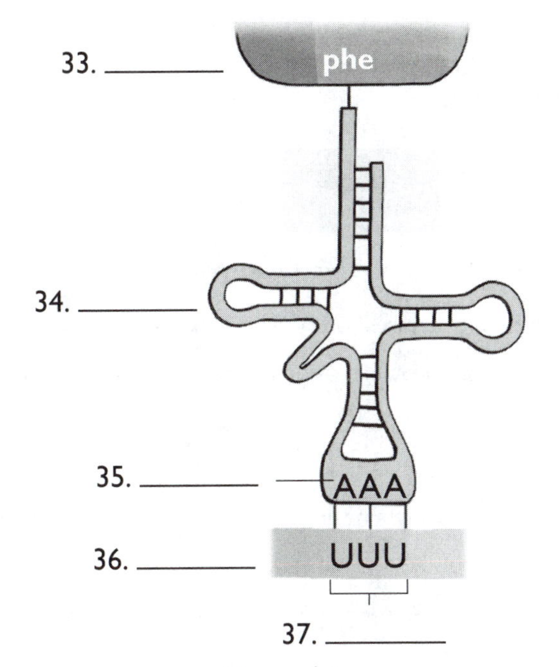

Figure 4.2

Roots to Remember

Use your knowledge of the root words presented in this chapter to answer the following questions.

38. Which root in the word *ribose* means "sugar"?
39. Glutamic and aspartic are two adjectives that could describe what?
40. Does the literal translation of *chromosome* provide a good description of a chromosome? Why or why not?
41. What does *recombinant* mean?

Crossword Puzzle

Across

42. "generally recognized as safe" by the FDA
43. an organism that is not transgenic
44. component of a nucleotide
45. circular DNA found in a bacterium
46. produced from AGU or AGC
47. one example is RNA polymerase

Down

48. RNA complement to TTACTGGAC
49. unpaired bases left by restriction enzymes
50. GM rice with added nutrients

Word Choice

Circle the word or phrase that correctly completes each sentence.

51. Enzymes called (nucleases/polymerases) break down RNA molecules in the cytoplasm of cells.
52. A point mutation that results in the substitution of one amino acid for another is called a (missense/nonsense) mutation.
53. An anticodon is found on (mRNA/tRNA).
54. (Transcription/Translation) occurs when the code for a gene is copied from DNA to RNA.
55. Genes are composed of (nucleotides/polypeptides).
56. A(n) (promoter/activator) sequence is located at the beginning of each gene.

Sequencing

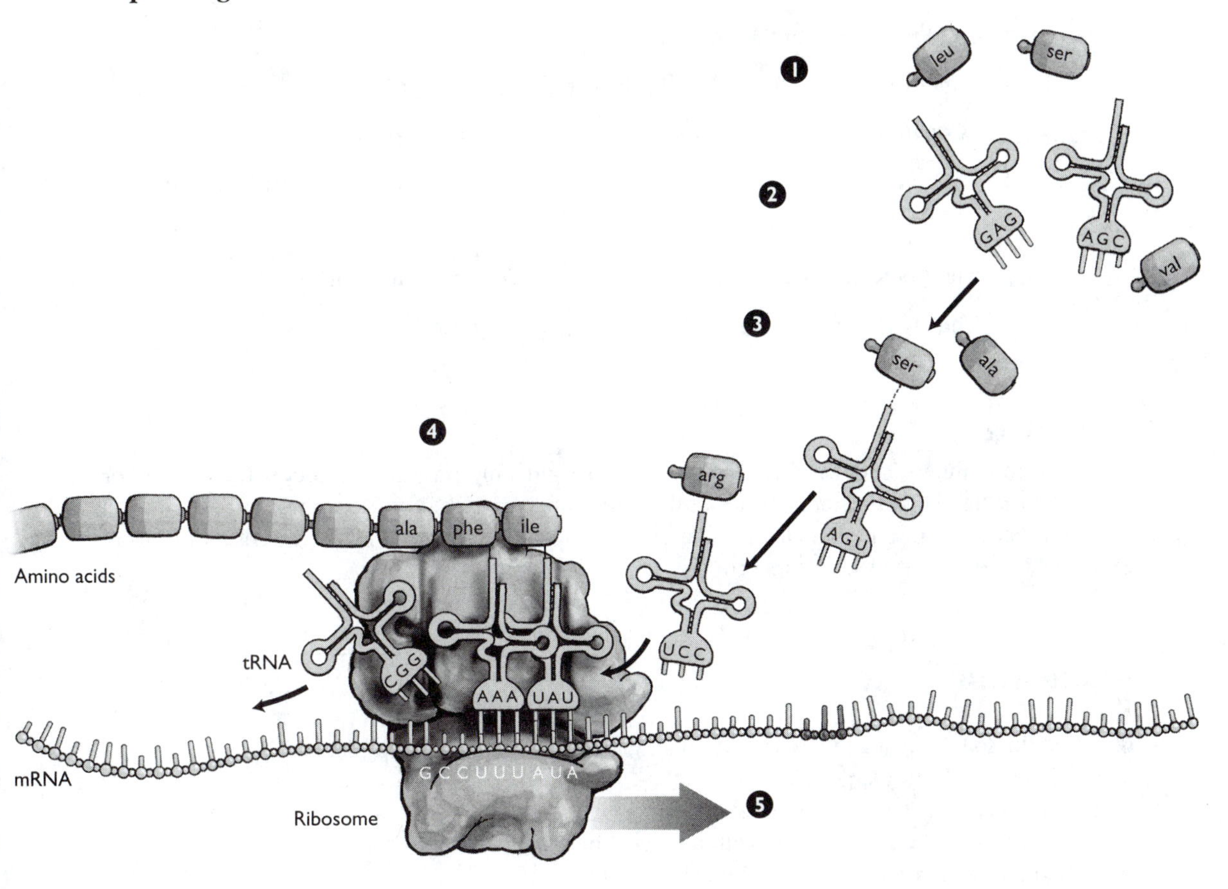

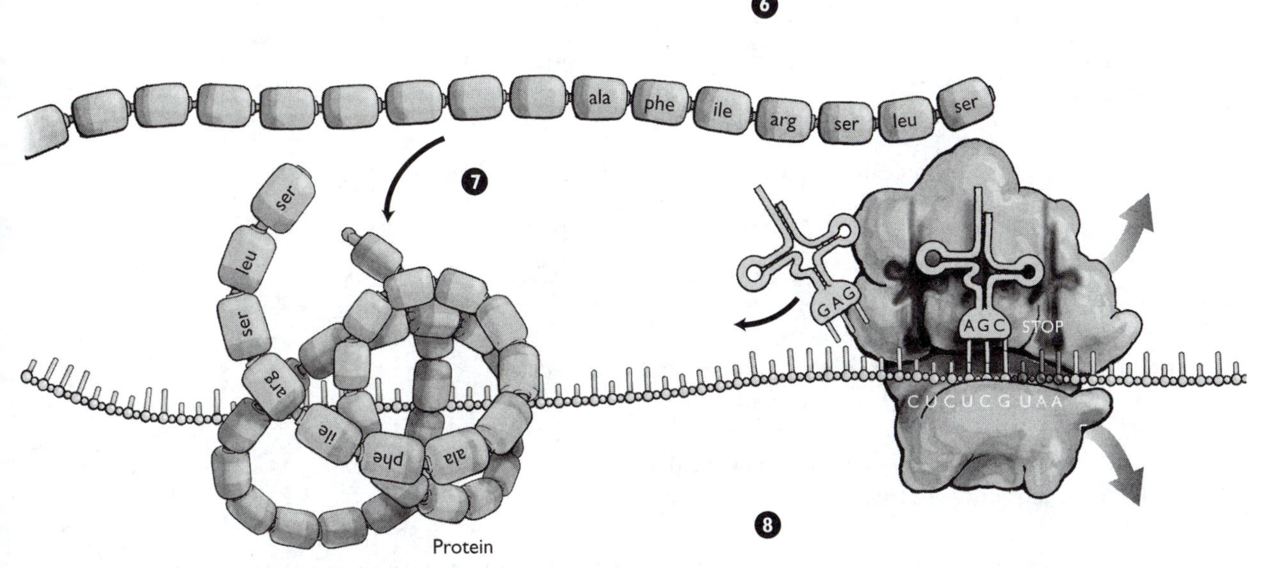

The Process of Translation

Use the diagram above to place these events related to translation in the correct sequence, with 1 being the first event and 8 being the last event.

57. ____________ Amino acids link together to form a polypeptide.

58. ____________ Enzymes facilitate binding of a tRNA to its appropriate amino acid.

59. ____________ When the ribosome reaches the stop codon, no tRNA can pair with the codon on the mRNA. RNA and the new protein are released.

60. ____________ Amino acids and tRNAs float in the cytoplasm.

61. ____________ A tRNA docks when its complementary RNA codon is present.

62. ____________ Subunits of the ribosome separate but can reassemble to translate another mRNA.

63. ____________ The ribosome moves to the next codon to receive the next tRNA.

64. ____________ The chain of amino acids folds into its globular form.

Critical Thinking

65. There are four mRNA codons that code for leucine and only one mRNA codon that codes for tryptophan. What might this indicate about the importance of each of these amino acids?
66. Gene flow is the transfer of alleles from one population to another. Why might gene flow from genetically modified crops be a concern?

Practice Test

1. Which of the following is *not* a reason to plant genetically modified crop plants?
   A. GM crops can have a larger yield.
   B. GM crops can have a higher nutritional value.
   C. GM crops can produce products with a longer shelf life.
   D. GM crops can increase hazards to beneficial insects.

2. What is the mRNA complement to AUU?
   A. TAA
   B. UAA
   C. UTT
   D. CGG

3. Which type of mutation can produce an abbreviated version of a protein?
   A. neutral mutation
   B. nonsense mutation
   C. frameshift mutation
   D. missense mutation

4. Which of the following is true about codons and amino acids?
   A. A codon can only encode for one specific amino acid.
   B. Each amino acid is encoded for by only one specific codon.
   C. Every codon can encode for an amino acid.
   D. All of the above are true.

5. A drug company wants to use recombinant gene technology to produce large quantities of *r*BGH. Which of the following is not a step in this process?
   A. Insert the *BGH* gene into a cow chromosome.
   B. Use restriction enzymes to remove the *BGH* gene from cow DNA.

C. Insert a recombinant plasmid into a bacterial cell.
D. Use restriction enzymes to insert the *BGH* gene into a plasmid.

6. An activator is a protein that
   A. turns a gene on in the presence of certain chemicals.
   B. decondenses chromosomes so their genes can be transcribed.
   C. helps RNA polymerase to bind to the promoter on a gene.
   D. slows down the activities of enzymes that can break down RNA.

7. Which molecule could be considered a blueprint that contains all of the instructions for building cell proteins?
   A. mRNA
   B. rRNA
   C. tRNA
   D. DNA

8. Sequences of DNA that do not code for proteins are called
   A. introns.
   B. exons.
   C. codons.
   D. anticodons.

9. Which of the following represents the largest unit of DNA?
   A. nucleotide
   B. chromosome
   C. genome
   D. gene

10. Which of the following is not directly involved in the process of translation?
    A. ribosomes
    B. DNA
    C. tRNA
    D. mRNA

11. A protein is
    A. an amino acid chain.
    B. a sequence of nucleotides.
    C. a nucleic acid chain.
    D. a sequence of codons.

12. A tRNA anticodon binds to its complementary mRNA codon through
    A. dehydration synthesis.
    B. hydrogen bonding.
    C. polypeptide synthesis.
    D. polymerase promotion.

13. Which of the following statements is true about ribosomes?
    A. Ribosomes are composed of one large unit and one small unit.
    B. Ribosomes are composed of rRNA and protein.
    C. Ribosomes can bind tRNA.
    D. All statements above are true.

Use the table below to answer questions 14 and 15.

First base	Second base: U	Second base: C	Second base: A	Second base: G	Third base
U	UUU, UUC – Phenyl-alanine (phe) UUA, UUG – Leucine (leu)	UCU, UCC, UCA, UCG – Serine (ser)	UAU, UAC – Tyrosine (tyr) UAA **Stop codon** UAG **Stop codon**	UGU, UGC – Cysteine (cys) UGA **Stop codon** UGG Tryptophan (trp)	U C A G
C	CUU, CUC, CUA, CUG – Leucine (leu)	CCU, CCC, CCA, CCG – Proline (pro)	CAU, CAC – Histidine (his) CAA, CAG – Glutamine (gln)	CGU, CGC, CGA, CGG – Arginine (arg)	U C A G
A	AUU, AUC, AUA – Isoleucine (ile) AUG Methionine (met) **Start codon**	ACU, ACC, ACA, ACG – Threonine (thr)	AAU, AAC – Asparagine (asn) AAA, AAG – Lysine (lys)	AGU, AGC – Serine (ser) AGA, AGG – Arginine (arg)	U C A G
G	GUU, GUC, GUA, GUG – Valine (val)	GCU, GCC, GCA, GCG – Alanine (ala)	GAU, GAC – Aspartic acid (asp) GAA, GAG – Glutamic acid (glu)	GGU, GGC, GGA, GGG – Glycine (gly)	U C A G

14. A gene undergoes the following mutation:
    Old gene: TACATTGCGCGAATTTCG
    Mutation: TACATTGCCCGAATTTCG
    What type of mutation has occurred?
    A. frameshift mutation
    B. missense mutation
    C. neutral mutation
    D. nonsense mutation

15. Which tRNA molecule would *not* add the amino acid arginine (arg) to a polypeptide?
    A. a tRNA molecule with the anticodon GCU
    B. a tRNA molecule with the anticodon UCC
    C. a tRNA molecule with the anticodon GCA
    D. a tRNA molecule with the anticodon CGU

Answer Key

Matching

1. g; **2.** a; **3.** n; **4.** e; **5.** i; **6.** d; **7.** o; **8.** k; **9.** h; **10.** j; **11.** b; **12.** c; **13.** f; **14.** l; **15.** m

Fill-in-the-Blank

16. RNA polymerase; **17.** Genes; chromosomes; **18.** transgenic; modified; **19.** neutral; **20.** UCG; **21.** stop codon; ribosome; **22.** mutation; codons; frameshift; **23.** mRNA; **24.** three; **25.** Transfer; messenger; **26.** decondense; polymerase; **27.** cuts; sticky ends; **28.** bacteria

Labeling

29. promoter; **30.** mRNA; **31.** RNA polymerase; **32.** RNA nucleotides; **33.** amino acid; **34.** tRNA; **35.** anticodon; **36.** mRNA; **37.** codon

Roots to Remember

38. *-ose*; **39.** acids; **40.** Not really, because a chromosome's literal translation is "colored body." A chromosome could be considered a small body. However, it does not have a distinct color; **41.** to combine again, or to combine in a different way

Crossword Puzzle

Across

42. GRAS; **43.** unmodified; **44.** phosphate group; **45.** plasmid; **46.** serine; **47.** enzyme

Down

48. AAUGACCUG; **49.** sticky; **50.** golden

Word Choice

51. nucleases; **52.** missense; **53.** tRNA; **54.** Transcription; **55.** nucleotides; **56.** promoter

Sequencing

57. 4; **58.** 2; **59.** 6; **60.** 1; **61.** 3; **62.** 8; **63.** 5; **64.** 7

Critical Thinking

65. Leucine may be part of more proteins produced by the human body than tryptophan is. Leucine may have also developed earlier in human evolution than tryptophan, so more codons developed to produce it.

66. Genetically modified alleles could be introduced into unmodified populations to introduce unwanted traits. For example, alleles from a population of cotton that is genetically modified to produce infertile seeds could spread to an unmodified population and reduce the fertility of the cotton seeds produced.

Practice Test

1. D; **2.** B; **3.** B; **4.** A; **5.** A; **6.** C; **7.** D; **8.** A; **9.** C; **10.** B; **11.** A; **12.** A; **13.** D; **14.** C; **15.** D

CHAPTER 5

TISSUES, ORGANS, AND ORGAN SYSTEMS: WORK OUT!

Learning Goals

1. Compare and contrast the structure and function of the epithelia.
2. Compare and contrast exocrine and endocrine glands.
3. Identify the structure and function of tight junctions, gap junctions, and adhesion junctions.
4. List the major differences between the different types of connective tissues.
5. Describe how the three separate types of muscle tissue differ from each other.
6. Describe how neurotransmitters help cells to communicate.
7. List the major organs found in ventral and dorsal body cavities.
8. List the five types of body membranes covered in this chapter.
9. Define the term *organ*.
10. Give an example of how two different organ systems work together.

Chapter Outline

I. Tissues and Their Functions
  A. Overview
    1. A tissue is composed of similar cell types that perform a common function.
    2. The main types of tissue in the human body are epithelial, connective, muscle, and nervous.
  B. Epithelial Tissue
    1. Epithelial tissue is composed of tightly packed sheets of cells.
      a) Some epithelial tissue covers outer surfaces.
      b) Some epithelial tissue lines organs, vessels, and body cavities.
    2. Epithelial tissue is anchored on one surface.
    3. The free or apical surface may be exposed to air or fluids.
    4. Epithelial tissue functions in protection, absorption, and secretion.
      a) Epithelial tissue in the skin protects the body from injury, UV light, water loss, and diseases.
      b) Epithelial tissues that line blood vessels and intestines absorb water and nutrients.
      c) Glands are specialized epithelial tissues that secrete substances such as mucus, oil, and sweat.
        (1) Endocrine glands secrete hormones into the bloodstream.
        (2) Exocrine glands secrete products into organs or ducts.
    5. Epithelial tissues are classified by shape and number of cell layers.
      a) Classification by Number of Cell Layers
        (1) Simple tissue types are composed of one cell layer.
        (2) Stratified tissue types are composed of more than one cell layer.
      b) Classification by Shape
        (1) Squamous epithelium
          (a) This type consists of flattened cells.
          (b) It provides protection and nutrient exchange.
          (c) Capillaries are simple squamous epithelium.
          (d) Stratified squamous epithelium is found in skin.
        (2) Cuboidal epithelium
          (a) This type is composed of tightly packed cells that are neither flat nor tall.
          (b) It protects other cells and tissues and secretes and absorbs water and small molecules.
          (c) Most glands are made of cuboidal epithelium.

(3) Columnar epithelium
(a) This type is composed of tall, column-shaped cells.
(b) It secretes and absorbs substances.
(c) Columnar epithelium lines parts of the digestive and reproductive systems and most of the respiratory system.
6. A layer of noncellular material called basement membrane is found below epithelial tissue.
a) Basement membrane provides structural support.
b) Polysaccharides and proteins that make up the basement membrane are secreted by epithelial cells.
c) Basement membrane provides chemicals and growth factors that heal damage to the epithelium.
7. Epithelial tissue constantly sheds dead cells.
8. Epithelial tissue is highly regenerative.
C. Cell Junctions
1. Cells are joined together by different types of intercellular junctions.
a) Tight junctions form tight, impermeable barriers between adjacent plasma membranes.
(1) Tight junctions limit the flow of chemicals between tissues.
(2) Tight junctions are found in the digestive system.
b) Adhesion junctions or desmosomes are composed of filaments that join cells together.
(1) Filaments in adhesion junctions allow some movement.
(2) Adhesion junctions are found in the skin epithelium.
c) Gap junctions are channels made of protein that join adjacent plasma membranes.
(1) Gap junctions permit the flow of substances between cells.
(2) Gap junctions in the heart allow substances that cause contraction to flow to cells.
D. Connective Tissue
1. Overview
a) Connective tissue binds organs and tissues to each other.
b) Cells are not tightly packed.
c) Spaces between cells are filled with a matrix of noncellular ground substance and protein fibers.
(1) Connective tissue such as blood has a liquid matrix.
(2) Connective tissue such as adipose tissue has a viscous matrix.
(3) Connective tissue such as bone has a solid matrix.
2. Loose Connective Tissue
a) Also called areolar tissue, this is the most widespread connective tissue.
(1) It connects epithelia to underlying tissues.
(2) It holds organs in place.
(3) It acts as padding under the skin.
b) Loose connective tissue is a loose weave of fibers.
c) Cells called fibroblasts synthesize and secrete proteins into the matrix.
d) Collagen and elastin proteins are the major components of the matrix.
(1) Collagen gives tissue tensile strength.
(2) Elastin gives tissue elasticity.
e) Aging, exposure to sunlight, and smoking are factors that can cause degradation of loose connective tissue.
(1) Elastin breaks down when exposed to these factors, causing wrinkles in skin.
(2) Aging causes fibroblasts to secrete less collagen.
(3) Exercise improves circulation and may lessen the effects of aging.
3. Adipose Tissue
a) Adipose, or fat, tissue is found under the skin and around the kidneys and heart.
b) It insulates and protects organs.
c) Adipose cells, or adipocytes, are the main component of this tissue.
(1) Adipocytes synthesize and store lipids.
(2) A fat droplet shrinks or swells in these cells.
(3) Adipocytes also produce adipokines that can trigger diseases, such as heart disease, stroke, and type 2 diabetes.
4. Blood
a) Blood is a connective tissue with a liquid matrix.
(1) Blood cells are suspended in plasma.
(a) Red blood cells carry oxygen.

(b) White blood cells fight infection.
(c) Platelets function in clotting.
b) Blood transports oxygen and nutrients to cells and removes wastes.
c) Exercise can increase blood volume and the number of red blood cells.
5. Dense Fibrous Connective Tissue
a) Dense irregular fibrous tissue contains randomly oriented collagen fibers (e.g., tissue in the layer of skin called the dermis).
b) Dense regular fibrous tissue contains collagen fibers that are parallel.
(1) This tissue is found in tendons, which connect muscle to bone.
(a) Tendonitis is inflammation of the tendon due to repetitive motion.
(b) Tendonitis often occurs at the heel or elbow.
(2) This tissue is found in ligaments, which connect bone to bone.
(a) Ligaments can be torn when direction is changed quickly.
(b) Ligaments can be torn by physical trauma.
6. Cartilage
a) Cartilage is composed of cells called chondrocytes, which secrete a matrix of collagen and other proteins.
b) Cartilage provides flexible support and shock absorption.
c) There are three types of cartilage.
(1) Hyaline cartilage has a matrix composed only of collagen.
(a) This cartilage is found in the nose, at the ends of long bones, and in the ribs.
(b) The fetal skeleton is composed of hyaline cartilage which is later replaced by bone.
(2) Elastic cartilage also contains elastic fibers and is more flexible.
(3) Fibrocartilage has a matrix of collagen fibers.
(a) This cartilage acts as a cushion between vertebrae.
(b) It also cushions the knee joint.
7. Bones
a) Bones provide support.
(1) They provide a skeletal framework.
(2) They protect and support tissues and organs.
b) Bone marrow produces blood cells.
c) Bone is a rigid connective tissue composed of osteocytes.
(1) Osteocytes are branched cells.
(2) Osteocytes secrete substances that harden into a solid matrix of collagen, calcium, and other minerals.
(a) Calcium and other minerals can be taken from bone if the body needs them.
(b) Calcium is needed by cells for nerve impulses, muscle contractions, and other functions.
d) Canals in bone house nerves and blood vessels.
e) Bone mass is improved by weight-bearing activities.
(1) Weight-bearing activities include running and weight lifting.
(2) These activities can help prevent osteoporosis.
E. Muscle Tissue
1. Overview
a) Muscle tissue contracts to help the body move.
b) Muscle tissue is composed of long, cylindrical cells called muscle fibers.
(1) Muscle fibers are grouped into bundles.
(2) Muscle fibers contain actin and myosin proteins that cause the cells to contract.
c) Muscle tissue movement is voluntary or involuntary.
d) Muscle tissue is striated or smooth.
(1) The banded pattern of striated tissue is formed by actin and myosin deposits.
(2) Smooth muscle contains actin and myosin that are not deposited in a banded pattern.
2. Cardiac Muscle
a) This muscle is found only in the wall of the heart.
b) It is involuntary and striated.
c) Cardiac muscle cells are branched and interwoven.
(1) They have gap junctions that allow contraction signals to be sent to many cells at once.
(2) They do not reproduce.
d) Aerobic exercise can strengthen cardiac muscle.

3. Skeletal Muscle
   a) This muscle is attached to bones.
   b) It is voluntary and striated.
   c) Skeletal muscle cells do not reproduce.
      (1) They can be enlarged through exercise (hypertrophy).
      (2) They can waste away due to lack of movement (atrophy).
4. Smooth Muscle
   a) This muscle is found in internal organs, blood vessels, and the digestive system.
   b) It is involuntary and smooth.
   c) Contractions of smooth muscle can be held for long periods of time.

F. Nervous Tissue
1. Nervous tissue helps the body to sense and respond to stimuli.
2. Cells called neurons conduct and transmit electrical impulses.
3. Cells called neuroglia take up most of the volume of nervous tissue.
   a) Astrocytes deliver nutrients to neurons.
   b) Microglia cells remove foreign substances.
   c) Oligodendrocytes and Schwann cells synthesize myelin that insulates neurons.
4. Neurotransmitters are chemicals that send signals to nerve cells.
   a) Exercise produces endorphins, which are neurotransmitters that produce a feeling of well-being.
   b) Exercise may also produce other neurotransmitters that fight depression.

II. Body Cavities and Membranes

A. Body Cavities
1. The Ventral Cavity
   a) This cavity is at the front of the body.
   b) It is composed of the thoracic and abdominal cavities.
      (1) The thoracic cavity contains the heart and lungs.
      (2) The abdominal cavity contains the abdominal organs.
      (3) The diaphragm separates the thoracic and abdominal cavities.
2. The Dorsal Cavity
   a) This cavity is composed of the cranial cavity and vertebral canal.
      (1) The cranial cavity houses the brain.
      (2) The vertebral canal houses the spinal cord.

B. Body Membranes
1. Epithelial Membranes
   a) Mucous Membranes
      (1) These moist membranes line the tubes of the digestive, respiratory, reproductive, and urinary systems.
      (2) They can absorb and secrete substances.
   b) Serous Membranes
      (1) These are double-layered membranes that line the thoracic and abdominal cavities, and their respective organs.
      (2) They are composed of epithelium and loose connective tissue.
      (3) They secrete watery fluid between the two membranes.
      (4) Serous membranes are named for their location.
         (a) The pleural membrane lines the thoracic cavity and lungs.
         (b) The pericardium surrounds the heart.
         (c) The peritoneum lines the abdominal cavity and covers its organs.
   c) Cutaneous membranes comprise hard, dry skin.
2. Meningeal Membranes
   a) These membranes are composed of connective tissue.
   b) They cover the brain and spinal cord.
3. Synovial Membranes
   a) These membranes are composed of connective tissue.
   b) They line synovial joints and secrete synovial fluid to lubricate these joints.

III. Organs and Organ Systems

A. Levels of Organization
1. Organs are structures composed of two or more tissues that work together.

2. Organs that work together are called an organ system.
3. All organ systems work together to make a functional organism.

B. Interdependence of Organ Systems
1. Different organ systems rely on each other.
2. Some systems are so closely integrated that they may be called by one name, such as the cardiorespiratory system.

IV. Homeostasis

A. Overview
1. Homeostasis is the ability to maintain constant internal conditions, even under environmental stress.
2. Every level of the organism works to maintain homeostasis.
3. Humans must maintain a relatively constant heart rate, blood pressure, water and mineral balance, temperature, and blood glucose level to survive.
4. Feedback mechanisms help to maintain homeostasis.

B. Negative Feedback
1. Negative feedback is when the product of a process helps to inhibit the process.
2. Thermoregulation occurs through negative feedback.
   a) When body temperature increases, a sensor in the brain is activated.
   b) The sensor sends signals to dilate blood vessels and activate sweat glands.
   c) When body temperature returns to its set point, the sensor sends signals to stop vessel dilation and sweating.
   d) If body temperature drops, the control center triggers muscles to shiver.

C. Positive Feedback
1. Positive feedback is when the product of a process intensifies the process.
2. Positive feedback triggers blood clotting, childbirth, production of breast milk, and the urge to urinate.

D. Improving Fitness
1. Exercising can improve overall fitness.
2. Daily aerobic exercise and frequent muscle strengthening exercises are recommended.
3. You should work within a safe range of heart rate.

Practice Questions

Matching

1. homeostasis
2. adipose tissue
3. atrophy
4. synovial fluid
5. fibroblast
6. collagen
7. gap junction
8. peritoneum
9. tight junction
10. microglia
11. diaphragm
12. serous membrane
13. meningeal membrane
14. pericardium
15. neuroglia

a. where plasma membranes are joined by channels of proteins
b. double-layered membrane, with a watery fluid between the layers
c. sheet of muscle separating the thoracic cavity from the abdominal cavity
d. helps to lubricate joints
e. maintenance of constant internal conditions
f. impermeable barrier between adjacent plasma membranes
g. support cells found in nervous tissue
h. membrane that lines the abdominal cavity
i. sac that surrounds the heart
j. tissue that stores lipids
k. fibers that provide tensile strength
l. cells that remove foreign substances from nervous tissue
m. wasting away of tissue
n. covering that protects the brain and spinal cord
o. cell that secretes proteins into the matrix of loose connective tissue

Fill-in-the-Blank

16. Body temperature is regulated through ____________ feedback.
17. ____________ muscle is involuntary and striated, and ____________ muscle is involuntary and smooth.
18. Bone is a ____________ tissue that is composed of branched cells called ____________ and a matrix that contains collagen, ____________, and other minerals.
19. Serous membranes are composed of epithelium and ____________.
20. An infection of the pleural membrane would affect the thoracic cavity or the ____________.
21. The ____________ cavity contains the brain and is part of the body's ____________ cavity.
22. ____________ tissue is at a lower level of organization than the brain, but at a higher level of organization than nerve cells or ____________.
23. The neuromuscular system encompasses both the ____________ system and the ____________ system.
24. The knee joint contains a ____________ membrane and ____________ that connect leg bones together.
25. A bodybuilder shows signs of ____________ in his or her ____________ muscles.
26. ____________ cartilage in a fetus is later replaced by ____________.
27. ____________ tissue and a sac called the ____________ surround the heart.
28. ____________ tissue in blood vessels is anchored on one side and exposed to blood on the other.

Labeling

Use the terms below to label Figure 5.1.

abdominal cavity
cranial cavity
diaphragm
dorsal cavity
thoracic cavity
ventral cavity
vertebral canal

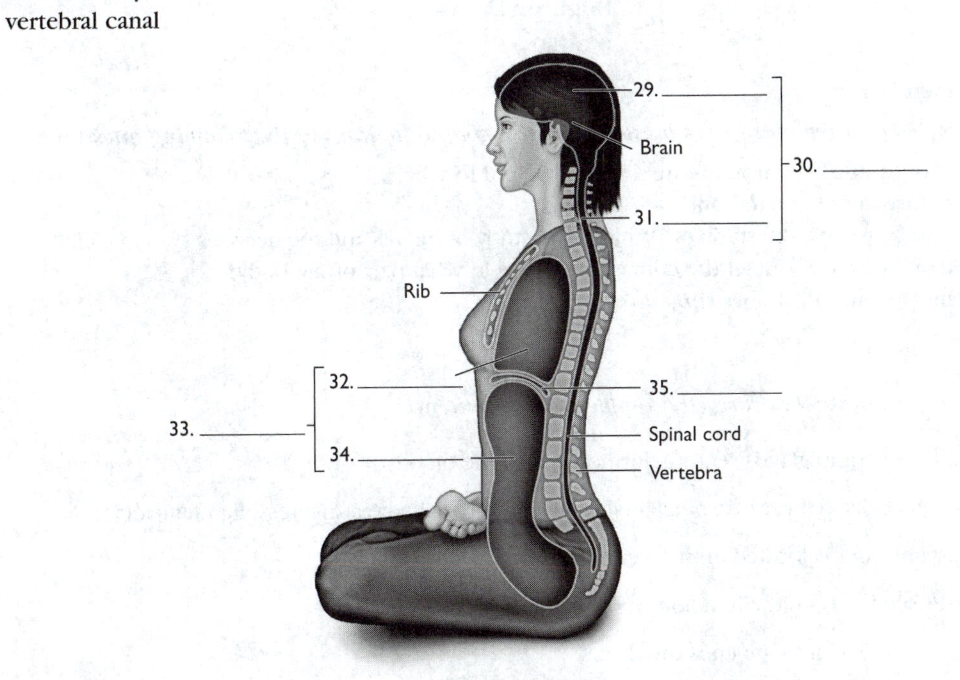

Figure 5.1

Use the terms below to label Figure 5.2.

astrocyte
microglia
neuron
oligodendrocyte

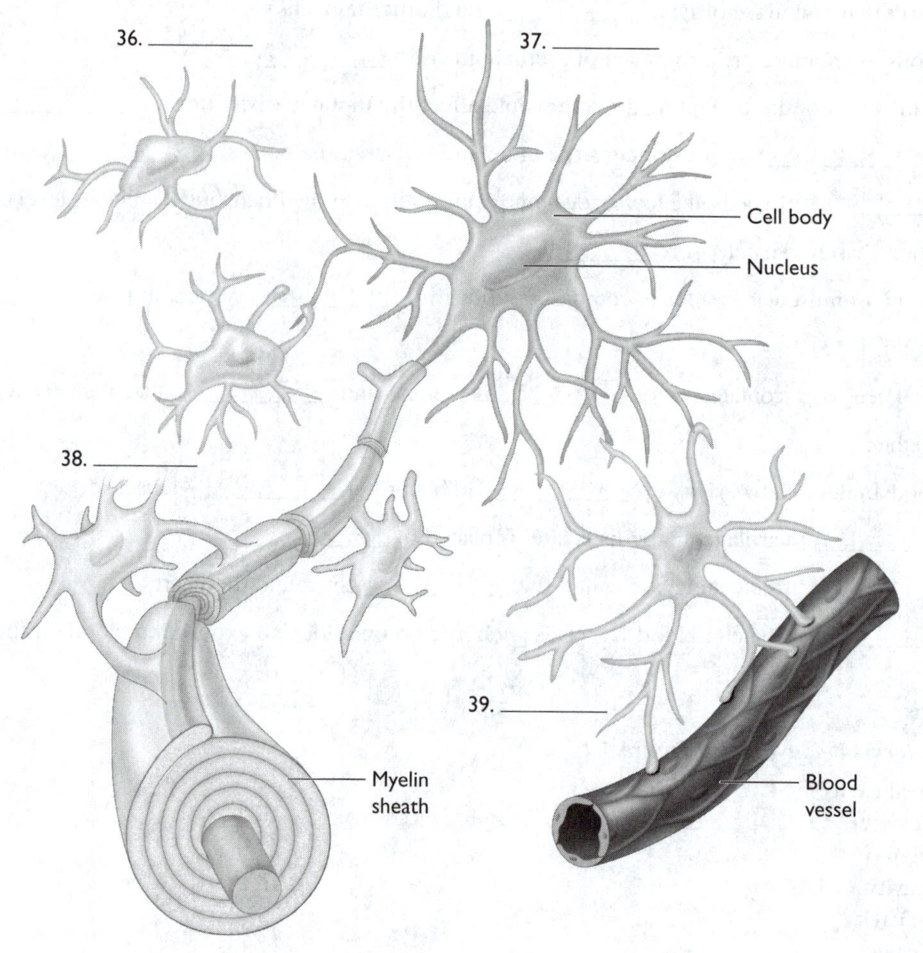

Figure 5.2

Roots to Remember

Use your knowledge of the root words presented in this chapter to answer the following questions.

40. Osteodystrophy and osteomyelitis are diseases related to what?
41. What structure joins bone to bone?
42. The root *-ology* means "the study of." What is the study of nerves and the nervous system called?
43. Periosteal disease would affect the connective tissue in what part of the body?
44. *Sub-* means "below." What does *subcutaneous* mean?

Word Choice

Circle the word or phrase that correctly completes each sentence.

45. (Endocrine/Exocrine) glands secrete hormones into the bloodstream.
46. The neuroglia cells that provide nutrients to neurons are called (astrocytes/oligodendrocytes).
47. The urinary bladder is located in the (ventral/dorsal) cavity.
48. (Cuboidal/Columnar) epithelium lines the small intestine.
49. A (mucus/serous) membrane lines the throat.
50. (Adhesion/Gap) junctions provide flexibility to the skin.

Table Completion

Name each component of connective tissue described in the table.

Component	Description
51.	a cell found in cartilage
52.	cells that carry oxygen
53.	cells that synthesize the matrix in loose connective tissue
54.	bundles of this protein are found in fibrocartilage
55.	the matrix in blood
56.	cell fragments that promote clotting
57.	fibers that allow tissue to stretch
58.	a cell that stores fat

Paragraph Completion

Complete the paragraph using terms from the chapter.

Epithelial tissue is a large component of your skin. This type of tissue protects your body from injury and disease, and helps the body to avoid (59)____________ loss. The epithelial tissue that makes up the outer layer of your skin is called the (60)____________. This layer is anchored on one side and exposed to the (61)____________ on its free surface, which is also called the (62)____________ surface. Specialized epithelial cells in the skin secrete mucus, oils, and sweat. These cells are called (63)____________ glands. Sweat glands help the body to maintain (64)____________.

Critical Thinking

65. Endorphins are natural opioids that provide a feeling of well-being. Recent studies have also shown that endorphins protect against heart attacks. How can you get these benefits from endorphins?
66. How do the muscle cells aid the transport of nutrients around the body?
67. Why isn't the production of breast milk controlled by negative feedback?

Practice Test

1. Which of the following is regulated by negative feedback?
   A. blood clotting
   B. body temperature
   C. childbirth
   D. urge to urinate

2. The digestive system is at a lower organizational level than
   A. the stomach.
   B. the human organism.
   C. smooth muscle in the intestines.
   D. adipocytes.

3. Skeletal muscle tissue is
   A. voluntary and striated.
   B. voluntary and smooth.
   C. involuntary and striated.
   D. involuntary and smooth.

4. Where would you expect to find chondrocytes?
   A. in cartilage
   B. in bone tissue
   C. within muscle tissue
   D. in the blood

5. Which of the following would *not* be found in a matrix of connective tissue?
   A. collagen fibers
   B. elastin fibers
   C. actin proteins
   D. calcium

6. All of the following contain smooth muscle *except*
   A. the gall bladder.
   B. blood vessels.
   C. the leg muscles.
   D. the small intestine.

7. Which of the following membranes is composed only of connective tissue?
   A. meningeal membrane
   B. pleural membrane
   C. peritoneum
   D. mucus membrane

8. How can you increase the number of skeletal muscle cells in your body?
   A. You can increase muscle cells by lifting weights.
   B. You can increase muscle cells by practicing yoga.
   C. You can increase muscle cells by running in marathons.
   D. You cannot increase muscle cells after you are born.

9. Which of the following does *not* contain collagen?
   A. tendons
   B. dermis
   C. ligaments
   D. cardiac muscle

10. Which organ(s) would be affected by disease in the pericardium?
    A. the lungs
    B. the heart
    C. the brain
    D. the skin

11. A myelin sheath
    A. insulates neurons.
    B. connects plasma membranes.
    C. adds strength to muscle cells.
    D. lines the thoracic cavity.

12. Bone contains the following components *except*
    A. collagen.
    B. osteocytes.
    C. calcium.
    D. chondrocytes.

13. The basement membrane
    A. is found on the surface of organs.
    B. is composed of elastin and collagen.
    C. is not composed of cells.
    D. is a connective tissue.

14. Muscle cells are generally
    A. round and elastic.
    B. flat and banded.
    C. long and thin.
    D. branched.

15. Which of the following may shrink after weeks of exercise?
    A. osteocytes
    B. adipocytes
    C. muscle fibers
    D. chondrocytes

Answer Key

Matching

1. e; **2.** j; **3.** m; **4.** d; **5.** o; **6.** k; **7.** a; **8.** h; **9.** f; **10.** l; **11.** c; **12.** b; **13.** n; **14.** i; **15.** g

Fill-in-the-Blank

16. negative; **17.** Cardiac; smooth; **18.** connective; osteocytes; calcium; **19.** loose connective tissue; **20.** lungs; **21.** cranial; dorsal; **22.** Nervous; neurons; **23.** nervous; muscular; **24.** synovial; ligaments; **25.** hypertrophy; skeletal; **26.** Hyaline; bone; **27.** Adipose; pericardium; **28.** Epithelial

Labeling

29. cranial cavity; **30.** dorsal cavity; **31.** vertebral cavity; **32.** thoracic cavity; **33.** ventral cavity; **34.** abdominal cavity; **35.** diaphragm; **36.** microglia; **37.** neuron; **38.** oligodendrocyte; **39.** astrocyte

Roots to Remember

40. bones; **41.** ligament; **42.** neurology; **43.** around the bone; **44.** below the skin

Word Choice

45. Endocrine; **46.** astrocytes; **47.** ventral; **48.** Columnar; **49.** mucus; **50.** Adhesion

Table Completion

51. chondrocyte; **52.** red blood cells; **53.** fibroblasts; **54.** collagen; **55.** plasma; **56.** platelets; **57.** elastin; **58.** adipocyte

Paragraph Completion

59. water; **60.** epidermis; **61.** air; **62.** apical; **63.** exocrine; **64.** homeostasis

Critical Thinking

65. exercise
66. Smooth muscle in the blood vessels contracts to move nutrients in the bloodstream to cells around the body.
67. With negative feedback, one action stops the activity of another action. If production of breast milk were controlled by negative feedback, the suckling of the infant would shut off the production of milk. This would not allow the infant to receive the proper nutrition.

Practice Test

1. B; **2.** B; **3.** A; **4.** A; **5.** C; **6.** C; **7.** A; **8.** D; **9.** D; **10.** B; **11.** A; **12.** D; **13.** C; **14.** C; **15.** B

CHAPTER 6

THE SKELETAL, MUSCULAR, AND INTEGUMENTARY SYSTEMS: SEX DIFFERENCES IN ATHLETICISM

Learning Goals

1. List the types of tissues found in the skeletal system.
2. Describe the structure and function of bones.
3. Compare the activities of osteoclasts and osteoblasts.
4. List the bones that compose the pelvic girdle.
5. Describe how osteocytes, which are embedded in solid bone, receive nutrients.
6. Describe the structure and function of muscles.
7. Describe how antagonistic muscle pairs work.
8. Describe the mechanism of muscular contraction.
9. List the functions of the integumentary system.
10. Describe the structure and function of the skin.

Chapter Outline

I. The Skeletal System
   A. Overview
      1. The skeletal system is an internal framework that provides structure and aids in body movement.
      2. The bones in the skeletal system store minerals and produce blood cells.
      3. Cartilage protects the ends of bones from degradation.
      4. Ligaments connect bone to bone.
      5. Tendons connect bone to muscle.
   B. Bones of the Skeleton
      1. Bone Types
         a) Long bones are longer than they are wide.
         b) Short bones are cube shaped.
         c) Flat bones are platelike.
         d) Round bones are circular in shape.
         e) Irregular bones have many different shapes.
      2. Bone Composition
         a) Compact Bone
            (1) Compact bone forms the hard outer shell of bones.
            (2) Compact bone is composed of cylindrical structures called osteons.
               (a) Bone cells called osteocytes are located within lacunae in osteons.
               (b) Lacunae are arranged in concentric circles around a central canal.
                  (i) Canaliculi connect lacunae to each other and to the central canal.
                  (ii) Blood vessels running through the central canal provide nutrients to osteocytes.
                  (iii) Osteocytes that are too far away from the central canal obtain nutrients passed through gap junctions.
         b) Spongy Bone
            (1) Spongy bone is loosely packed and porous.

(a) Support structures called trabeculae add strength to spongy bone.
(b) In some bones, spaces between trabeculae contain red bone marrow, which produces blood cells.
(2) Spongy bone is found on the inner surfaces of bone and at the ends of long bone.

3. Bone Structures
a) The epiphysis is the spongy end of a long bone that is covered with cartilage.
b) The diaphysis is the shaft of a long bone.
(1) The walls of the diaphysis are composed of compact bone.
(2) This compact bone surrounds the medullary cavity, which is filled with yellow bone marrow.
c) The periosteum is a fibrous layer of connective tissue rich in blood vessels and nerves that surrounds bone.
d) Articular cartilage covers bone surfaces at joints.

C. Bone Development, Growth, Remodeling, and Repair

1. Bone Development
a) Long bones begin to form when an embryo is 6 weeks old.
b) Long bones are composed of hyaline cartilage.
c) Hyaline cartilage is replaced with bone through the process of ossification.
(1) Chondroblasts that produce cartilage die, and cartilage begins to dissolve.
(2) The periosteum develops and produces bone-forming cells called osteoblasts.
(3) Osteoblasts secrete collagen and enzymes that promote the crystallization of mineral salts to produce hydroxyapatite.
(4) Hydroxyapatite secretions trap osteoblasts.
(5) Trapped osteoblasts eventually form lacunae.

2. Bone Growth
a) Bones lengthen throughout childhood and into early adulthood.
b) A large proportion of growth occurs in the extremities.
c) Bones grow from a growth plate or epiphyseal plate.
(1) An epiphyseal plate has four layers.
(a) The resting zone attaches the plate to the epiphysis.
(b) The proliferating zone undergoes cell division to produce new chondroblasts.
(c) Cartilage cells die in the degenerating zone.
(d) Bone forms in the ossification layer.
(2) The epiphyseal plate is replaced with bone in early adulthood, as osteoblasts grow faster than chondroblasts.
d) In adolescence, sex hormones produce a growth spurt by stimulating osteoblasts and chondroblasts.
e) The skeletons of males and females develop differences as bones grow.

3. Bone Remodeling
a) Bone is continually being broken down and built up.
(1) Osteoclasts break down bone.
(2) Osteoblasts help to regenerate bone.
b) Weight-bearing exercises and calcium intake help maintain bone strength.
c) Hormones work together to regulate blood calcium levels.
(1) When blood calcium is too low:
(a) Parathyroid hormone signals the kidneys to decrease the amount of calcium secreted in urine.
(b) Parathyroid hormone also stimulates the release of calcium from bone.
(2) When blood calcium is too high:
(a) The calcitonin hormone produced by the thyroid gland decreases the concentration of blood calcium.
d) Osteoclasts and osteoblasts oppose each other to maintain blood calcium levels.
e) Estrogen interferes with osteoclast activity.
(1) When estrogen declines in older women, osteoclast activity increases.
(2) Osteoporosis is a result of more osteoclast activity and less osteoblast activity.

4. Bone Repair
a) A fractured bone undergoes repair.
(1) A mass of clotted blood, called a hematoma, fills the space between broken bones.
(2) Fibroblasts migrate to the area and produce cartilage.

(3) A fibrocartilaginous callus forms between the broken bones.
(4) Osteoclasts arrive.
(a) They break down broken and dead fragments.
(b) They remove the hematoma.
(5) Osteoblasts lay down bony callus.
(6) Osteoblasts and osteoclasts remodel the bony callus to resemble the original bone.
b) A fractured bone can be identified on an X-ray, even after it has healed.

D. Axial Skeleton: The Central Structure
1. The Skull
a) The skull protects the brain and gives the head structure.
b) The cranium is the braincase that forms the upper skull.
(1) The frontal bone creates the shape of the forehead.
(2) The sphenoid extends across the floor of the cranium and attaches to the following bones:
(a) two temporal bones that are located near the ears;
(b) the occipital bone that forms the base of the skull;
(c) two parietal bones that form the side walls of the cranium; and
(d) the ethmoid bone that forms the orbits of the eyes and the nasal septum.
2. The Facial Bones
a) Two maxillae form the upper jaws.
b) The lower jaw is the mandible.
c) The nasal bones are found above the maxillae.
d) The zygomatic bones form cheekbones.
e) Air pockets in the bones of the face form sinuses.
f) Facial bones are different sizes in males and females.
3. Hyoid Bone
a) The hyoid bone is located at the root of the tongue.
b) It attaches to the larynx by a membrane and to temporal bones by muscles and ligaments.
c) The hyoid bone does not form a joint.
d) It anchors the tongue and attaches muscles involved in swallowing.
4. Vertebral Column
a) The vertebral column is composed of 33 vertebrae.
b) This column protects the spinal cord and anchors muscles and ribs.
c) Intervertebral disks of fibrocartilage provide cushioning between individual vertebrae.
d) Vertebrae are classified by their location.
(1) Cervical vertebrae are in the neck region.
(2) Thoracic vertebrae are in the upper back region.
(3) Lumbar vertebrae are in the small of the back.
5. Rib Cage
a) The rib cage consists of thoracic vertebrae, 24 curved ribs, and the sternum or breastbone.
b) Costal cartilages connect the first seven pairs of ribs to the breastbone.
c) Costal cartilages connect the next three pairs of ribs to cartilage above them.
d) Cartilages that end in the muscle of the abdominal wall support two pairs of floating ribs.

E. Appendicular Skeleton
1. Pelvic Girdle
a) Pelvic Girdle Function
(1) The pelvic girdle connects the trunk of the body to the legs.
(2) The pelvic girdle provides support for the body.
b) Pelvic Girdle Structure
(1) The sacrum consists of the five fused vertebrae at the end of the spine.
(2) The paired hip bones have three bones each.
(a) The ilium is the blade-shaped hip bone that produces the width of the hips.
(b) The ischium is the bone at the back of the hips.
(i) Ischial bones have protuberances.
(ii) These are called the "sit bones."
(c) The pubis is the bone at the front of the hips.
(3) The hip bones attach to the sacrum and form part of the acetabulum, or socket, that forms the hip joint.
(4) The pelvis forms a ring shape called the pelvic inlet, or birth canal, in females.

(5) Pelvises are different shapes in males and females.
(a) Differences in pelvic shape can cause a difference in the Q angle between kneecap and femur.
(b) An increased Q angle in female athletes may lead to knee injuries.
c) Legs and Feet
(1) Upper Leg
(a) The femur is attached to the thigh and buttock muscles and hip flexors by the greater and lesser trochanters.
(b) The femur has medial and lateral condyles that articulate with the lower leg.
(2) Lower Leg
(a) The tibia is the shinbone.
(b) The fibula is a non-weight-bearing bone.
(c) The patella is the kneecap.
(3) Foot
(a) Seven tarsal bones form the ankle and heel.
(b) Five metatarsal bones form the arch.
(c) Phalanges form the feet.
2. Pectoral Girdle
a) Pectoral Girdle Structure
(1) The shoulder blade (scapula) and collarbone (clavicle) are part of the pectoral girdle.
(2) The glenoid cavity is a depression in the scapula that holds the humerus.
b) Pectoral Girdle Motion
(1) The range of motion in the pectoral girdle is greater than the range of motion in the pelvic girdle.
(2) There is a greater risk of dislocation and injury.
(a) Rotator cuff injuries are rips in the tendons that connect the humerus to muscles attached to the scapula.
(b) Rotator cuff injuries are common in athletes.
c) Arms and Hands
(1) Upper Arm
(a) The humerus is the bone of the upper arm.
(b) Protuberances of the humerus called the trochlea and capitulum articulate with the bones of the forearm.
(2) Forearm
(a) The radius is shorter than the ulna and has a disk-shaped head.
(b) The ulna has a large projection called the olecranon that articulates with the humerus.
(3) Hands
(a) Eight carpal bones form the wrist.
(b) Metacarpals and phalanges form the palm and hands.
(c) The thumb is opposable and has a wide range of movement.
F. Joints and Movement
1. Immovable Joints
a) These joints have limited movement.
b) The sutures in the skull are immovable joints.
2. Synovial Joints
a) Structure
(1) Ligaments form a capsule around the joint.
(2) The capsule is lined with a synovial membrane that secretes synovial fluid to lubricate the joint.
(3) Articular surfaces in these joints are covered with cartilage.
b) Hinge Joints
(1) These synovial joints allow back-and-forth movement.
(2) The knee and elbow are hinge joints.
c) Pivot Joints
(1) This synovial joint allows the head to turn.
(2) The neck has a pivot joint.

d) Ball-and-Socket Joints
(1) These synovial joints enable arms and legs to move in three dimensions.
(2) Hips and shoulders are ball-and-socket joints.
e) Movement of Synovial Joints
(1) Flexion decreases the joint angle.
(2) Extension increases the joint angle.
(3) Adduction is movement of a body part toward the midline.
(4) Abduction is movement of a body part away from the midline.
(5) Rotation is movement of a body part around its own axis.
(6) Circumduction is movement of a body part in a wide circle.

— II. The Muscular System
A. Overview
1. Skeletal muscle helps the body to move.
a) The origin of a skeletal muscle is on a stationary bone.
b) The insertion is on the bone that moves.
2. Blood vessels supply oxygen and nutrients to skeletal muscle.
3. Nerves control muscle contraction.
4. Skeletal muscle works in antagonistic pairs.
a) One muscle relaxes while the other contracts.
b) Biceps and triceps are an example of an antagonistic pair.
B. Names and Actions of Skeletal Muscles
1. Names of muscles are based on characteristics such as size, shape, and location.
2. Names are also based on number of attachments or muscle action.
C. Skeletal Muscle Structure
1. Muscles are composed of bundles of parallel muscle fibers.
a) Each bundle is called a fascicle.
b) Each fiber in the bundle is a muscle cell.
(1) Each muscle cell contains parallel filaments called myofibrils.
(a) Each myofibril contains units of contraction called sarcomeres.
(i) A sarcomere contains thin filaments of actin and thick filaments of myosin.
(ii) A sarcomere occupies the region between dark lines, called Z discs, in the myofibril.
(b) A myofibril may contain thousands of sarcomeres.
(2) The sarcoplasm is the cytoplasm of a muscle cell.
(a) Mitochondria and other organelles are found in the sarcoplasm.
(b) The sarcoplasm contains glycogen, a sugar required for muscle contraction.
(3) Muscle cells are rich in myoglobin, a red pigment that binds oxygen.
2. Muscles are covered with connective tissue called fascia that becomes the tendon anchoring the muscle to bone.
D. Skeletal Muscle Contraction
1. Nerves activate skeletal muscles.
a) Motor neurons release a neurotransmitter called acetylcholine.
b) Acetylcholine diffuses across a neuromuscular junction to the muscle cell.
c) Acetylcholine binds with the cell membrane, and the muscle cell transmits an impulse down the sarcolemma (plasma membrane of a muscle cell).
d) T tubules in the sarcolemma penetrate the muscle cell and contact the sarcoplasmic reticulum.
e) The sarcoplasmic reticulum releases calcium ions.
f) Calcium ions associate with troponin in the myofibril.
(1) Troponin is attached to tropomyosin.
(2) Tropomyosin threads are wound around the actin filament in a myofibril.
g) The troponin changes shape and shifts position.
h) This shape change causes tropomyosin threads to shift position, exposing binding sites for myosin.
2. Sliding filaments shorten the sarcomere.
a) The heads of myosin molecules attach to binding sites on actin molecules.
b) ATP molecules attach to myosin heads, causing them to detach from the binding site.

c) The breakdown of ATP provides energy to the myosin head, and the head changes shape.
d) The head binds to another binding site on an actin molecule that is farther along on the sarcomere.
e) This process contracts the Z discs and shortens the sarcomere.

E. Energy Inputs for Muscle Contraction
1. Skeletal muscles can synthesize ATP from creatine phosphate.
a) Creatine phosphate has a high-energy phosphate group.
b) This group can be transferred to ADP to produce ATP.
c) Hydrolysis of creatine phosphate is used during intense exercise.
(1) The process continues as long as oxygen is available.
(2) When oxygen is unavailable, fermentation takes place.
2. Skeletal muscle cells contain more mitochondria than other cells.
a) People who exercise have more mitochondria in muscle cells than do sedentary people.
b) More mitochondria allow the muscle to produce ATP through cellular respiration without relying as heavily on creatine phosphate breakdown and fermentation.

III. The Integumentary System
A. Overview
1. The integumentary system is composed of skin and accessory organs.
a) Skin is composed of two regions.
(1) The superficial layer is the epidermis.
(2) The inner layer is the dermis.
(3) Glands are found in these regions.
b) The subcutaneous layer, or hypodermis, is located between the skin and underlying muscle or bone.
2. The integumentary system provides protection.
a) It protects the body from pathogens, chemicals, and ultraviolet (UV) light.
b) It prevents water loss and helps to regulate body temperature.

B. Epidermis
1. The epidermis is the outer layer of skin.
2. It is nonvascular.
3. It is composed of stratified squamous epithelial cells.
a) The outer layers of these cells are dead.
b) Living cells get pushed to the surface.
(1) These cells die as they get farther away from the blood supply.
(2) Epidermal cells flatten as they migrate.
(3) These cells produce the waterproof protein called keratin.
(4) Melanocytes in epidermis produce melanin, which absorbs UV light.

C. Dermis
1. The dermis is located beneath the epidermis.
2. It is vascular.
3. It is composed of dense, fibrous connective tissue that is rich in collagen and elastin.
4. The dermis contains sensory receptors, sweat glands, and hair follicles.

D. Accessory Structures of the Skin
1. Nails
a) Nails cover the tips of fingers and toes.
b) They grow from epithelial cells at the base of each nail.
(1) These cells produce keratin as they grow.
(2) These cells are covered at the base by a fold of skin called the cuticle.
2. Hair
a) Hair grows from a root in the dermis up through the epidermis.
b) Hair is composed of keratin.
c) Arrector pili muscles cause hair to stand straight up.
3. Sebaceous Glands
a) Sebaceous glands grow from the epithelium that surrounds a hair.
b) These glands produce oils to lubricate hair and skin.
4. Sweat Glands
a) Sweat glands are located in all regions of the skin.
b) They regulate body temperature.
c) Mammary glands are specialized sweat glands.

E. Subcutaneous Layer
   1. This layer, called the hypodermis, is below the skin.
   2. It is composed of loose connective tissue and adipose tissue.
   3. Due to a higher concentration of fat, this layer is thicker in women than in men.

Practice Questions

Matching

1. lacunae	a. protein that interacts with actin
2. fibula	b. connective tissue that covers muscles
3. myosin	c. bone that gives shape to the forehead
4. cranium	d. secretes neurotransmitters
5. fascia	e. surround osteocytes
6. skull	f. parallel filaments in a muscle fiber
7. hypodermis	g. breastbone
8. metacarpal	h. layer of tissue below the skin
9. motor neuron	i. braincase
10. sarcolemma	j. a bone in the foot
11. sternum	k. bone located in the lower leg
12. frontal bone	l. bone in the hand
13. fascicle	m. underlying structure of the head
14. metatarsal	n. bundle of muscle fibers
15. myofibril	o. specialized plasma membrane

Fill-in-the-Blank

16. Each ____________ in a muscle fiber contains thick filaments of ____________ protein and thin filaments of ____________ protein.
17. The ____________ joint of the knee is a type of ____________ joint.
18. Your biceps and triceps muscles are called a(n) ____________.
19. The movement produced when you straighten your leg is called ____________.
20. Goosebumps are caused by the ____________ of the ____________ muscles.
21. During intense exercise, skeletal muscles can synthesize ____________ for energy from ____________, a compound that contains a high-energy phosphate group.
22. Nails are made of ____________ that is produced by ____________ cells.
23. The layer of skin that does not contain blood vessels is called the ____________.
24. Teeth are located on the lower jaw, or ____________, and the upper jaw, or ____________.
25. The ____________ gland produces hormones when blood calcium is low.

Labeling

Use the terms below to label Figure 6.1.

articular cartilage
compact bone
diaphysis
epiphysis
medullary cavity
spongy bone

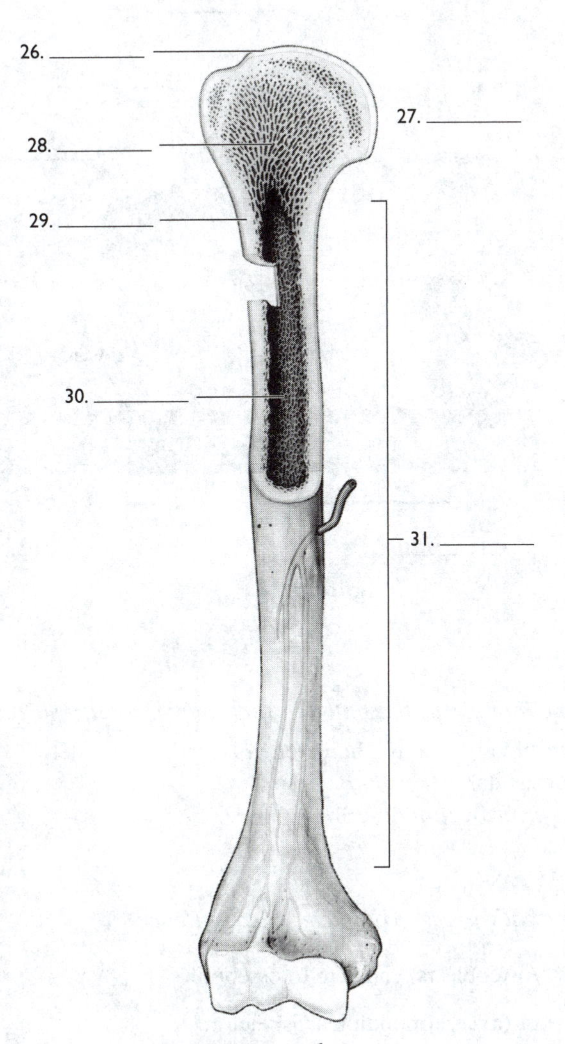

Figure 6.1

Use the terms below to label Figure 6.2.

fascia
fascicle
muscle fiber
myofibril
sarcomere
Z disc

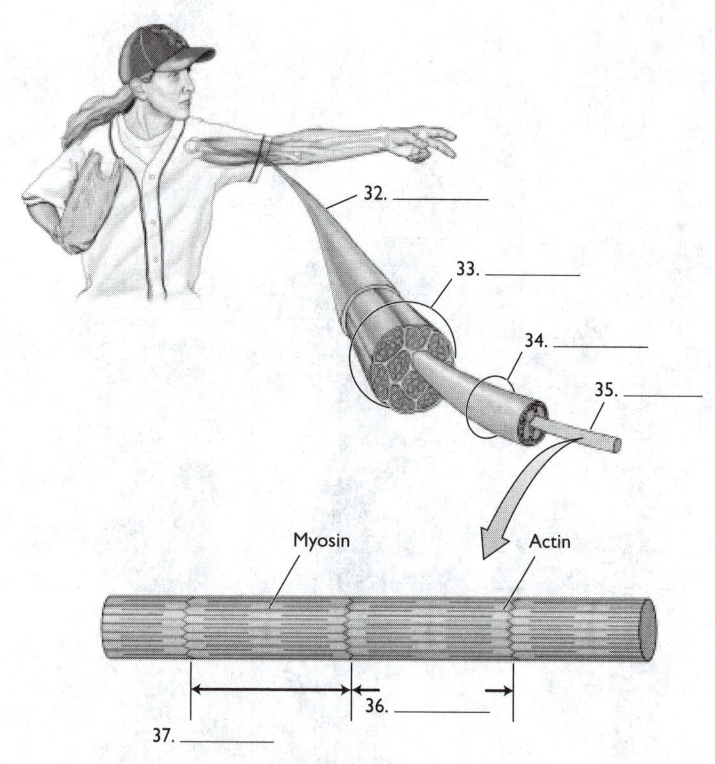

Figure 6.2

Roots to Remember

Use your knowledge of the root words presented in this chapter to answer the following questions.

38. The root *hema-* means blood. What is a hematoma?
39. What would an iconoclast do?
40. What would the term *myocardium* describe?

Word Choice

Circle the word or phrase that correctly completes each sentence.

41. Overactive (osteoclasts/osteoblasts) can lead to osteoporosis.
42. Your ribs are part of your (axial/appendicular) skeleton.
43. The (humerus/hyoid) bone does not form a joint with another bone.
44. Only (five/seven) pairs of ribs are attached directly to bone by costal cartilage.
45. The (greater/lesser) Q angle in women may make them more likely to experience knee injury.
46. The (Z discs/T tubules) in a muscle cell help transmit nerve impulses.
47. Muscle cells can generate ATP using (creatine phosphate/fermentation) in the presence of oxygen.
48. (Melanin/Keratin) is a substance found in skin that absorbs UV light.
49. The (diaphysis/epiphysis) protects the ends of long bone.
50. When calcium ions are released by the sarcoplasmic reticulum, they associate with (troponin/tropomyosin) to change the shape of that protein.

Table Completion

Name each component of bone described in the table.

Component	Description
51.	cells that produce cartilage
52.	small canals that connect lacunae
53.	substance that produces blood cells
54.	main mineral salt found in bone
55.	structures that add strength to spongy bone
56.	cells that secrete collagen and enzymes
57.	connective tissue surrounding bone that contains blood vessels
58.	a cylindrical structure in compact bone

Sequencing

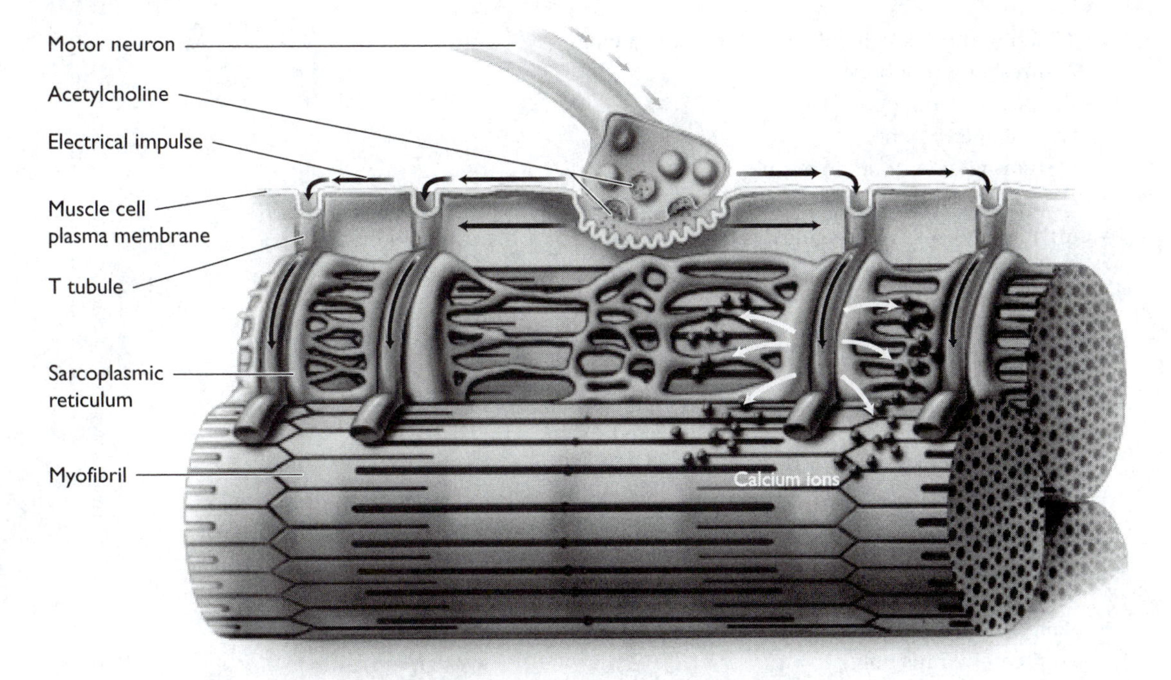

Use the diagram above to place these events related to translation in the correct sequence, with 1 being the first event and 7 being the last event.

59. ____________ Acetylcholine diffuses across a neuromuscular junction to the muscle cell.

60. ____________ The sarcoplasmic reticulum releases calcium ions.

61. ____________ T tubules in the sarcolemma penetrate the muscle cell and contact the sarcoplasmic reticulum.

62. ____________ The muscle cell transmits an impulse down the sarcolemma.

63. ____________ Motor neurons release a neurotransmitter called acetylcholine.

64. ____________ Acetylcholine binds with the cell membrane.

65. ____________ Calcium ions associate with troponin in the myofibril.

Critical Thinking

66. Why is a broken leg potentially more serious in an adolescent than it is in an adult?
67. Some medications to prevent osteoporosis inhibit the activity of osteoclasts. How might these drugs affect homeostasis in the body?
68. What is the purpose of the fibrocartilaginous callus that is created by the body after a fracture?

Practice Test

1. What kind of bone is the patella?
   A. flat bone
   B. round bone
   C. long bone
   D. short bone

2. The space between a motor neuron and skeletal muscle is called the
   A. gap junction.
   B. neuromuscular junction.
   C. musculoskeletal junction.
   D. tight junction.

3. Which of the following does *not* require calcium?
   A. breakdown of bone
   B. muscle contraction
   C. building of bone
   D. transmission of nerve impulses

4. An athlete bruises his zygomatic bone. Which part of his body did he injure?
   A. his leg
   B. his arm
   C. his back
   D. his face

5. The socket that forms your hip joint is called the
   A. lateral condyle.
   B. glenoid cavity.
   C. acetabulum.
   D. trochlea.

6. Hair grows through
   A. the epidermis only.
   B. the dermis and epidermis only.
   C. the hypodermis and dermis only.
   D. the hypodermis, dermis, and epidermis.

7. What is the sarcolemma?
   A. the organelle in a muscle cell that stores calcium
   B. the cytoplasm in a muscle cell
   C. the plasma membrane of a muscle cell
   D. the unit of contraction of a muscle fiber

8. Which of the following is the first step of bone repair?
   A. Fibroblasts lay down cartilage.
   B. Osteoclasts remove broken or dead bone.
   C. Osteoblasts lay down new bone.
   D. Osteoclasts remove bony callus.

9. Which is the smallest component in the following list?
   A. fascicle
   B. sarcomere
   C. myofibril
   D. Z disc

10. A motor unit contains
   A. a motor neuron.
   B. a skeletal muscle cell.
   C. a neuromuscular junction.
   D. all of the above.

11. A dancer folds her arms in close to her body and then flings them up overhead. The movement she made when she folded her arms in could be described as
   A. abduction.
   B. adduction.
   C. extension.
   D. circumduction.

12. Which of the following muscles is found in the lower back?
   A. pectoralis major
   B. adductor longus
   C. deltoid
   D. latissimus dorsi

13. A myofibril contains all of the components below *except*
   A. myosin.
   B. sarcomeres.
   C. collagen.
   D. actin.

14. Which joint allows the most movement in all directions?
   A. ball-and-socket
   B. hinge
   C. pivot
   D. synovial

15. Which of the following bones is found in the pectoral girdle?
   A. fibula
   B. ilium
   C. scapula
   D. pubis

Answer Key

Matching

1. e; **2.** k; **3.** a; **4.** i; **5.** b; **6.** m; **7.** h; **8.** l; **9.** d; **10.** o; **11.** g; **12.** c; **13.** n; **14.** j; **15.** f

Fill-in-the-Blank

16. sarcomere; myosin; actin; **17.** hinge; synovial; **18.** antagonistic muscle pair; **19.** extension; **20.** contraction; arrector pili; **21.** ATP; creatine phosphate; **22.** keratin; epithelial; **23.** epidermis; **24.** mandible; maxillae; **25.** parathyroid

Labeling

26. articular cartilage; **27.** epiphysis; **28.** spongy bone; **29.** compact bone; **30.** medullary cavity; **31.** diaphysis; **32.** fascia; **33.** fascicle; **34.** muscle fiber; **35.** myofibril; **36.** sarcomere; **37.** Z disc

Roots to Remember

38. a swelling or mass of blood; **39.** break down iconic figures; **40.** heart muscle

Word Choice

41. osteoclasts; **42.** axial; **43.** hyoid; **44.** seven; **45.** greater; **46.** T tubules; **47.** creatine phosphate; **48.** Melanin; **49.** epiphysis; **50.** troponin

Table Completion

51. chondroblasts; **52.** canaliculi; **53.** red bone marrow; **54.** hydroxyapatite; **55.** trabeculae; **56.** osteoblasts; **57.** periosteum; **58.** osteon

Sequencing

59. 2; **60.** 6; **61.** 5; **62.** 4; **63.** 1; **64.** 3; **65.** 7

Critical Thinking

66. The leg bones of an adolescent are growing very quickly, and the leg bones of an adult are finished growing. If the break damages the epiphyseal growth plate in the adolescent, it can interfere with the growth in that leg.

67. By inhibiting osteoclast activity, these drugs may cause blood calcium levels to drop. This could inhibit cell activities that require calcium.

68. The fibrocartilaginous callus provides temporary stability and support until osteoblasts can rebuild the bone.

Practice Test

1. B; **2.** B; **3.** A; **4.** D; **5.** C; **6.** B; **7.** C; **8.** A; **9.** D; **10.** D; **11.** B; **12.** D; **13.** C; **14.** A; **15.** C

CHAPTER 7

THE DIGESTIVE SYSTEM: WEIGHT-LOSS SURGERY

Learning Goals

1. List the organs of the digestive tract and describe the functions of each.
2. List the different types of teeth and indicate their function.
3. Describe how and where peristalsis occurs.
4. Discuss the mechanism by which food is swallowed, and describe how it is prevented from entering the trachea.
5. Identify the secretions produced by the stomach, and explain how they aid in digestion.
6. Explain why the small intestine is considered to be the major site of digestion and absorption.
7. Describe the structure and function of the liver.
8. Explain how the gallbladder aids digestion.
9. Explain how the pancreas functions as both an exocrine gland and an endocrine gland.
10. Describe how homeostasis of blood glucose levels is maintained.

Chapter Outline

I. The Digestive Tract
  A. Overview
    1. The digestive system contains organs and glands that break down food into its component parts.
    2. Digestion occurs as the food travels through the alimentary canal, or digestive tract.
    3. Digestion involves both mechanical and chemical processes.
      a) Mechanical digestion involves crushing, grinding, and churning of food.
      b) Chemical digestion involves chemical action from acids and enzymes.
  B. The Wall of the Digestive Tract
    1. The digestive tract is composed of multiple layers.
      a) The mucosal layer, or mucosa, lines the lumen (central cavity).
        (1) This layer is a mucus membrane composed of epithelia surrounded by connective tissue, which is surrounded by smooth muscle.
        (2) Epithelial cells secrete digestive enzymes.
        (3) Goblet cells secrete mucus.
      b) The submucosa is a layer of connective tissue containing blood vessels that surrounds the mucosa.
        (1) Peyer's patches are lymph nodes in this layer.
        (2) Lymph nodes prevent infection.
      c) The muscularis surrounds the submucosa and is composed of two layers of muscle.
        (1) The inner layer is circular muscle that surrounds the gut.
        (2) The outer layer is longitudinally oriented.
      d) The serosa layer surrounds the muscularis.
        (1) This layer is composed of squamous epithelium.
        (2) It secretes fluids that keep organs moist and able to slide past each other.
  C. The Mouth: The Actions of the Teeth and Saliva
    1. An anterior hard palate and posterior soft palate separate the oral cavity from the nasal cavity.

2. The tongue is a skeletal muscle covered with taste buds.
   a) The tongue helps to shape food into a bolus.
   b) The tongue moves the bolus toward the throat.
3. Tonsils in the throat protect the body from infection.
4. Teeth mechanically digest food.
   a) Types of Teeth
      (1) Incisors bite off pieces of food.
      (2) Canines rip food.
      (3) Premolars tear food.
      (4) Molars grind food.
   b) Tooth Structure
      (1) A tooth has a visible crown and a root below the gums.
      (2) An outer shell of enamel is composed of calcium and phosphate compounds.
      (3) Dentin is a living layer below the enamel.
      (4) The pulp cavity contains nerves and blood vessels.
      (5) The periodontal membrane holds the tooth in its bony socket.
      (6) The gums surround the upper tooth.
5. Saliva chemically digests food.
   a) Salivary glands are exocrine glands in the mouth that produce enzymes.
      (1) The parotids are located in the upper cheeks.
      (2) The sublinguals are located below the tongue.
      (3) The submandibular glands are below the floor of the oral cavity.
   b) Saliva moves from the glands to the mouth through salivary ducts.
   c) Saliva contains salivary amylase, which breaks down starch, and lingual lipase, which breaks down fats.

D. The Pharynx and the Esophagus: Transport to the Stomach
1. The pharynx branches into the trachea and esophagus.
2. The epiglottis is a flap of cartilage that keeps food from entering the trachea.
3. The esophagus is a long, muscular tube.
   a) No chemical digestion occurs in the esophagus.
   b) Food is moved by peristalsis from the muscularis layer.
4. The gastroesophageal sphincter is located between the esophagus and stomach.
   a) This sphincter prevents acid from backing up into the esophagus.
   b) Damage to the sphincter can cause gastroesophageal reflux disease (GERD).
      (1) GERD can be treated with antacids.
      (2) Untreated GERD may cause esophageal ulcers.

E. The Stomach: Digestion in an Acid Bath
1. The stomach is located beneath the diaphragm.
2. The stomach stores and digests, but does not absorb, most food.
   a) Stretching of the stomach signals peristalsis in the esophagus to intensify.
   b) Stomach contractions mix food.
   c) Cells in the gastric glands produce gastric juice.
      (1) Hydrochloric acid (HCl) lowers pH and kills bacteria in food.
      (2) Pepsinogen is modified to form the enzyme pepsin, which digests proteins.
   d) Columnar epithelial cells secrete thick mucus to protect the stomach lining.
3. Chyme produced by digestion in the stomach is delivered to the small intestine by the pyloric sphincter.
4. Pancreatic secretions raise the pH of chyme.

F. The Small Intestine: Where Most Digestion Happens
1. The small intestine consists of the duodenum, jejunum, and ileum.
   a) The duodenum is the site of most digestion.
   b) The ileum is the longest region.
2. The small intestine is the main site of chemical digestion.
   a) The small intestine has a huge surface area, due to folding.
      (1) Folds exist in the intestinal wall itself.
      (2) These folds are covered by fingerlike projections called villi.
      (3) The plasma membrane of each cell in the villi has small projections called microvilli.
   b) A large surface area increases the absorption of nutrients.

3. Once nutrients are digested by the intestine, they move into the circulatory and lymphatic systems.
   a) Capillaries from the circulatory system are located within villi.
   b) Lacteal (lymph capillaries) from the lymphatic system are located within villi.
4. Subunits of nutrients enter cells and are used as building blocks for cells or as sources of energy.

G. Regulation of Digestive Secretions
1. The stomach produces the gastrin hormone after a meal, which stimulates the production of gastric juices in the stomach.
2. The duodenum produces the hormones secretin and cholecystokinin (CCK).
   a) HCl stimulates secretin production.
   b) Partially digested proteins and fats stimulate CCK production.
   c) Secretin and CCK signal the pancreas and gallbladder to increase output of digestive juices.

H. The Large Intestine: Absorption and Elimination
1. The large intestine is shorter and wider than the small intestine.
2. Its primary functions are absorption of water, salts, and some vitamins, and the formation, storage, and lubrication of fecal matter.
   a) Fecal matter (feces) is composed of water, fiber, bacteria, and other indigestible materials.
   b) Fecal matter is expelled after it travels through the large intestine.
3. The large intestine is composed of the cecum, colon, rectum, and anus.
   a) A vestigial projection called the appendix hangs from the cecum.
   b) The colon is divided into the ascending colon, transverse colon, descending colon, and sigmoid colon.
      (1) Polyps in the colon may develop into colon cancer.
      (2) Inflammatory bowel diseases are caused by inflammation in the colon.
      (3) Diverticulitis can occur when fecal matter gets trapped in pouches in the colon and becomes infected.
   c) The rectum is the terminal portion of the large intestine.
      (1) Nerves in the rectal wall are stimulated when feces build up.
      (2) These nerves produce a desire to defecate.
   d) The anus is the sphincter through which fecal matter exits the body.
      (1) Constipation is caused by hard, dry stools.
         (a) Constipation makes it difficult to defecate.
         (b) Exercise, eating high-fiber foods, and drinking water can prevent constipation.
      (2) Diarrhea is loose, watery stool.
         (a) Diarrhea may be caused by pathogenic bacteria, viruses, or other parasites in the intestine.
         (b) When the intestines are irritated, they increase peristalsis.
         (c) Increased peristalsis moves food through quickly, so less water is absorbed into the intestines.

I. Gastric Bypass Surgery: Scaling Back Digestion
1. Gastric bypass surgery decreases stomach size to reduce food intake.
2. Gastric bypass surgery allows food to bypass the major sites of digestion and absorption.
   a) A small pouch is made from the upper stomach.
   b) The duodenum is bypassed.
   c) The small pouch is attached to the jejunum.
   d) The stomach and duodenum are sewn shut and remain in the abdomen.
3. Gastric bypass surgery can lead to a number of complications, including death.

II. Three Accessory Organs of the Digestive System

A. Overview
1. Accessory organs are organs located outside the alimentary canal that aid in digestion.
2. The liver, gallbladder, and pancreas are accessory organs.
3. Ducts transport secretions from these organs to the duodenum.
   a) The common hepatic duct transports bile from the liver to the gallbladder.
   b) The common bile duct transports bile from the gallbladder to the duodenum.
   c) The pancreatic duct transports digestive enzymes from the pancreas to the duodenum.

B. The Liver
   1. The liver is a reddish-brown organ that is located below the diaphragm.
   2. The liver gets its color from the blood that it filters.
   3. The four lobes of the liver are divided into lobules.
      a) Liver cells in lobules secrete substances.
         (1) Liver cells produce greenish-yellow bile.
            (a) Bile contains salts that help dissolve fats.
            (b) Bile gets its color from bilirubin—a substance produced by the breakdown of hemoglobin.
         (2) Liver cells produce cholesterol for cell membranes.
         (3) Liver cells produce clotting factors and other blood proteins.
         (4) Liver cells produce immune proteins that fight infection.
      b) Liver cells in lobules remove substances from blood.
         (1) Liver cells remove toxins, dead cells, pathogens, drugs, and alcohol.
         (2) Liver cells convert fats to fatty acids and carbohydrates.
         (3) Liver cells convert blood glucose to glycogen for storage.

C. The Gallbladder
   1. The gallbladder is a greenish sac attached at the bottom of the liver.
   2. The gallbladder stores and concentrates bile.
   3. It releases bile into the common bile duct for transport into the duodenum.
   4. Salts can crystallize and form gallstones that can block bile ducts.

D. The Pancreas
   1. The pancreas is a gland that is located behind the stomach.
   2. The pancreas produces many digestive enzymes.
      a) Pancreatic amylase breaks down starches.
      b) Lipase breaks down lipids.
      c) Trypsin breaks down proteins.
   3. Pancreatic cells release digestive enzymes, water, and sodium bicarbonate to neutralize chyme in the small intestine.
   4. The pancreas secretes hormones to regulate blood-glucose levels.
      a) Insulin is released when blood glucose is too high.
         (1) The release of insulin triggers the liver to store glucose as glycogen.
         (2) Blood-glucose level falls.
      b) Glucagon is released when blood glucose is too low.
         (1) The release of glucagon triggers the liver to release glucose from glycogen.
         (2) Blood-glucose level rises.
   5. Diabetes mellitus occurs when blood-glucose level is no longer regulated.
      a) Type 1, or insulin-dependent diabetes mellitus, usually begins in childhood.
         (1) In this disease, the immune system attacks and destroys the person's own pancreatic cells.
         (2) Not enough insulin is produced to regulate blood glucose.
         (3) Daily insulin shots are needed.
         (4) Type 1 diabetes mellitus is not correlated with obesity.
      b) Type 2, or non-insulin-dependent diabetes mellitus, usually begins after 40.
         (1) In this disease, the pancreas may decrease secretions of insulin.
         (2) Target cells may also become resistant to insulin.
         (3) Sugars accumulate in the bloodstream.
         (4) Type 2 diabetes mellitus is strongly correlated with obesity.
      c) Diabetes can cause many illnesses.
         (1) High blood sugar can cause weakness, confusion, convulsions, coma, or death.
         (2) Nerve damage can put the diabetic at risk for infections in the limbs.
         (3) Damage to the kidneys can result in a need for dialysis.
         (4) Deterioration of the blood vessels to the retina can cause blindness.

III. Weighing the Risks of Gastric Bypass Surgery
   A. Gastric bypass surgery may help people to lose weight and keep the weight off.
   B. This surgery may alleviate health problems associated with diabetes.
   C. This surgery has a high risk of death.
   D. The risk of death from complication due to obesity is higher than the risk of death from the surgery.

Practice Questions

Matching

1. bile	a. hormone that stimulates the liver to break down glycogen
2. glucagon	b. ends in the uvula
3. ileum	c. enzyme in gastric juice
4. cecum	d. a flap of cartilage that controls the movement of food as you swallow
5. bilirubin	e. helps to dissolve fats
6. pepsin	f. hormone that stimulates the conversion of glucose to glycogen
7. soft palate	g. longest portion of the small intestine
8. pharynx	h. also called a bicuspid
9. premolar	i. first part of the large intestine
10. microvilli	j. chamber just above the esophagus
11. epiglottis	k. the site of the most digestion
12. villi	l. produced when hemoglobin breaks down
13. lobules	m. multicellular projections in the small intestine
14. insulin	n. projections of the plasma membrane
15. duodenum	o. portions of the liver

Fill-in-the-Blank

16. Bile moves from the liver into the ____________ duct and then into the ____________.

17. The polymer ____________ contains multiple glucose ____________.

18. Cells in the liver are called ____________.

19. Each tooth contains an outer layer of ____________ on the crown and an inner layer of ____________.

20. In the mouth, the ____________ mechanically digest food and ____________ chemically digest food.

21. Acid that flows backward into the esophagus through the gastroesophageal ____________ can cause ____________ in the esophagus.

22. Cells in the stomach secrete ____________ acid and ____________.

23. The ____________ intestine contains folds, villi, and ____________, which increase the intestine's surface area.

24. The large intestine is composed of the cecum, ____________, rectum, and anus.

25. The ____________ in the throat help protect the body from diseases.

Labeling

Use the terms below to label Figure 7.1.

canine
incisors
molars
premolars
salivary duct
salivary glands

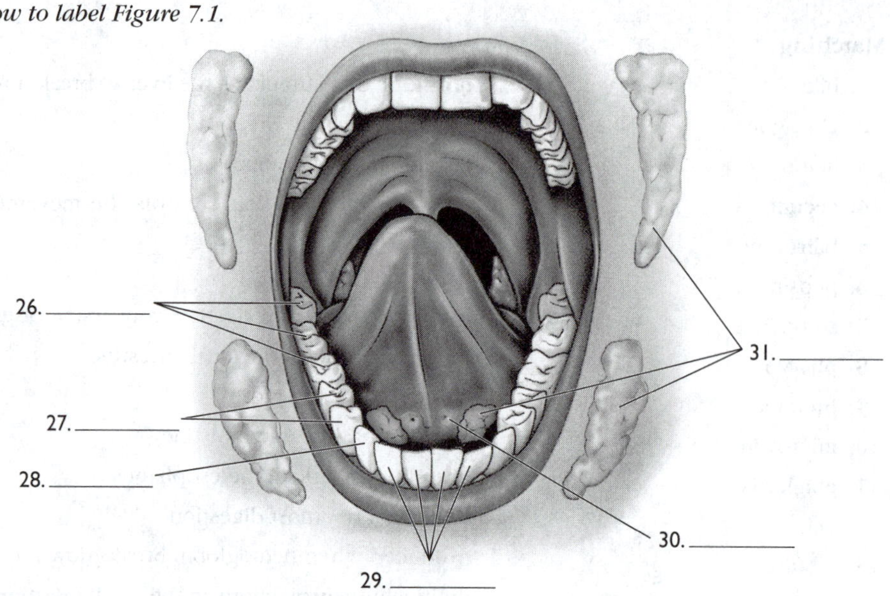

26. ________
27. ________
28. ________
29. ________
30. ________
31. ________

Figure 7.1

Use the terms below to label Figure 7.2.

common bile duct
common hepatic duct
gallbladder
liver
pancreas
pancreatic duct

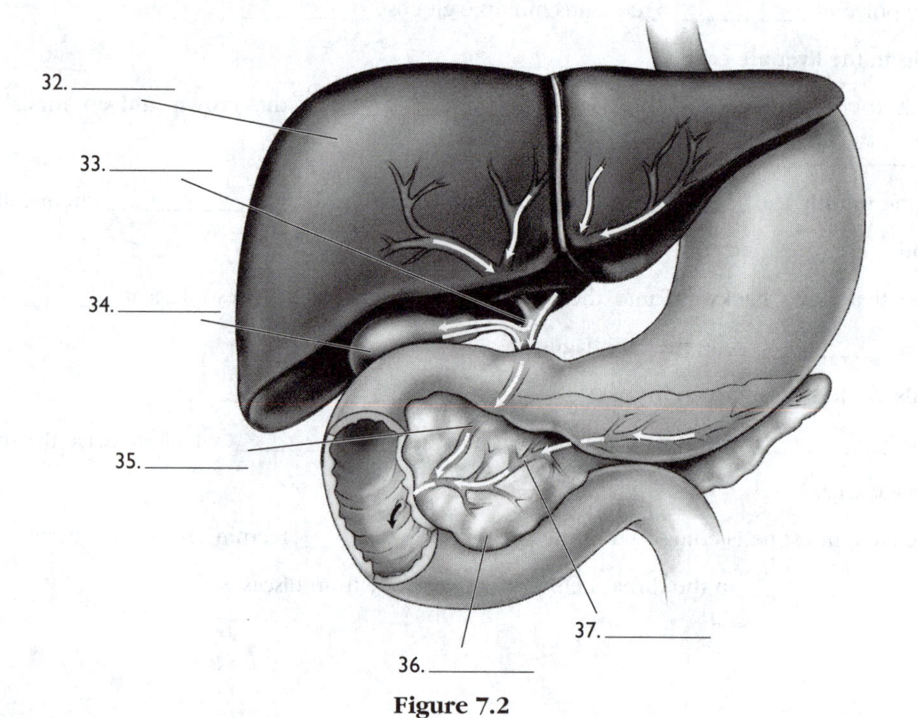

Figure 7.2

Roots to Remember

Use your knowledge of the root words presented in this chapter to answer the following questions.

38. According to the word's roots, what is hepatitis?
39. What is the name of an enzyme that can break down fat?
40. The root *pep-* is used in words that relate to what process?

Word Choice

Circle the word or phrase that correctly completes each sentence.

41. Liver (lobules/lipases) filter substances from the blood.
42. The (gallbladder/pancreas) is an accessory organ that holds bile.
43. Chyme is produced through digestion in the (stomach/duodenum).
44. The (incisors/molars) help to grind food.
45. When the concentration of blood sugar gets too low, the pancreas produces (insulin/glucagon).
46. Bile produced in the liver empties into the (stomach/duodenum) to aid digestion of fats.
47. Chemical digestion does *not* occur in the (mouth/esophagus).

Crossword Puzzle

Across

48. this amylase breaks down starch
49. you have both small and large
50. intolerance to milk products
51. yellowish-green substance
52. secreted by the pancreas

Down

53. rhythmic movement
54. infection in a vestigial organ
55. lacking essential nutrients
56. root meaning *tooth*
57. gastrin, secretin, and CCK are examples

Sequencing

Place the following organs of the digestive system in the correct sequence, with 1 being the first organ that encounters food that is being broken down and 8 being the last.

58. ____________ cecum

59. ____________ small intestine

60. ____________ rectum

61. ____________ tongue

62. ____________ esophagus

63. ____________ colon

64. ____________ stomach

65. ____________ anus

Critical Thinking

66. Laparoscopic gastric banding is a procedure that places a belt around the stomach to produce a small pouch. This band can be adjusted as the patient loses weight, and then removed completely. What are the potential advantages and disadvantages of the gastric band versus gastric bypass surgery?
67. Parasites and other pathogens can cause diarrhea. Why would diarrhea be an adaptation to the presence of a pathogen?

Practice Test

1. Which type of cell secretes mucus in the digestive tract?
   A. goblet
   B. mucosa
   C. squamous epithelial
   D. submucosa

2. After bile leaves the gallbladder, it moves into
   A. the duodenum.
   B. the common bile duct.
   C. the common hepatic duct.
   D. the stomach.

3. Which organ helps to break down wine or beer?
   A. the duodenum
   B. the stomach
   C. the liver
   D. the pancreas

4. Each accessory organ to the digestive system
   A. produces enzymes that are used by the system.
   B. participates in mechanical digestion.
   C. contributes substances that help with chemical digestion.
   D. filters toxins out of digested material.

5. Where is trypsin produced?
   A. in the liver
   B. in the pancreas
   C. in the stomach
   D. in the colon

6. Where does the digestion of protein begin?
   A. in the mouth
   B. in the stomach
   C. in the small intestine
   D. in the liver

7. Which of the following is *not* a major function of the large intestine?
   A. absorption of water
   B. lubrication of fecal matter
   C. absorption of nutrients
   D. storage of fecal matter

8. Which enzyme stimulates the production of gastric juices?
   A. cholecystokinin (CCK)
   B. pepsin
   C. secretin
   D. gastrin

9. Which of the following does *not* contribute to chemical digestion in the mouth?
   A. periodontal membrane
   B. sublinguals
   C. parotids
   D. submandibular glands

10. Air and food both naturally move through which structure?
    A. epiglottis
    B. trachea
    C. esophagus
    D. pharynx

11. Which part of the colon is located across the body and right below the stomach?
    A. ascending colon
    B. transverse colon
    C. descending colon
    D. sigmoid colon

12. After gastric bypass surgery, which of the following loses function?
    A. the stomach only
    B. the duodenum only
    C. the stomach and duodenum
    D. All organs retain functionality.

13. What happens in the body to cause type 1 diabetes mellitus?
    A. The pancreas starts to produce too much insulin.
    B. The pancreas stops being able to detect blood sugar levels.
    C. Immune cells mistakenly destroy pancreatic cells.
    D. Target cells become resistant to insulin.

14. Which structure brings oxygen-rich blood to the lobules of the liver?
    A. a branch of the hepatic artery
    B. the central vein
    C. a branch of the hepatic portal vein
    D. the bile duct

15. Which type of tissue is found in the serosa layer of the digestive tract?
    A. muscle
    B. connective
    C. nervous
    D. epithelial

Answer Key

Matching

1. e; **2.** a; **3.** g; **4.** i; **5.** l; **6.** c; **7.** b; **8.** j; **9.** h; **10.** n; **11.** d; **12.** m; **13.** o; **14.** f; **15.** k

Fill-in-the-Blank

16. common hepatic; gallbladder; **17.** glucagon; monomers; **18.** hepatocytes; **19.** enamel; dentin; **20.** teeth; enzymes; **21.** sphincter; ulcers; **22.** hydrochloric; pepsinogen; **23.** small; microvilli; **24.** colon; **25.** tonsils

Labeling

26. molars; **27.** premolars; **28.** canine; **29.** incisors; **30.** salivary duct; **31.** salivary glands; **32.** liver; **33.** common hepatic duct; **34.** gallbladder; **35.** common bile duct; **36.** pancreas; **37.** pancreatic duct

Roots to Remember

38. an inflammation of the liver; **39.** lipase; **40.** digestion

Word Choice

41. lobules; **42.** gallbladder; **43.** stomach; **44.** molars; **45.** glucagon; **46.** duodenum; **47.** esophagus

Crossword Puzzle

Across

48. pancreatic; **49.** intestines; **50.** lactose; **51.** bile; **52.** insulin

Down

53. peristalsis; **54.** appendicitis; **55.** deficient; **56.** dent; **57.** hormones

Sequencing

58. 5; **59.** 4; **60.** 7; **61.** 1; **62.** 2; **63.** 6; **64.** 3; **65.** 8

Critical Thinking

66. Advantages: The gastric band surgery would not require such major surgery, so risks would be lower and recovery would be faster. The band could be removed, but the gastric bypass is permanent. Disadvantages: The patient may regain weight after the band is removed.

67. Diarrhea is a way to flush the system quickly. This prevents the pathogen from remaining in the body and causing further damage to body systems. Therefore, diarrhea would be an adaptation to protect the long-term health of the body with short-term disease.

Practice Test

1. A; **2.** B; **3.** C; **4.** C; **5.** B; **6.** B; **7.** C; **8.** D; **9.** A; **10.** D; **11.** B; **12.** C; **13.** C; **14.** A; **15.** D

CHAPTER 8

THE BLOOD: MALARIA—A DEADLY BITE

Learning Goals

1. Summarize the major functions of blood in the body.
2. List the constituents of both the liquid and solid portions of the blood and describe their functions.
3. Explain how homeostasis of the oxygen-carrying capacity of the blood is maintained.
4. Describe the ABO and Rh blood type systems and explain the consequences of a blood type mismatch between donors and recipients (or between a mother and a developing fetus).
5. Describe the fate of red blood cells when they die.
6. List the techniques employed by the circulatory system to minimize blood loss from a damaged vessel and summarize the process of blood clotting.

Chapter Outline

I. The Constituents of Blood
  A. Overview
    1. The circulatory system ties all the systems of the body together.
    2. Blood is the principle connective tissue in this system.
      a) Blood transports nutrients and oxygen to cells.
      b) Blood removes metabolic wastes and carbon dioxide from cells to organs that can eliminate these wastes.
      c) Blood promotes homeostasis through regulation of body temperature, water volume, pH, and constituents that fight infection.
    3. Malaria is a blood-borne disease caused by a parasite.
      a) Malaria is spread by *Anopheles* mosquitoes.
      b) These mosquitoes inject *Plasmodium* parasites into human hosts.
  B. Plasma
    1. Plasma is approximately 90% water.
    2. The remaining 10% includes dissolved nutrients, gases, salts, and proteins.
      a) 60% of the proteins in plasma are albumins.
        (1) Albumins serve as transport proteins for substances that are not water soluble.
        (2) Albumins maintain tissue water balance by increasing the solute concentration in blood.
      b) Globulins are proteins that are also found in blood.
        (1) Globulins transport insoluble or small molecules throughout the body.
        (2) Antibodies that attack foreign proteins are globulins.
      c) The remaining plasma proteins are clotting factors.
    3. Plasma transports hormones and drugs throughout the body.
    4. Plasma minus its clotting factor proteins is called serum.
  C. Formed Elements: The Cellular Portion of Blood
    1. Red Blood Cells
      a) Red blood cells, or erythrocytes, carry oxygen from the lungs to body cells.
      b) An erythrocyte is a disk-shaped membrane filled with proteins.
        (1) The cell does not have a nucleus or organelles.
        (2) A lack of internal structure allows erythrocytes to fit through tight spaces.
        (3) The cell's pinched shape increases the surface area that is available for oxygen exchange.

c) Red blood cells make up the majority of cells in blood.
d) The red color of these cells is due to hemoglobin proteins.
  (1) Hemoglobin is made of four chains; each chain contains an iron atom in a heme group.
  (2) Each iron atom can bind to one oxygen molecule.
  (3) Oxygen-rich hemoglobin is called oxyhemoglobin and is bright red.
e) Erythrocytes turn purple in the absence of oxygen.
  (1) When hemoglobin releases its oxygen molecules, it becomes deoxyhemoglobin.
  (2) Oxygen is released to metabolically active tissues.
  (3) It is also released in conditions of low pH, such as when tissues are producing waste acids.
  (4) Deoxyhemoglobin is common in veins and gives veins a blue cast.
f) Heme groups on erythrocytes can bind to carbon dioxide.
  (1) Binding of carbon dioxide occurs under conditions of low oxygen and low pH.
  (2) Hemoglobin transports around 25% of the waste carbon dioxide in blood.
g) Malaria causes a dramatic decrease in red blood cells, which can reduce the oxygen-carrying capacity of blood.

2. White Blood Cells
  a) White blood cells, or leukocytes, attack invading organisms and remove toxins, wastes, and damaged cells from blood.
  b) Leukocytes are larger than red blood cells and contain organelles.
  c) These cells make up about 1% of total blood volume.
  d) Leukocytes can be classified by their appearance.
    (1) Granular leukocytes have small, grainy vesicles.
    (2) Agranular leukocytes contain small vesicles that are not visible.
  e) Leukocytes can be classified by their function.
    (1) Some leukocytes participate in a nonspecific immune response.
      (a) Eosinophils defend against large pathogens, such as tapeworms and flukes.
      (b) Basophils secrete histamine in response to injury, to trigger inflammation.
      (c) Neutrophils surround and engulf small, common invaders (such as bacteria).
      (d) Monocytes that have differentiated into phagocytic macrophages engulf large invaders and damaged cells.
    (2) Some leukocytes respond to specific pathogens.
      (a) Lymphocytes respond to unique chemical markers on invaders.
        (i) B lymphocytes produce protein antibodies that attach to and disable invaders.
        (ii) T lymphocytes differentiate into killer cells that target specific pathogens.
      (b) Cytokines produced by monocytes trigger activity of lymphocytes.

3. Platelets
  a) Platelets are membrane-bounded cell fragments without nuclei.
    (1) They are derived from megakaryocytes in bone marrow.
    (2) They last only three to five days.
  b) They play a role in blood clotting.

II. Malaria and the Blood

A. Overview
  1. Malaria may be one of the oldest human diseases.
  2. A few centuries ago, it likely caused 10% of human deaths.
  3. Malaria now causes up to 3 million deaths each year.
  4. Children in malaria-prone areas are most affected.

B. Malaria Infection
  1. Several species of *Plasmodium* can cause malaria.
  2. The *Plasmodium* parasite has a complicated life cycle.
    a) Infection by *Plasmodium falciparum*
      (1) *P. falciparum* exists in the form of sporozoites within the salivary gland of a mosquito.
      (2) Sporozoites are injected into the bloodstream through a mosquito bite.

(3) Sporozoites move into human liver cells and produce offspring called merozoites.
(4) Liver cells burst and release merozoites into the bloodstream.

b) Immune System Reaction and Parasite Response
(1) Monocytes release cytokines to trigger an immune response.
(2) Merozoites enter red blood cells to avoid immune system cells.
(3) The immune response decreases.
(4) Merozoites reproduce in red blood cells.
(5) Infected red blood cells burst and release more merozoites.
(6) Merozoites again stimulate the immune response.
(7) The cycle of reaction and response continues unless:
(a) the immune system kills the parasite;
(b) drugs are used to kill the parasite; or
(c) the infected person dies from the disease.

c) Disease Transmission
(1) Some merozoites develop into gametocytes, the sexual form of *Plasmodium*.
(2) An *Anopheles* mosquito picks up gametocytes when biting an infected person.
(3) Gametocytes produce sporozoites in the mosquito's gut.
(4) Sporozoites migrate to the mosquito's salivary gland and are injected when the mosquito bites another person.

C. Anemia and Blood Cell Production
1. Malaria can cause anemia.
2. Anemia reduces oxygen levels in the blood.
a) Low oxygen levels cause symptoms of fatigue and weakness.
b) Low oxygen levels trigger the kidneys to release erythropoietin.
(1) Erythropoietin stimulates the production of erythrocytes.
(2) This process may not keep up with the continuing destruction of red blood cells.

D. Blood Types and Transfusions
1. Severe anemia can be treated with transfusion of whole blood.
2. Blood types must match between donors and recipients.
a) ABO System
(1) Three alleles determine blood type in the ABO system.
(a) One allele codes for an enzyme that produces the A sugar.
(b) One allele codes for an enzyme that produces the B sugar.
(c) One allele (O) does not code for sugar production.
(d) Each person has a combination of two of these alleles.
(2) Individuals produce antibodies to the sugars they do not carry.
(a) Antibodies bind to foreign sugars (antigens) on blood cells.
(b) Agglutination of these cells occurs.
(c) Tagged cells are also broken apart by immune system cells.

b) Rh Factor
(1) Rh^+ people have Rh protein on red blood cells.
(2) A transfusion of Rh^+ blood into an Rh^- recipient can result in agglutination.
(3) The blood of an Rh^- mother may produce antibodies to the blood of an Rh^+ fetus.

c) Duffy negative
(1) Some people in West Africa lack the Duffy antigen on the surface of red blood cells.
(2) The Duffy antigen is required for *Plasmodium vivax* invasion.
(3) Duffy-negative individuals are immune to malaria caused by *P. vivax*.

E. Recycling Red Blood Cells
1. Damaged or old erythrocytes are sent to the liver and spleen for recycling.
a) Macrophages engulf the cells.
b) Components of the cells are recycled.
(1) Iron is sent to bone marrow for incorporation into new hemoglobin.
(2) Proteins and lipids are disassembled and carried to sites of protein or fat synthesis.
(3) Heme groups are transported to the liver for processing and excretion.
(a) Bilirubin is incorporated into bile.
(b) Bilirubin is modified by bacteria in the intestines.
(c) Modified bilirubin is defecated in feces.

2. Erythrocytes are recycled at a high rate during malarial infection.
    a) Increased recycling activity causes enlargement of the spleen.
    b) Bilirubin production increases and accumulates in blood plasma.
        (1) High bilirubin causes jaundice, or a yellowing of the skin, mucous membranes, and eyes.
        (2) Jaundice is also a sign of serious liver disease.

III. Blood Clotting
  A. Overview
    1. A vast network of blood vessels brings nutrients and oxygen to cells throughout the body.
    2. These vessels are susceptible to damage.
    3. Components in blood limit the loss of fluids due to vessel injury.
  B. The Clotting Cascade
    1. Hemostasis is the process that stems the flow of blood out of damaged blood vessels.
    2. Hemostasis involves a response from many components of blood vessels and the bloodstream.
        a) When a rupture occurs, muscles in the vessel constrict to slow the flow of blood.
        b) Collagen proteins trigger changes in platelets, so they swell and become sticky.
        c) Sticky platelets attach to damaged tissue, forming a plug.
        d) Nearby cells release the clotting factor thromboplastin, beginning the complex pathway that forms fibrin.
            (1) Thromboplastin sparks a series of reactions that produce an enzyme called prothrombin activator.
            (2) Prothrombin activator converts the plasma protein prothrombin into the enzyme thrombin.
            (3) Thrombin catalyzes the conversion of fibrogen into fibrin.
        e) Fibrin threads form a blood clot.
        f) As the clot solidifies, platelets within it contract, pulling vessel walls closer together.
        g) Cell division eventually seals the cut.
        h) The fibrin patch dissolves.
  C. Clotting Disorders
    1. Thrombocytopenia is a deficiency of platelets.
        a) Low platelet counts can lead to hemorrhage.
        b) Hemorrhage can lead to shock and/or death.
    2. Hemophilia is an inherited condition that leads to reduced or nonexistent blood clot formation.
    3. Malaria can cause blood cells to stick together.
        a) Accumulations of blood cells become clots that reduce gas and nutrient exchange.
        b) In the brain, these effects can lead to cerebral malaria, with symptoms that include dizziness, confusion, convulsions, coma, and death.

IV. Ending Malaria
  A. Drugs to Fight Malaria
    1. Chloroquine is the primary drug for treating malaria.
        a) It is cheap and widely available.
        b) In some parts of Africa and Southeast Asia, *P. falciparum* has developed resistance to this drug.
        c) After this treatment has been stopped for a few decades, in some areas, it may become effective again.
    2. Alternatives to chloroquine are expensive or have severe side effects.
  B. Immunization
    1. A vaccine has not yet been developed.
    2. New money is available, which has spurred research.
    3. An experimental vaccine reduced malaria incidence by 58% in immunized children.
  C. Prevention
    1. Draining standing water or filling small ponds can reduce mosquito breeding areas.
    2. Treating the interiors of houses with insecticides can reduce the risk of mosquito bites.
    3. Using insecticide-coated bed nets can reduce the risk of malaria by 25% or more.

Practice Questions

Matching

1. hemorrhage
2. anemia
3. macrophage
4. plasma
5. basophil
6. platelet
7. hemoglobin
8. leukocyte
9. fibrin
10. jaundice
11. erythrocyte
12. fibrinogen
13. serum
14. albumin
15. immunization

a. excessive blood flow
b. engulfs large foreign particles in the bloodstream
c. caused by high concentrations of bilirubin
d. white blood cell
e. blood protein that maintains tissue water balance
f. secretes histamine
g. liquid that contains water, nutrients, gases, salts, proteins, and clotting factors
h. treatment that stimulates the development of antibodies
i. creates a fibrous net to stop blood flow
j. caused by a reduction in red blood cells
k. liquid without clotting factors
l. protein in blood cells that transports oxygen
m. plasma protein that is activated by thrombin
n. cell fragment that contributes to clotting
o. disk-shaped cell with no nucleus

Fill-in-the-Blank

16. Someone who produces antibodies to both A and B antigens would have type ____________ blood.
17. Leukocytes called ____________ target small foreign cells, whereas monocytes that have differentiated into phagocytic ____________ target larger foreign cells.
18. During malaria infection, ____________ infect liver cells and produce ____________.
19. ____________ in a heme group of a ____________ molecule can bind to an oxygen molecule.
20. Leukocytes that contain large vesicles can be classified as ____________ leukocytes.
21. Megakaryocytes in the ____________ produce ____________.
22. The release of ____________ from the kidneys stimulates the production of ____________ blood cells.
23. Anti-Rh ____________ in an Rh^- mother can trigger ____________ disease in a subsequent ____________ fetus.
24. When the *Plasmodium* parasite first infects a person, it avoids the human immune system by hiding in ____________ cells.
25. ____________ can cause fatigue and weakness, due to a decline in ____________, or the percentage of red blood cells in the blood.

Labeling

Use the terms below to label Figure 8.1.

erythrocyte
lymphocytes
megakaryocyte
myeloid stem cell
neutrophil
white blood cells

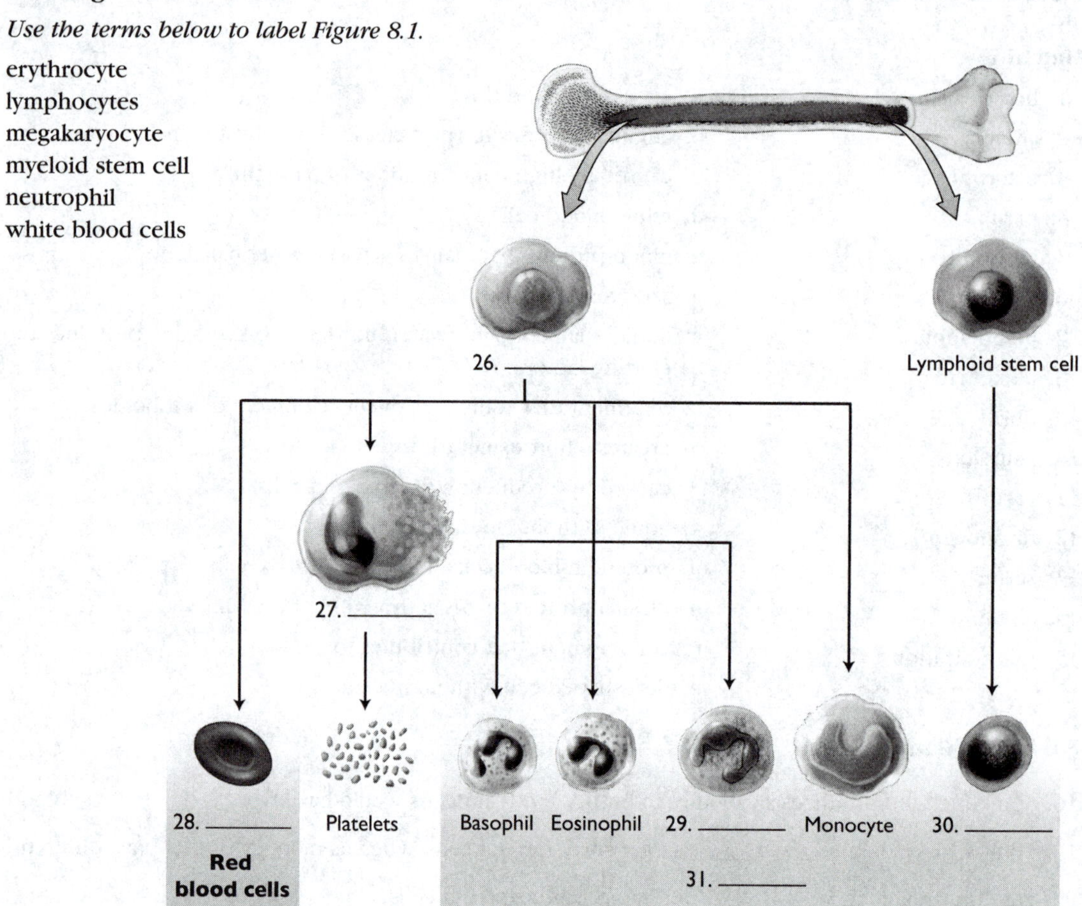

Figure 8.1

Use the terms below to label Figure 8.2.

fibrin
fibrinogen
prothrombin
thromboplastin

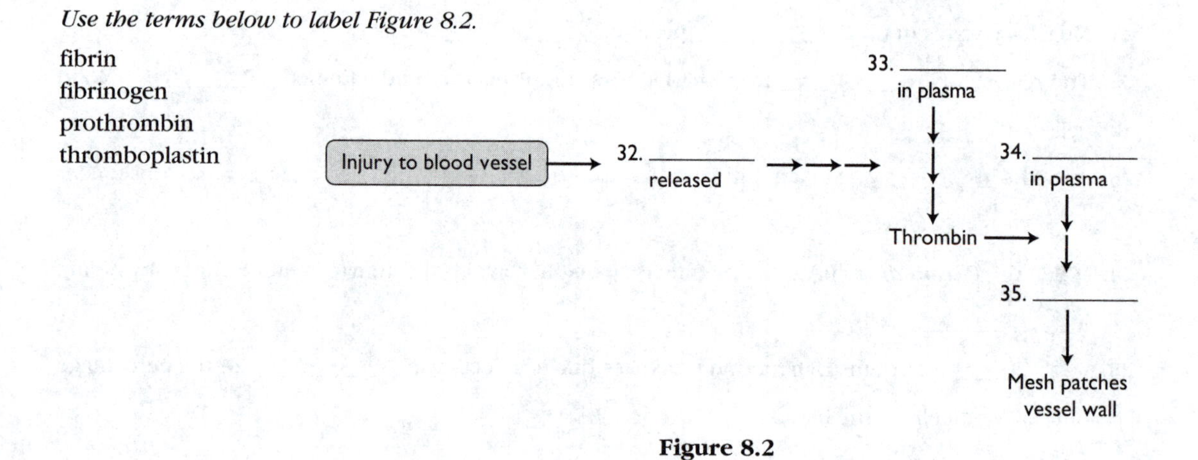

Figure 8.2

Roots to Remember

Use your knowledge of the root words presented in this chapter to answer the following questions.

36. What does *hemorrhage* literally mean, according to its roots?
37. What is a phagocyte?
38. What has happened when someone gets deep vein thrombosis (DVT)?
39. Using the roots in this chapter, name the disease that causes an abnormal increase in white blood cells.
40. Which roots in this chapter relate to blood?

Word Choice

Circle the word or phrase that correctly completes each sentence.

41. The liquid portion of blood that contains dissolved materials is called (plasma/serum).
42. A person with type B blood produces (anti-A/anti-B) antibodies.
43. Blood cells form in (red/yellow) bone marrow.
44. Homeostasis of blood oxygen is maintained with a (positive/negative) feedback loop.
45. *Plasmodium* sporozoites reproduce within (liver/red blood) cells.
46. (B/T) lymphocytes produce antibodies to specific antigens.
47. (Lymphoid/Myeloid) stem cells produce eosinophils.

Table Completion

Complete the following table showing the consequences of mixing different types of blood in the ABO blood system. Fill in each box with either "No Reaction" or "Agglutination" to indicate the outcome of each combination.

	Recipient Blood Type			
Donor Blood Type	**A**	**B**	**AB**	**O**
A	No Reaction	48.	49.	Agglutination
B	50.	51.	No Reaction	52.
AB	53.	54.	No Reaction	55.
O	No Reaction	56.	57.	No Reaction

Sequencing

Put the steps in the life cycle of Plasmodium falciparum *in the correct order, using the numbers 1 through 8. Step 1 begins just as the mosquito bites a person already infected with the malaria parasite.*

58. ____________ Sporozoites in liver cells produce offspring called merozoites.
59. ____________ Gametocytes reproduce in the gut of the *Anopheles* mosquito.
60. ____________ Sporozoites infect liver cells.
61. ____________ Gametocytes produced by merozoites enter the *Anopheles* mosquito.
62. ____________ Merozoites leave liver cells and infect erythrocytes.
63. ____________ Erythrocytes burst and release merozoites, which are attacked by the immune system.
64. ____________ Sporozoites are injected into a person by an infected mosquito.
65. ____________ Sporozoites move into the salivary gland of the *Anopheles* mosquito.

Critical Thinking

66. How is global warming likely to affect the distribution of malaria?
67. Sporozoites hide from the immune system in liver cells. Merozoites hide from the immune system in erythrocytes. What must be missing from the surfaces of these cells in order for the malarial parasites to avoid being attacked by the immune system?
68. Blood thinners are anticoagulant medications that are used to prevent heart attack and stroke. What are the risks of these medications?

Practice Test

1. A technician takes a sample of blood and counts the cells in the sample. Which type of cell is likely to be most common in the sample?
   A. erythrocyte
   B. eosinophil
   C. monocyte
   D. neutrophil

2. How many O_2 molecules can be bound to one heme group?
   A. one
   B. two
   C. three
   D. four

3. *Plasmodium* damages cells in which human organ(s)?
   A. the heart
   B. the stomach
   C. the liver
   D. the kidneys

4. Which of the following leukocytes is able to specifically target *Plasmodium* cells in the body?
   A. basophil
   B. monocyte
   C. eosinophil
   D. T lymphocyte

5. Erythropoietin is most similar to
   A. erythrocytes.
   B. testosterone.
   C. plasma.
   D. bone marrow.

6. As malaria changes and destroys red blood cells, it can cause all of the following *except*
   A. anemia.
   B. jaundice.
   C. cirrhosis.
   D. convulsions.

7. Which of the following is *not* found in blood serum?
   A. water
   B. thromboplastin
   C. albumin
   D. globulins

8. A person with AB^- blood would produce
   A. anti-A antibodies.
   B. anti-B antibodies.
   C. anti-A and anti-B antibodies.
   D. neither anti-A nor anti-B antibodies.

9. An Rh^- mother delivers a healthy Rh^+ baby with no complications. Which of the following is most likely to be true?
   A. The mother had already given birth to an Rh^+ baby.
   B. The mother had antibodies to the Rh antigen.
   C. The fetus and the mother had the same ABO system blood type.
   D. No fetal blood had entered the mother's system before birth.

10. What component of blood is the first line of defense after a blood vessel is damaged?
    A. plasma
    B. fibrin
    C. platelets
    D. macrophages

11. Oxyhemoglobin is
    A. bright red and oxygen rich.
    B. bright red and oxygen poor.
    C. purple and oxygen rich.
    D. purple and oxygen poor.

12. Where are sporozoites of the *Plasmodium* parasite produced?
    A. in the gut of the *Anopheles* mosquito
    B. in human liver cells
    C. in the salivary gland of the *Anopheles* mosquito
    D. in human red blood cells

13. Which leukocyte defends the body against tapeworms?
    A. basophil
    B. eosinophil
    C. monocyte
    D. neutrophil

14. Globulins are involved in
    A. the transport of oxygen and carbon dioxide.
    B. the immune response against antigens.
    C. the maintenance of tissue water balance.
    D. both the transport of gases and the maintenance of tissue water balance.

15. Which of the following conditions can promote the spread of malaria?
    A. presence of running water
    B. use of insecticide-coated bed nets
    C. immunization against the *Plasmodium* parasite
    D. long-term use of the chloroquine drug

Answer Key

Matching

1. a; **2.** j; **3.** b; **4.** g; **5.** f; **6.** n; **7.** l; **8.** d; **9.** i; **10.** c; **11.** o; **12.** m; **13.** k; **14.** e; **15.** h

Fill-in-the-Blank

16. O; **17.** neutrophils; macrophages; **18.** sporozoites; merozoites; **19.** Iron; hemoglobin; **20.** granular; **21.** bone marrow; platelets; **22.** erythropoietin; red; **23.** antibodies; hemolytic; Rh^+; **24.** liver; **25.** Anemia; hemocrit

Labeling

26. myeloid stem cell; **27.** megakaryocyte; **28.** erythrocyte; **29.** neutrophil; **30.** lymphocytes; **31.** white blood cells; **32.** thromboplastin; **33.** prothrombin; **34.** fibrinogen; **35.** fibrin

Roots to Remember

36. blood that bursts or flows excessively; **37.** a cell that "eats"; **38.** a clot has formed in a deep vein; **39.** leukemia; **40.** *-emia, hema-, hemato-, thromb-,* and *thrombo-*

Word Choice

41. plasma; **42.** anti-A; **43.** red; **44.** negative; **45.** liver; **46.** B; **47.** Myeloid

Table Completion

48. Agglutination; **49.** No Reaction; **50.** Agglutination; **51.** No Reaction; **52.** Agglutination; **53.** Agglutination; **54.** Agglutination; **55.** Agglutination; **56.** No Reaction; **57.** No Reaction

Sequencing

58. 6; **59.** 2; **60.** 5; **61.** 1; **62.** 7; **63.** 8; **64.** 4; **65.** 3

Critical Thinking

66. Global warming is likely to increase malaria distribution as temperate areas become more tropical.
67. Markers or antigens that identify the parasite as an invader must be missing.
68. If a person is injured, uncontrolled bleeding may occur because the medication would interfere with the clotting response.

Practice Test

1. A; **2.** A; **3.** C; **4.** D; **5.** B; **6.** C; **7.** B; **8.** D; **9.** D; **10.** C; **11.** A; **12.** A; **13.** B; **14.** B; **15.** D

CHAPTER 9

THE CARDIOVASCULAR SYSTEM: CAN WE STOP THE NUMBER-ONE KILLER?

Learning Goals

1. List the primary components of the circulatory system and describe the function of each.
2. Compare and contrast the structure and function of arteries, arterioles, capillaries, and veins.
3. List the materials that are exchanged between body and blood in a capillary bed.
4. Summarize how blood pressure is regulated in the body.
5. Describe the structure of the heart, including the four chambers and the valves.
6. Describe how blood moves through the double circulation system.
7. Describe the heart's electrical system, including the structures involved and the path and timing of the electrical signal.
8. Illustrate the cardiac cycle.
9. Describe coronary circulation.
10. Summarize the various ways that components of the cardiovascular system can fail, and describe the medical tools for repairing those failures.
11. List the five steps for maintaining good heart health.

Chapter Outline

I. Blood and Lymphatic Vessels: The Circulation Pipes
   A. Overview
      1. Systems
         a) The vascular system is the system of tubes that carry blood in the body.
         b) The cardiovascular system includes the vascular system, heart, and blood.
         c) The circulatory system includes the cardiovascular system and the lymphatic system.
      2. Blood Vessels
         a) Arteries carry blood away from the heart.
         b) Veins carry blood to the heart.
         c) Capillaries are tiny vessels that create a net of channels between veins and arteries.
   B. Arteries and Arterioles
      1. Large arteries have thick, elastic walls of smooth muscle and connective tissue.
         a) Walls balloon out as blood flows through.
         b) Walls snap back after blood passes.
         c) The wave of blood is called a pulse.
      2. Arterioles are surrounded by smooth muscle.
         a) Vasoconstriction reduces arteriole diameter and increases blood pressure.
         b) Vasodilation increases arteriole diameter and decreases blood pressure.
   C. Capillaries: The Distribution Network
      1. Capillaries have a single thin endothelial wall.
         a) The wall is composed of simple squamous epithelial cells.
         b) A thin wall allows diffusion of gases, nutrients, and other substances.
         c) Macrophages can move between epithelial cells.
         d) Gaps or pores between cells in some capillaries allow for exchange of larger molecules.
      2. Capillaries are about the diameter of a single red blood cell.
      3. Exchange of substances occurs in capillary beds.
         a) Materials in the capillaries are forced out near the arterial end of the bed.

b) Wastes and water move down the concentration gradient into capillaries on the venous end of the bed.
c) Precapillary sphincters regulate blood flow in the bed.

D. Veins: The Path Back to the Heart
1. Venules move blood from the venous capillary bed to the veins.
2. Veins have thin walls that are less elastic than arteries.
a) Blood pressure is lower in veins.
b) Most blood volume is stored in veins.
3. Other structures help blood to move through veins to the heart.
a) Skeletal muscles squeeze veins.
b) Veins have one-way valves to keep blood from moving backward.
c) The respiratory pump forces blood to the heart.
d) Veins can constrict or dilate to control blood flow.

E. The Lymphatic System: Draining the Tissues
1. Fluids accumulate in tissues around capillary beds.
2. The lymphatic system collects excess fluid from tissues.
a) Fluid and white blood cells are called lymph.
b) Lymph drains into lymphatic capillaries.
c) Lymphatic capillaries drain into larger lymphatic vessels.
(1) Lymphatic vessels are similar to veins.
(2) Lymphatic vessels have valves like veins.
d) Lymphatic vessels empty into lymph ducts.
e) Lymph ducts return lymph to chest veins.

F. Control of Blood Pressure
1. Blood pressure is created by
a) the force of blood moving from the heart, and
b) the diameter of arterioles.
2. The body responds when blood pressure is too high.
a) Artery walls stretch.
b) Stretching of artery walls signals a decrease in water retention by the kidneys.
c) The cardiovascular center in the brain causes vasodilation and a decrease in heart rate.
d) Endothelial cells in arterioles that are damaged by the pressure release nitric acid, which causes the smooth muscle cells in the arterioles to dilate.
3. The body responds when blood pressure is too low.
a) The cardiovascular center increases heart rate and causes vasoconstriction.
b) The kidneys retain more water.

G. Fixing the Pipes
1. Hypertension is chronically high blood pressure.
2. Causes of hypertension include
a) stress,
b) high salt intake,
c) obesity,
d) genetic susceptibility, and
e) atherosclerosis.
(1) Atherosclerosis is the accumulation of fatty plaque in arteries.
(2) Genetics, aging, high-fat diet, and smoking can cause atherosclerosis.
3. A sphygmomanometer measures blood pressure.
a) Systolic pressure is the highest pressure in the artery.
b) Diastolic pressure is the lowest pressure in the artery.
c) Normal blood pressure is 120 mm Hg systolic and 80 mm Hg diastolic pressure.
4. Hypertension can trigger disease.
a) Risk of aneurysm increases.
b) Development of atherosclerosis can also increase.
(1) Plaques can break up and clog vessels.
(2) Blockage of blood vessels near the brain can cause stroke.
(3) Blockage of blood vessels near the heart can cause heart attack.
5. Hypertension can be treated with lifestyle and diet changes or drugs.

II. The Mechanical Heart
 A. Structure of the Heart
  1. The heart has side-by-side muscular pumps.
   a) The right pump sends oxygen-poor blood from the body to the lungs.
   b) The left pump sends oxygen-rich blood from the lungs to the body.
   c) The pumps are divided by the septum.
  2. Each pump has two chambers.
   a) The atrium is thin walled and elastic.
   b) The ventricle is thick walled.
  3. The heart is surrounded by a membranous sac called the pericardium.
  4. Valves prevent backflow of blood in the heart.
   a) Atrioventricular (AV) valves are found between atria and ventricles.
    (1) The tricuspid valve is in the right pump.
    (2) The bicuspid or mitral valve is in the left pump.
   b) Semilunar valves are found between ventricles and associated blood vessels.
    (1) The pulmonary valve is between the right ventricle and pulmonary artery.
    (2) The aortic valve is between the left ventricle and aorta.
 B. The Cardiovascular Pathway
  1. The cardiovascular system has two main circuits.
   a) The pulmonary circuit moves blood between the lungs and heart.
    (1) Deoxygenated blood from the body moves from the right ventricle to the pulmonary artery.
    (2) Blood travels to the capillary bed in the lungs.
    (3) Oxygenated blood moves to the heart through the pulmonary veins.
   b) The systemic circuit moves blood between the heart and the rest of the body.
    (1) Blood moves from the left ventricle of the heart into the aorta.
    (2) Oxygenated blood moves through arteries, arterioles, and into capillary beds.
    (3) Deoxygenated blood moves from capillary beds to venules and veins.
    (4) The superior vena cava carries deoxygenated blood from the head, neck, and arms to the heart.
    (5) The inferior vena cava carries deoxygenated blood from the abdomen and legs to the heart.
  2. There are two smaller circuits.
   a) Coronary circulation provides blood to heart muscle.
   b) The hepatic portal system connects most organs to the digestive system and liver.
 C. Repairing the Pump
  1. A hole in the heart results from incomplete formation of the septum.
   a) This disease causes oxygenated blood to mix with deoxygenated blood.
   b) This disease can be repaired by surgery.
  2. Valve damage or deformation can cause bad circulation or blood clots.
   a) Valves can be damaged by rheumatic fever.
   b) Damaged valves create a swishing sound called a heart murmur.
   c) Valves can be replaced.
  3. Heart attacks can lead to heart failure.
   a) In congestive heart failure, fluid backs up into the lungs.
   b) Heart failure can often be treated with drugs.
   c) A damaged heart can sometimes be replaced with a donor heart or artificial heart.

III. The Electrical Heart
 A. The Cardiac Cycle
  1. The myocardium, or heart muscle, contains intercalated discs.
   a) Intercalated discs are areas where the membranes of muscle cells are interconnected by gap junctions.
   b) Intercalated discs allow communication between heart muscle cells.
  2. The sinoatrial (SA) node is the heart's pacemaker.
   a) It is located in the right atrium wall.
   b) It sends electrical signals that trigger atrial contraction.

3. The atrioventricular (AV) node triggers contraction of the ventricles.
   a) It receives its signal from the SA node.
   b) The delay between atrial and ventricular contractions allows the ventricles to fill with blood.
4. The process of filling the heart with blood and pumping blood out of the heart is called the cardiac cycle.
   a) Diastole is the relaxed period that allows blood to fill the heart.
   b) Systole is the contraction phase of the cardiac cycle.
      (1) Atrial systole is when atria contract.
      (2) Ventricular systole is when ventricles contract.
   c) AV valves close during ventricular systole.
   d) Semilunar valves close at the beginning of diastole.
5. Heart rate is the speed of the cardiac cycle.
   a) Negative feedback regulates heart rate and stroke volume (the amount of blood pumped per beat).
   b) Hormones, stress, fear, and drugs can affect heart rate.

B. Steadying the Heartbeat
1. Cardiac arrest is when the heart suddenly stops beating.
2. Cardiac arrest often occurs when electrical signals become too fast, too slow, or disorganized.
   a) Fibrillation is chaotic contraction of the heart.
      (1) Atrial fibrillation can increase clot formation.
         (a) Blood thinners can reduce the risk of clots.
         (b) Filling an indentation in the left atrium can also reduce the risk of clots.
      (2) Ventricular fibrillation stops blood from leaving the heart and can cause death.
         (a) Cardiopulmonary resuscitation can keep blood flowing to essential organs.
         (b) A defibrillator can shock the SA node back into normal rhythm.
3. Underlying heart conditions such as arrhythmia can cause cardiac arrest.
   a) An artificial pacemaker may be used to override a faulty SA node.
   b) A pacemaker sends an electrical pulse if heart rate deviates from normal.

IV. Power for the Heart

A. Coronary Blood Vessels
1. Coronary blood vessels are on the outer surface of the heart.
   a) Coronary arteries receive blood from the aorta.
   b) Coronary arteries split into arterioles and capillaries.
   c) Venules drain into cardiac veins.
   d) Cardiac veins drain into the right atrium.
2. Coronary blood vessels are under high pressure.
   a) They are more prone to damage.
   b) Cholesterol can accumulate on damaged surfaces within these vessels.

B. Maintaining the Heart's Energy Supply
1. Myocardial infarction (MI) or heart attack occurs when the coronary artery is blocked.
2. MI causes heart muscle to become oxygen starved and die.
3. Angina signals the beginning of blockage.
   a) Angina causes chest discomfort.
   b) Angina can be diagnosed with a stress test.
4. An angiogram or echocardiogram can show blockage.
   a) Blockage is a condition called coronary heart disease.
   b) Blockage can be treated.
      (1) In angioplasty, a stent is implanted to prop open blood vessels.
      (2) In coronary bypass surgery, new blood vessels are attached to the heart to bypass clogged vessels.

C. The Healthy Heart
1. Some genes are associated with heart disease.
2. You can reduce the risk of heart disease by making good lifestyle choices.
   a) Don't smoke or use tobacco.
   b) Exercise.
   c) Eat a diet high in fruits, vegetables, whole grains, and low-fat dairy.
   d) Maintain a healthy weight.
   e) Have regular checkups.

Practice Questions

Matching

1. pericardium	a. heart muscle
2. venules	b. vessels directly upstream of capillaries
3. myocardial infarction	c. sac surrounding the heart
4. aorta	d. chaotic heart contractions
5. myocardium	e. relaxation phase of the cardiac cycle
6. systole	f. caused by blockage to blood vessels around the brain
7. fibrillation	g. artery that delivers blood from the heart to the body
8. ventricle	h. vessels directly downstream of capillaries
9. lymph	i. white blood cells and fluid from tissues
10. stroke	j. thin-walled heart chamber
11. arterioles	k. thick-walled heart chamber
12. defibrillator	l. when blockage stops blood flow to heart muscle
13. atrium	m. a device that restores regular heart rhythm
14. heart murmur	n. contraction phase of the cardiac cycle
15. diastole	o. caused by valve malfunction

Fill-in-the-Blank

16. Ventricles are separated by the ____________.

17. The two vena cavae carry blood from the ____________ to the ____________.

18. The ____________ valve is a(n) ____________ valve, which is also called the bicuspid valve.

19. ____________ carry blood that brings oxygen and nutrients to heart muscle.

20. The ____________ valves are forced open during ____________ systole.

21. Heart rate can be increased by the release of the stress hormone ____________, which is also called adrenaline.

22. The ____________ ventricle sends blood to the body, and the ____________ ventricle sends blood to the lungs.

23. The ____________ is an organ in the lymphatic system that stores and ____________ blood.

24. Arteries have ____________ layers of tissue, and veins have ____________ layers.

25. ____________ within veins keep blood from moving ____________ the heart.

Labeling

Use the terms below to label Figure 9.1.

aorta
aortic valve
left atrium
left ventricle
mitral (bicuspid) valve
pulmonary artery
pulmonary valve
pulmonary vein
right atrium
right ventricle
superior vena cava
tricuspid valve

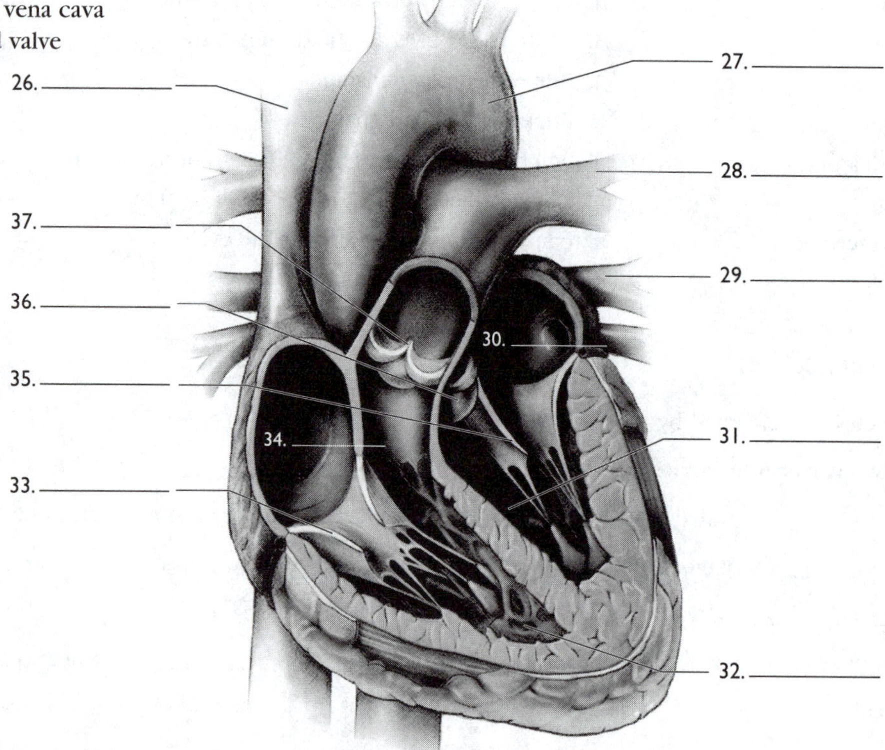

Figure 9.1

Roots to Remember

Use your knowledge of the root words presented in this chapter to answer the following questions.

38. What is the meaning of *angioplasty*, according to its word roots?
39. A pulmonary embolism is a floating blood clot that affects what organs?
40. What is a cardio workout?

Word Choice

Circle the word or phrase that correctly completes each sentence.

41. An (angiogram/echocardiogram) is an ultrasound image of the heart.
42. Blockage in a (coronary/cardiac) artery can result in a heart attack.
43. An artificial pacemaker sends an electronic pulse to override the (SA/AV) node.
44. The (bicuspid/tricuspid) valve lies between the right atrium and ventricle.
45. Chest pain, or (angina/arrhythmia), is generally diagnosed with a stress test.
46. A bulge in a blood vessel is called (an aneurysm/plaque).
47. Blood pressure is lowest during (diastole/systole).

Table Completion

Complete the following table on cardiovascular conditions and diseases.

Disease or Condition	Description	Treatment or Prevention
48.	Blood vessel bulges or ruptures	Surgery to bypass or reinforce the weakened vessel
Arrhythmia	49.	Drugs or pacemaker
Coronary heart disease	50.	51.
52.	High blood pressure	53.
Atherosclerosis	54.	55.
56.	Heart murmur	57.

Sequencing

Use the diagram to trace the pathway of blood within the cardiovascular system. Write each number in the diagram next to its matching statement below.

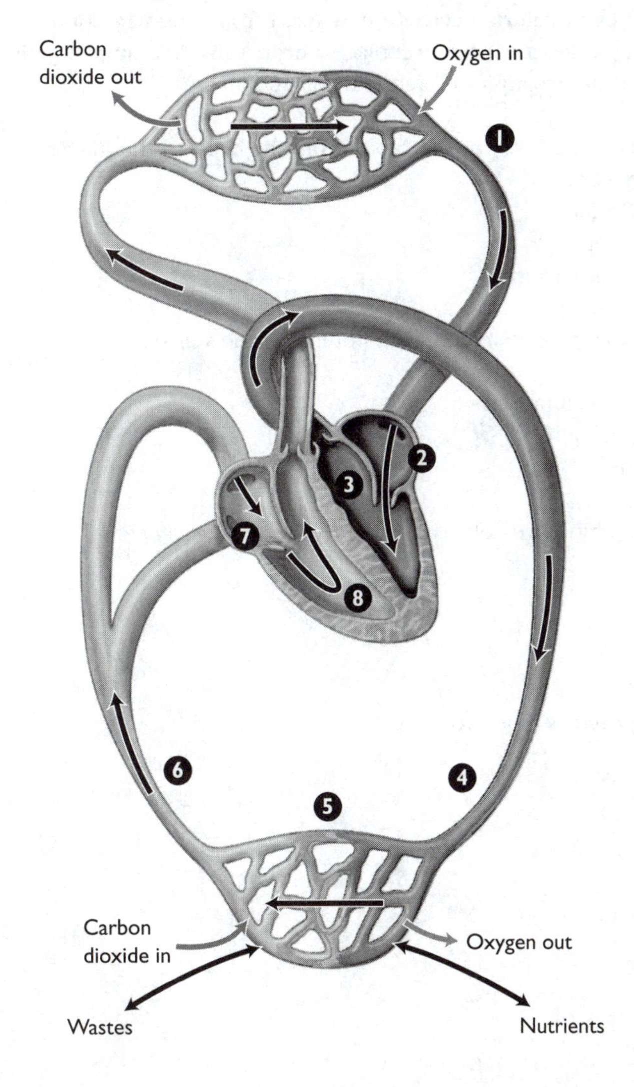

58. ____________ Contraction moves blood from the right atrium into the right ventricle.

59. ____________ Blood moves from the arterioles into capillary beds.

60. ____________ Blood from the lungs travels to the heart.

61. ____________ Gas exchange occurs in capillary beds.

62. ____________ Contraction moves blood from the left atrium into the left ventricle.

63. ____________ Contraction moves blood from the right ventricle to the lungs.

64. ____________ Contraction moves blood from the left ventricle to the body.

65. ____________ Blood moves through the venules and veins.

Critical Thinking

66. A study at the University of California, Irvine, found that isolated systolic hypertension is the dominant form of uncontrolled high blood pressure for people over age 50. Why could a high systolic pressure reading lead to health problems?
67. Why might an injury that severs an artery be more deadly than an injury that severs a vein?
68. A person confined to a hospital bed may wear a device that inflates and applies pressure to the calves. Why would such a device be prescribed?

Practice Test

1. Which of the following is an example of xenotransplantation?
   A. An LVAD device is implanted into a patient with heart failure.
   B. A patient with heart failure receives a new heart from a human donor.
   C. A valve from a pig heart is used to replace a heart valve in a human patient.
   D. All of the above are examples of xenotransplantation.

2. Which cardiovascular pathway moves blood from the heart to the lungs?
   A. the pulmonary circuit
   B. the systemic circuit
   C. the coronary circuit
   D. the hepatic portal system

3. Which of the following keeps blood from pooling in the veins?
   A. elastic walls
   B. squeezing of skeletal muscles
   C. valves within the veins
   D. dilation of veins

4. All of the following blood vessels have three tissue layers *except*
   A. arterioles.
   B. venules.
   C. arteries.
   D. veins.

5. The renal artery provides blood to the
   A. liver.
   B. kidneys.
   C. legs.
   D. abdomen.

6. Rheumatic fever can damage
   A. heart valves.
   B. heart muscle.
   C. arterial walls.
   D. the pericardium.

7. Which of the following happens right after the SA node fires in a healthy heart?
   A. The AV node fires.
   B. The ventricles fill with blood.
   C. The AV valves open.
   D. The atria contract.

8. The blood thinner *warfarin* is commonly used to prevent
   A. hypertension.
   B. stroke.
   C. arrhythmia.
   D. atherosclerosis.

9. Which of the following is *not* always true about veins?
   A. Veins carry blood toward the heart.
   B. Vein walls are composed of three layers of tissue.
   C. Veins carry deoxygenated blood.
   D. Venules drain into veins.

10. Place the following steps of angioplasty in the proper sequence:
    1. Stent expands to flatten plaque.
    2. Catheter is inserted.
    3. Balloon is deflated and removed.
    4. Balloon is inflated.

    A. 4, 1, 2, 3
    B. 1, 2, 4, 3
    C. 2, 4, 1, 3
    D. 4, 2, 1, 3

11. What does a sphygmomanometer measure?
    A. heart rate
    B. electrical signals in the heart
    C. blood pressure
    D. rate of respiration

12. What does a beta blocker do?
    A. It blocks the nerve receptors that trigger an increase in heart rate.
    B. It blocks the formation of blood clots in blood vessels.
    C. It blocks the production of LDL cholesterol.
    D. It blocks sodium from entering heart muscles, so heart rate is more stable.

13. The hepatic portal system
    A. carries blood from the liver to the digestive organs.
    B. is a main part of the systemic circuit.
    C. carries waste materials from the digestive organs to the liver.
    D. only operates right after you eat a meal.

14. A middle-aged woman is slightly overweight and a smoker. Which of the following lifestyle changes is likely to cause the greatest reduction in her risk for heart disease?
    A. starting an exercise program
    B. losing a few pounds
    C. switching to a low-fat diet
    D. quitting smoking

15. A physician listening to a patient's heart hears the following sound: "lub dupp whoosh, lub dupp whoosh." What might the physician conclude?
    A. The patient's heart is operating normally.
    B. The patient may be suffering from arrhythmia.
    C. The patient may have a faulty heart valve.
    D. The patient may need an artificial pacemaker.

Answer Key

Matching

1. c; **2.** h; **3.** l; **4.** g; **5.** a; **6.** n; **7.** d; **8.** k; **9.** i; **10.** f; **11.** b; **12.** m; **13.** j; **14.** o; **15.** e

Fill-in-the-Blank

16. septum; **17.** body; heart; **18.** mitral; atrioventricular (AV); **19.** Coronary arteries; **20.** semilunar; ventricular; **21.** epinephrine; **22.** left; right; **23.** spleen; purifies; **24.** three; three; **25.** Valves; away from

Labeling

26. superior vena cava; **27.** aorta; **28.** pulmonary artery; **29.** pulmonary vein; **30.** left atrium; **31.** left ventricle; **32.** right ventricle; **33.** tricuspid valve; **34.** right atrium; **35.** mitral (bicuspid) valve; **36.** aortic valve; **37.** pulmonary valve

Roots to Remember

38. forming or molding of a blood vessel; **39.** the lungs; **40.** a series of exercises that get the heart pumping faster

Word Choice

41. echocardiogram; **42.** coronary; **43.** SA; **44.** tricuspid; **45.** angina; **46.** an aneurysm; **47.** diastole

Table Completion

48. *Aneurysm*; **49.** Irregular or unusual heart rhythm; **50.** Blockage of coronary arteries; **51.** Angioplasty or coronary bypass surgery; **52.** *Hypertension*; **53.** Stress reduction, drugs, reduction in salt intake; **54.** Accumulation of plaque in arterial walls; **55.** Drugs, reduction in dietary fat, smoking cessation; **56.** *Heart valve malfunction*; **57.** Replacement with synthetic or biological valve

Sequencing

58. 7; **59.** 4; **60.** 1; **61.** 5; **62.** 2; **63.** 8; **64.** 3; **65.** 6

Critical Thinking

66. The systolic pressure is the highest pressure of the artery. Having a high systolic pressure indicates that pressure on the blood vessels goes too high and puts the blood vessels at risk for aneurysm.

67. Arteries are under more pressure, so blood can be released more quickly from a severed artery.

68. The device presses on the veins to maintain blood flow back to the heart. The device might be needed when the person is unable to walk around and exercise the calf muscles.

Practice Test

1. C; **2.** A; **3.** B; **4.** A; **5.** B; **6.** A; **7.** D; **8.** B; **9.** C; **10.** C; **11.** C; **12.** A; **13.** C; **14.** D; **15.** C

CHAPTER 10

THE RESPIRATORY SYSTEM: SECONDHAND SMOKE

Learning Goals

1. Describe the structure and function of the respiratory system.
2. Describe the path of air into the body.
3. Describe the muscles involved in breathing and explain how their movements facilitate air movement into and out of the lungs.
4. Explain the principle of partial pressure, and detail how differences in partial pressure cause gas exchange across the respiratory membrane in the lungs.
5. Explain the role of hemoglobin in gas exchange.
6. List the effects of smoking on the various structures and functions of the respiratory tract.
7. Briefly explain the reasons why smoking is powerfully addictive.

Chapter Outline

I. Respiratory System Anatomy: The Path of Smoke into the Lungs
  A. Upper Respiratory Tract
    1. Consists of the structures in the head and neck.
      a) Air enters through the nose and mouth.
      b) Air moves from the nostrils to the nasal cavity.
      c) The nasal cavity connects to Eustachian tubes and sinuses.
      d) The pharynx, epiglottis, glottis, and larynx are found in the neck.
    2. Filters and conditions air and shapes sounds.
      a) Air entering the nasal cavity is filtered, warmed, and humidified.
      b) Cilia on the epithelial cells of the mucous membranes trap particles and move them out of the upper respiratory system.
      c) The epiglottis covers the glottis and prevents food from entering the lungs.
      d) Sounds produced by vocal cords in the larynx are modified by structures in the mouth and muscles in the throat.
  B. Lower Respiratory Tract
    1. Consists of the larynx down to the lungs.
      a) Air moves down the trachea into bronchi.
      b) Air moves from the bronchi into the bronchioles of the lungs.
      c) Bronchioles end in small sacs called alveoli.
      d) Lungs are encased in a pleural membrane.
    2. Conducts gas exchange, expels foreign particles, and produces vocal sounds.
      a) Vocal cords in the larynx produce sound waves.
      b) Coughing sends a burst of air to clear the bronchi and trachea.
      c) Cilia in bronchi move foreign particles up and out of the body.
      d) Alveoli surrounded by capillaries facilitate gas exchange.

II. Tobacco Smoke and the Respiratory Tract
  A. The Composition of Tobacco Smoke
    1. Types of Smoke
      a) Mainstream smoke is inhaled from a cigarette, cigar, or pipe, or is exhaled by the smoker.
      b) Sidestream (secondhand) smoke is emitted from the lit end of a cigarette, cigar, or pipe.
    2. Tobacco smoke contains more than 4,500 chemicals.
      a) Dozens of these chemicals are carcinogens.

      b) Carbon monoxide is the main gas in secondhand smoke.
      c) Airborne particulates are also produced.
  B. Smoke Damages the Respiratory System
    1. Mucous membranes lining the nasal passages and sinuses can be damaged.
      a) When cilia are destroyed, the body cannot move particles out of the respiratory system quickly.
      b) Particles trapped in the respiratory system can cause sinusitis.
    2. Secondhand smoke can cause ear infection, tonsillitis, or asthma in children.
    3. Smoke can irritate vocal cords, causing laryngitis.
    4. Smoking can cause bronchitis or pneumonia.
    5. Smoking can cause head and neck cancers, or lung cancer.

III. Inhaling and Exhaling
  A. The Mechanics of Breathing
    1. The diaphragm and intercostal muscles between the ribs control inhalation.
      a) Muscle contraction is triggered by the respiratory center.
      b) Inhalation is triggered by contraction of the diaphragm.
      c) As air is inhaled, the intercostals increase the chest's diameter by moving the ribs up and outward.
      d) As the volume of the chest cavity increases, the volume of the lungs increases.
    2. Pressure changes during inhalation and exhalation.
      a) Air pressure in the lungs drops during inhalation.
      b) Air flows from higher pressure outside the body to the lower pressure inside the lungs.
      c) When stretch receptors signal the respiratory center to shut off signals, the diaphragm and intercostal muscles relax.
      d) Chest volume decreases, causing higher air pressure in the lungs.
      e) High pressure air flows out of the body, through exhalation.
    3. Exhalation can be passive or active.
      a) Resting exhalation is passive.
      b) Exhalation during exercise is active.
        (1) Active exhalation involves contraction of muscles in the chest wall and abdomen.
        (2) Active exhalation helps the body to get rid of carbon dioxide quickly.
  B. The Control of Breathing
    1. Breathing is automatic.
      a) The medulla oblongata contains a respiratory center that triggers the muscles controlling inhalation.
      b) Regular impulses are sent at a steady rate during rest.
    2. Sensors gauge carbon dioxide (CO_2) levels in body fluids.
      a) A drop in pH causes an increase in the respiratory cycle rate.
      b) An increase in respiratory cycle rate signals deeper or more frequent breathing.
    3. Receptors in arteries sense oxygen levels in blood.
      a) These sensors are not as sensitive as pH sensors in the brain.
      b) A decline in oxygen levels triggers an increase in the respiratory cycle rate.
    4. The cerebral cortex allows conscious control over breathing rate.
    5. Apnea can be caused by a malfunctioning respiratory center or obstruction of the upper respiratory tract.
  C. Smoking and Breathing
    1. Deep inhalation of smoke causes damage.
      a) Smoke damages alveoli.
      b) Damage to alveoli reduces the surface area for gas exchange.
      c) Bronchioles may collapse, causing trapped air to build up in lungs.
        (1) Damage to alveoli and bronchioles causes emphysema.
        (2) Patients with emphysema have trouble breathing.
    2. Emphysema and chronic bronchitis caused by smoking combine to form a condition called chronic obstructive pulmonary disease (COPD).

IV. Gas Exchange in the Lungs
  A. A Closer Look at Gas Exchange
    1. Breathing is called ventilation.
      a) Tidal volume is the amount you breathe in and out.

b) Vital capacity is the maximum amount of air you can exhale after a deep breath.
c) Residual volume is the amount of air that always remains in your lungs.

2. Gas exchange relies on changes in pressure.
   a) Partial pressure is the concentration of one gas in a gas mixture.
   b) The pressure difference between oxygen and CO_2 in blood is large.
      (1) The pressure difference allows diffusion of carbon dioxide from cells into the blood and from the blood into the lungs.
      (2) The pressure difference allows diffusion of oxygen from the lungs into the blood and from the blood into cells.
   c) Gases must dissolve before diffusing.
      (1) A soaplike surfactant coats the inner respiratory surface.
      (2) Surfactant can be negatively affected by tobacco smoke.
   d) Oxygen uptake is promoted by high partial pressure of oxygen, low temperature, and high pH.
      (1) Hemoglobin binds oxygen when oxygen partial pressure is high.
      (2) The lungs are a low-temperature, high-pH environment.
   e) Oxygen release is promoted by low partial pressure of oxygen, high temperature, and low pH.
      (1) Hard-working tissues produce heat and pH-lowering CO_2.
      (2) Hemoglobin releases oxygen to active tissue.
   f) CO_2 diffuses into the bloodstream to be removed.
      (1) 10% of this CO_2 dissolves into plasma.
      (2) 90% enters erythrocytes.
         (a) 20% of the CO_2 in erythrocytes binds with the globin in hemoglobin.
         (b) 80% combines with water in the erythrocyte and dissociates into hydrogen and bicarbonate ions.
            (i) Bicarbonate ions exit the cell and accumulate in plasma.
            (ii) Bicarbonate ions maintain a stable blood pH.
            (iii) Excess hydrogen ions bind with the protein in hemoglobin and produce reduced hemoglobin.
   g) Low partial pressure of CO_2 in the lungs leads to diffusion of CO_2 from blood into air.

B. Smoking and Gas Exchange
   1. The heme group in hemoglobin has a strong affinity for carbon monoxide.
   2. Large amounts of hemoglobin can be tied up with carbon monoxide instead of oxygen.
   3. Lack of oxygen can damage tissues.
   4. Smoking during pregnancy can deprive the fetus of oxygen and cause low birth weight and diminished brain function in the baby.

C. Nicotine: Why Tobacco is Habit Forming
   1. Nicotine is an alkaloid that affects the nervous system.
   2. Nicotine stimulates the brain's pleasure center.
   3. Lack of nicotine can cause depression, fatigue, and cravings.

V. Beyond the Lungs

A. The Effects of Smoke on Other Organ Systems
   1. Cardiovascular System
      a) Most deaths due to smoking result from cardiovascular disease.
      b) Nicotine increases production of LDL "bad" cholesterol and decreases production of HDL "good" cholesterol.
      c) Nicotine stimulates blood clot formation.
   2. Digestive and Excretory System
      a) Toxins from tobacco smoke increase the risk of stomach or pancreatic cancer.
      b) Smoke exposure increases the risk of stomach ulcers caused by *Helicobacter pylori* bacteria.
      c) Excreted toxins increase the risk of kidney and bladder cancer.
   3. Senses and Brain
      a) Hydrogen cyanide damages sensory receptors in the nose and mouth.
      b) Smoking increases the risk of cataracts.
      c) Smoking increases the risk of stroke.
      d) Smoking causes addiction.

B. Preventing Smoking-Related Illness
   1. There are many tools available to help people quit smoking.
   2. Societal factors that promote smoking can be reduced.

Practice Questions

Matching

1. septum	a. cavities in the skull
2. ventilation	b. divides the nasal cavity
3. sinusitis	c. inflammation in the lower respiratory system
4. lungs	d. partly caused by collapse of bronchioles
5. emphysema	e. infection in the upper respiratory system
6. larynx	f. breathing
7. trachea	g. opening to the lungs
8. glottis	h. the throat
9. sinuses	i. cone-shaped structures
10. apnea	j. branched structures with cartilage bands
11. epiglottis	k. grapelike
12. pharynx	l. the windpipe
13. bronchitis	m. flap that protects the lungs
14. alveoli	n. the location of the vocal cords
15. bronchi	o. can be caused by blockage of the upper respiratory system

Fill-in-the-Blank

16. The diaphragm muscle contracts and ____________ during inhalation.

17. When the partial pressure of oxygen is ____________, hemoglobin will bind oxygen molecules.

18. When ____________ leaves a cell, about 90% of it enters ____________ and about 10% dissolves in plasma.

19. ____________ is the addictive substance in cigarettes that stimulates the ____________ in the brain.

20. Both ____________ expiration and ____________ require muscle contraction.

21. Toxins from smoking can cause cancer of the kidneys or ____________ when they are ____________ from the body.

22. When carbon dioxide dissolves in water, it forms ____________, which can change the ____________ of body fluids.

23. The ____________ muscles are located between the ribs.

24. ____________ is a sticky substance that captures and ____________ particles from the respiratory system.

25. The vocal cords are composed of ____________ that ____________ when air passes over it.

Labeling

Use the terms below to label Figure 10.1.

epiglottis
glottis
larynx
nasal conchae
pharynx
sinuses

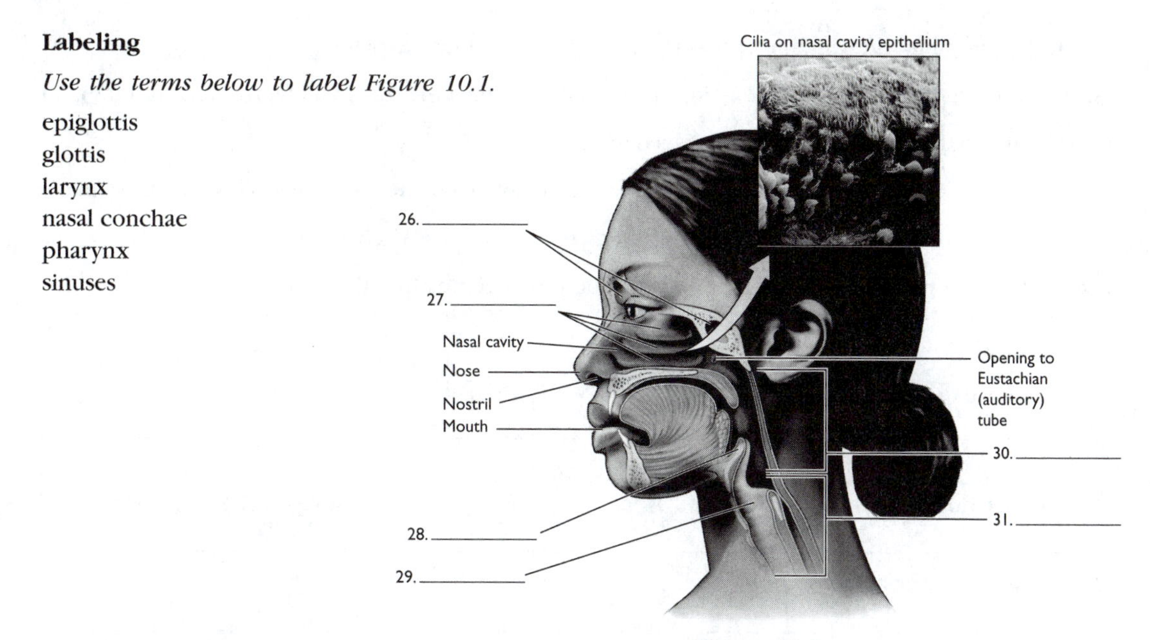

Figure 10.1

Use the terms below to label Figure 10.2. Any term listed twice will be used twice.

contract(s)
contract(s)
decrease(s) in volume
expand(s)
increase(s) in volume
relax(es)

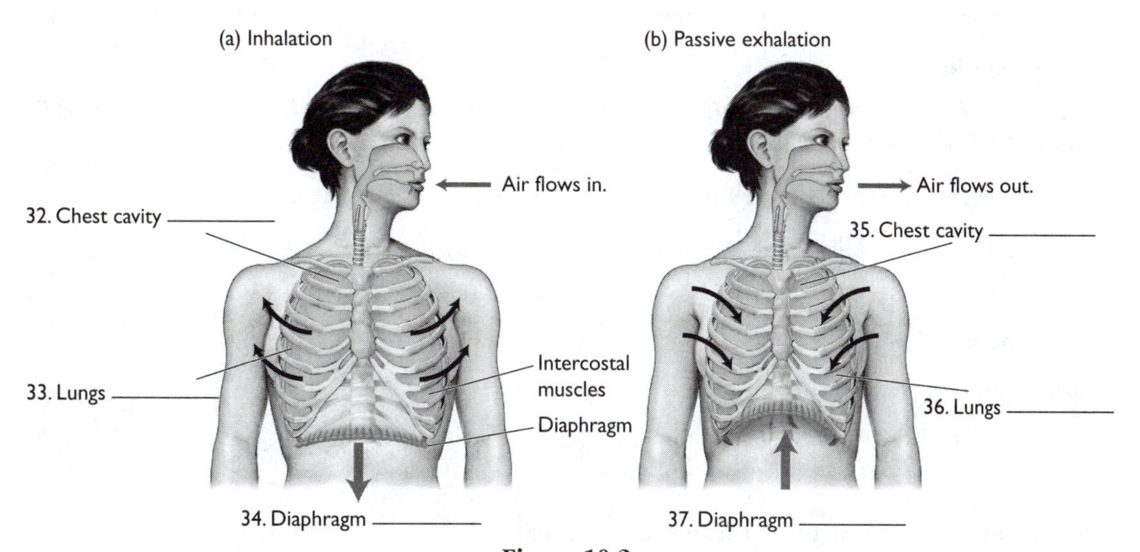

Figure 10.2

Roots to Remember

Use your knowledge of the root words presented in this chapter to answer the following questions.

38. A grasshopper has spiracles. What do these organs help the grasshopper do?
39. What is tendonitis?
40. A pneumatic tube system within a bank or hospital is used to transport containers. What is the likely material used in this system to move the containers?

Word Choice

Circle the word or phrase that correctly completes each sentence.

41. The (septum/sinus) separates the nose into two cavities.

42. During (inspiration/expiration), pressure decreases as the lungs expand.

43. The amount of air that always remains in your lungs is called the (tidal volume/residual volume).

44. (Plasma/erythrocytes) transport(s) oxygen in the blood.

45. Constriction of the bronchial tubes and overproduction of mucus are symptoms of (asthma/bronchitis).

46. The body moves oxygen into the lungs through (ventilation/gas exchange).

47. The (glottis/epiglottis) is a flap that prevents food from getting into the lungs.

Table Completion

Complete the following table on respiratory conditions and diseases.

Disease or Condition	Part(s) of Respiratory System Affected	Description/Symptoms
Emphysema	48.	49.
Chronic bronchitis	50.	51.
52.	Sinuses	53.
Laryngitis	54.	55.
56.	Tonsils	57.

Cause and Effect

For each action listed below, the body's response is partially filled in. Complete each sentence to fully explain the body's response.

58. **Action:** Stretch receptors sense that critical lung volume has been reached.

    Response: The respiratory center ____________ to the ____________.

59. **Action:** You start breathing quickly as you break into a run.

    Response: Your body expels ____________ and takes in ____________.

60. **Action:** Carbon dioxide levels increase in the blood.

    Response: The pH receptors in the ____________ signal the ____________ to increase its ____________.

61. **Action:** Blood traveling in the veins reaches the lungs.

    Response: Carbon dioxide diffuses from the ____________ and out of the ____________.

62. **Action:** Carbon dioxide in red blood cells combines with water to form carbonic acid.

    Response: Carbonic acid dissociates into ____________.

63. **Action:** Chest cavity volume decreases.

    Response: Air flows ____________.

64. **Action:** A smoker takes a drag on a cigarette.

    Response: Nicotine stimulates ____________.

65. **Action:** The diaphragm contracts.

    Response: The chest cavity expands and ____________.

Critical Thinking

66. Why is carbon monoxide particularly dangerous?
67. Even if we don't smoke, we inhale many airborne particulates over a lifetime. Why don't the lungs fill up with these materials?
68. Why is the respiratory center located in the most primitive part of the brain?

Practice Test

1. A cough clears debris from the
   A. alveoli.
   B. sinuses.
   C. bronchi.
   D. Eustachian tubes.

2. What is the purpose of surfactant in the lungs?
   A. It keeps lung tissue from sticking together.
   B. It allows gases to dissolve and gas exchange to occur.
   C. It keeps the respiratory surface moist.
   D. All of the above.

3. Oxygen uptake is promoted by
   A. low temperature and relatively low pH.
   B. low temperature and relatively high pH.
   C. high temperature and relatively low pH.
   D. high temperature and relatively high pH.

4. What is the main gas in environmental tobacco smoke?
   A. oxygen
   B. hydrogen cyanide
   C. carbon monoxide
   D. ammonia

5. Which structure in the throat is also called the voice box?
   A. larynx
   B. pharynx
   C. glottis
   D. epiglottis

6. Which of the following are most like the leaves on a tree?
   A. bronchi
   B. alveoli
   C. bronchioles
   D. trachea

7. A smoker is likely to be more successful at quitting smoking if he or she
   A. just cuts down to a few cigarettes a day.
   B. hangs out with friends who are smokers.
   C. takes up an exercise program.
   D. has quit smoking before.

8. Which of the following is found in the trachea?
   A. muscle tissue and cartilage only
   B. cartilage and mucus membrane only
   C. muscle tissue and mucous membrane only
   D. cartilage, muscle tissue, and mucous membrane

9. Most smoking-related deaths are due to
   A. lung cancer.
   B. cardiovascular disease.

C. emphysema.
D. asthma.

10. Where are the stretch receptors located?
A. in the ribs
B. in the abdomen
C. in the diaphragm
D. near the alveoli

11. The amount of air that you normally breath in and out is called the
A. tidal volume.
B. residual volume.
C. inspiration volume.
D. vital capacity.

12. Which of the following conditions affects the upper respiratory system?
A. tonsillitis
B. bronchitis
C. laryngitis
D. sinusitis

13. In order to speak in a soft voice, you need
A. lots of air passing quickly over your vocal cords.
B. very little air passing quickly over your vocal cords.
C. lots of air passing slowly over your vocal cords.
D. very little air passing slowly over your vocal cords.

14. The nasal cavity connects to the ears via the
A. Eustachian tubes.
B. sinuses.
C. nasal conchae.
D. pharynx.

15. Which of the following does *not* bind to hemoglobin?
A. bicarbonate
B. carbon dioxide
C. oxygen
D. carbon monoxide

Answer Key

Matching

1. b; **2.** f; **3.** e; **4.** i; **5.** d; **6.** n; **7.** l; **8.** g; **9.** a; **10.** o; **11.** m; **12.** h; **13.** c; **14.** k; **15.** j

Fill-in-the-Blank

16. flattens; **17.** high; **18.** carbon dioxide; erythrocytes; **19.** Nicotine; pleasure center; **20.** active; inspiration or inhalation; **21.** bladder; excreted; **22.** carbonic acid; pH; **23.** intercostal; **24.** Mucus; removes; **25.** cartilage; vibrates

Labeling

26. sinuses; **27.** nasal conchae; **28.** epiglottis; **29.** glottis; **30.** pharynx; **31.** larynx; **32.** expands; **33.** increase in volume; **34.** contracts; **35.** contracts; **36.** decrease in volume; **37.** relaxes

Roots to Remember

38. They help the grasshopper to breathe; **39.** inflammation of a tendon; **40.** pressurized air

Word Choice

41. septum; **42.** inspiration; **43.** residual volume; **44.** erythrocytes; **45.** asthma; **46.** ventilation; **47.** epiglottis

Table Completion

48. Alveoli and bronchioles; **49.** Air becomes trapped in the chest, causing the chest to become barrel shaped; ventilation becomes difficult and gas exchange decreases; **50.** Bronchial tubes; **51.** Bronchial tubes become inflamed; mucus builds up; patient experiences early morning cough; **52.** *Sinusitis*; **53.** Sinuses become infected; **54.** Vocal cords; **55.** Vocal cords swell; voice become husky or is lost temporarily; **56.** *Tonsillitis*; **57.** Tonsils become inflamed.

Cause and Effect

58. shuts off contraction signals; diaphragm and intercostal muscles; **59.** more carbon dioxide; more oxygen; **60.** brain; respiratory center; cycle rate; **61.** blood into the lungs; body; **62.** into hydrogen ions and bicarbonate ions; **63.** out of the body; **64.** the pleasure center in the brain; **65.** the lungs fill with air

Critical Thinking

66. The human body cannot detect it, but it is potentially deadly.

67. We remove larger particles when we cough, so they do not reach the lungs. The mucus escalator traps and moves smaller particles out of the bronchi and trachea, so these particles do not reach the lungs.

68. Breathing is an essential function that maintains life in many organisms, so that function would be found in the earliest brain forms. Nonessential functions, such as language, would be found in less-primitive parts of the brain.

Practice Test

1. C; **2.** D; **3.** B; **4.** C; **5.** A; **6.** B; **7.** C; **8.** D; **9.** B; **10.** D; **11.** A; **12.** D; **13.** D; **14.** A; **15.** A

CHAPTER 11

THE URINARY SYSTEM: SURVIVING THE IRONMAN

Learning Goals

1. Compare and contrast elimination and excretion, and describe the different methods by which wastes are eliminated from the body.
2. List the structures and functions of the urinary system.
3. Summarize the steps in the process of urination.
4. List the major components of urine and their sources.
5. List the steps in the process of excretion, and briefly describe what happens in each step.
6. Summarize the process of tubular reabsorption, explaining the role of sodium transport in the process.
7. Describe how countercurrent exchange can maximize water retention in the kidneys, and list the effects of hormones on the volume of water lost in urine.
8. Describe how blood pH is regulated in both the short and long term.
9. Summarize the role of the kidney in maintaining salt balance.
10. Explain how dialysis can replace kidney function in individuals with renal disease.

Chapter Outline

I. An Overview of the Urinary System
   A. Homeostasis and the Urinary System
      1. Waste products of metabolism must be removed from the body.
         a) Carbon dioxide and water waste can be removed via the respiratory system.
         b) Water and some metabolic wastes can be removed via the integumentary system, through sweat.
         c) Most metabolic wastes are removed via the urinary system, through excretion.
      2. The main organs of the urinary system, the kidneys, filter wastes from the bloodstream and help maintain homeostasis.
         a) Kidneys maintain water balance by regulating the amount of salt excreted.
         b) Kidneys maintain blood pH balance by adjusting ions in the blood.
         c) Kidneys influence other organ systems by releasing hormones.
   B. Structure of the Urinary System
      1. Kidneys filter and clean blood.
         a) Kidneys are paired, fist-shaped organs covered by a protective membrane called the renal capsule.
         b) They are located in the upper abdomen, behind the liver and stomach.
         c) Renal arteries bring blood to be filtered by the kidneys.
         d) Renal veins remove blood that has been filtered by the kidneys.
         e) Kidneys have three layers.
            (1) The outer layer is the renal cortex.
            (2) The inner layer is the renal medulla.
            (3) The hollow center is the renal pelvis.
         f) The renal cortex and medulla contain networks of nephrons.
            (1) Nephrons are the filters of the kidney.
            (2) Nephrons in the medulla are grouped in renal pyramids.
         g) Urine is the end product of filtration.
      2. Ureters are tubes that pass urine from the kidneys to the urinary bladder.

3. The urinary bladder receives and stores urine from the kidneys.
   a) The bladder is highly expandable.
   b) The detrusor muscle that makes up the bladder wall expels urine when it contracts.
4. The urethra is the tube that expels urine from the bladder to the outside of the body.
   a) Urethral sphincters control the movement of urine.
      (1) The internal urethral sphincter at the top of the urethra holds urine in the bladder.
      (2) The external urethral sphincter controls the flow of urine out of the body.
   b) Urethra form and function vary with gender.
      (1) The female urethra is 3–4 cm long and carries only urine.
      (2) The male urethra is 18–20 cm long and carries urine or sperm.

C. Urination
1. Urination is triggered when the bladder is full.
   a) Stretch receptors in the bladder wall trigger the micturition reflex.
      (1) The internal urethral sphincter relaxes.
      (2) The detrusor muscles contract.
   b) Urine flows into the urethra and causes a need-to-urinate feeling.
   c) The external urethral sphincter opens and urine flows out.
      (1) The muscles that open the external urethral sphincter are under conscious control.
      (2) The external urethral sphincter can be held closed until urination is appropriate.
2. Urination relies on a positive-feedback loop.
   a) Urine flow stimulates additional detrusor muscle contractions.
   b) All urine in the bladder is squeezed out.
   c) When urine flow stops, detrusor contractions stop.
3. Frequency of urination may be gender related.
   a) Females have smaller bladder capacity due to the uterus compressing the bladder.
   b) An enlarged prostate gland in males can cause frequent urination.

II. Excretion

A. The Composition of Urine
1. Water is a major component of urine.
2. Urea is a modified waste product of protein metabolism.
   a) Ammonia is produced when proteins are broken down during cellular respiration.
   b) Ammonia is toxic to cells.
   c) The liver breaks ammonia down into urea.
3. Uric acid is a product of DNA metabolism.
   a) Uric acid may build up in blood.
   b) When it crystallizes in joints, it can cause gout.
4. Excess amounts of ions such as sodium, chloride, potassium, calcium, and hydrogen are excreted in urine.
5. Toxins, drugs, and other metabolic wastes are secreted in urine.
6. Urobilins from the breakdown of bilirubin give urine its color.
7. Creatinine is a waste product produced by intense muscle use.
   a) When muscles are working hard, aerobic respiration cannot produce ATP quickly enough.
   b) Muscles can get additional ATP when creatine phosphate donates a phosphate group and becomes creatinine.

B. Urine Formation
1. Urine formation takes place in nephrons.
   a) The glomerular capsule, or Bowman's capsule, is the cup-shaped structure that receives plasma to filter.
   b) The capsule surrounds the glomerulus, which is a mass of capillaries.
   c) Fluid from the glomerular capsule flows into tubules.
      (1) The proximal tubule is in the renal cortex.
      (2) The nephron loop, or loop of Henle, is a long loop between the proximal and distal tubules.
         (a) In 20% of nephrons, the nephron loop extends into the renal medulla.
         (b) In 80% of nephrons, the nephron loop stays in the renal cortex.
      (3) The distal tubule is at the end of the nephron and is located in the renal cortex.
   d) The distal tubule of each nephron drains into a collecting duct.
   e) Each nephron is surrounded by a network of capillaries.

2. Step 1: Glomerular Filtration
   a) The afferent arteriole moves blood through glomerulus capillaries.
      (1) Blood pressure is high in the glomerulus.
      (2) High pressure forces most of the blood plasma through pores in the capillary walls.
      (3) Plasma moves into the capsule of the nephron.
   b) The efferent arteriole moves cells and blood proteins out of the glomerulus.
3. Step 2: Tubular Reabsorption
   a) Many substances are reabsorbed from filtrate back into blood.
      (1) Epithelial cells in the proximal tubule use ATP to actively transport sodium from within the cells into the renal cortex.
         (a) This sets up an electrical gradient that causes chloride to diffuse out of the tubule and into the renal cortex.
         (b) The presence of sodium and chloride ions in the renal cortex creates an osmotic gradient that causes water to flow out of the tubule and into the cortex.
         (c) The active transport of sodium out of the epithelial cells of the tubule sets up a sodium concentration gradient.
            (i) Sodium from filtrate inside the tubule moves down its concentration gradient and into the epithelial cells of the tubule.
            (ii) Amino acids and glucose are cotransported with sodium from filtrate in the tubule into epithelial cells, through protein transport molecules.
            (iii) Glucose and amino acids diffuse down their concentration gradient from epithelial cells into the renal cortex.
      (2) Substances in the renal cortex diffuse into adjacent capillaries and return to the blood.
   b) Microvilli increase the surface area of epithelial cells, thus increasing reabsorption.
4. Step 3: Tubular Secretion
   a) Blood can actively secrete some wastes into the proximal or distal tubules.
   b) Wastes such as creatinine, drugs, food additives, and inorganic chemicals are removed through secretion.
   c) Excess hydrogen ions that may lower blood pH are also removed by secretion.

III. Water, pH, and Salt Balance
   A. Hormones and Water Depletion
      1. High solute concentrations in blood stimulate nerve cells.
         a) These cells stimulate the pituitary gland.
         b) The pituitary gland releases antidiuretic hormone (ADH).
         c) Sites on the collecting ducts receive ADH.
         d) ADH allows increased water retention in the kidney.
         e) More water is returned to the blood, decreasing solute concentrations.
      2. Low blood pressure triggers the release of hormones that regulate salt concentration.
         a) Reduced blood pressure on juxtaglomerular cells allows them to expand and release the hormone renin.
         b) Renin activates the protein angiotensinogen.
         c) This produces the active regulatory hormone angiotensin II.
            (1) Angiotensin II narrows blood vessels.
            (2) It stimulates the brain to produce a feeling of thirst.
            (3) Angiotensin II stimulates the adrenal glands to produce the hormone aldosterone.
               (a) Aldosterone increases active transport of sodium from the distal tubule.
               (b) A high sodium concentration in the kidney causes water to move from tubules and collecting ducts into kidney tissue.
               (c) Water then moves back into blood.
   B. Countercurrent Exchange in the Kidney
      1. In countercurrent exchange, gas or fluid flows in opposite directions in tubes that are next to each other.
      2. Countercurrent exchange in the kidneys creates a concentration gradient within the renal medulla.
      3. The concentration gradient helps the body to recapture water from filtrate.
         a) When filtrate flows down the descending arm of the nephron loop,
            (1) the tubule is highly permeable to water, so water diffuses out, and
            (2) the tubule becomes relatively impermeable to salt ions.

b) When filtrate flows into the ascending arm of the nephron loop,
   (1) the tubule becomes impermeable to water, and
   (2) the tubule becomes permeable to sodium and chloride ions, so these ions diffuse out to maintain the solute-rich medulla.

c) Blood flowing through the vasa recta helps to maintain the concentration gradient in the medulla.
   (1) Sodium and chloride diffuse into the bloodstream in the descending capillaries, making the blood rich in solutes.
      (a) Water diffusing out of the descending nephron loop is taken up by solute-rich blood in the vasa recta.
      (b) This prevents dilution of the solute-rich medulla.
   (2) Sodium and chloride diffuse out of the ascending capillaries and into the medulla.

d) The concentration gradient is maintained by tubule structure.
   (1) The ascending tubule widens and becomes impermeable to water and most other solutes.
   (2) Sodium transporters actively excrete salt into the medulla from filtrate.

e) ADH increases the permeability of the collecting duct to water.
   (1) Water flows from the dilute interior to the solute-rich medulla.
   (2) This produces highly concentrated urine.

4. When solutes in blood are low, ADH production is inhibited.
   a) Collecting ducts become impermeable to water.
   b) Water is excreted in dilute urine.

5. When blood volume is high, cells in the heart sense heart muscle stretching.
   a) These cells produce atrial natriuretic peptide (ANP).
   b) ANP is a diuretic.
   c) ANP inhibits renin and aldosterone release, thus increasing excretion of sodium chloride and water.

C. Maintaining Blood pH

1. If hydrogen ions are allowed to build up in blood, the low pH that results can cause acidosis.
   a) Acidosis can denature enzymes.
   b) Enzyme-mediated processes in the body are stopped.

2. Buffers in the blood maintain pH.
   a) Buffers include the bicarbonate ion (HCO_3^-) and hemoglobin.
      (1) Buffers take up excess hydrogen ions when pH is low.
      (2) Buffers release ions when pH is high.
   b) Nephron epithelial cells generate bicarbonate.

3. Hydrogen ions are also secreted into the lumen and excreted by the kidneys.
   a) When enough bicarbonate is available,
      (1) hydrogen ions secreted into the lumen can combine with bicarbonate to form carbonic acid,
      (2) the enzyme carbonic anhydrase breaks down carbonic acid into carbon dioxide and water, and
      (3) water is excreted in urine.
   b) When pH levels are lower,
      (1) hydrogen ions secreted into the lumen can combine with hydrogen phosphate, and
      (2) this product is excreted in urine.
   c) When pH levels are even lower,
      (1) amino acids in nephron epithelia are metabolized to form ammonia,
      (2) ammonia picks up an extra hydrogen ion to form ammonium, and
      (3) ammonium is excreted in urine.

D. Salt Balance: The Right Amount of Sodium

1. Salt balance is maintained during normal activity.
2. As water and sodium are excreted through sweat during athletic events, salt balance is harder to maintain.
3. Loss of too much sodium can lead to hyponatremia (water intoxication).
   a) Blood becomes hypotonic.
   b) Water diffuses from the blood into tissues.
   c) Tissues swell from too much water.

  - d) Too much water in brain tissue can lead to slurred speech, disorientation, seizures, coma, or death.
4. Certain medications may also cause hyponatremia.

IV. When Kidneys Fail
- A. Many Causes of Kidney Malfunction
  1. Kidney stones can block urine flow out of the body.
     - a) Kidney stones are masses of calcium, phosphorus, uric acid, and protein crystals.
     - b) They can block the ureter.
     - c) If they remain trapped in the renal pelvis, urine can back up and destroy nephrons.
     - d) Small kidney stones may pass out in urine flow.
     - e) Large kidney stones must be removed surgically or broken up with ultrasound.
  2. Infections can move up the urethra and bladder and destroy nephrons.
  3. Chronic high blood pressure can damage the glomerulus and glomerular capsule.
     - a) Diabetes can increase the solute concentration of blood, leading to high blood pressure.
     - b) Nearly half of all kidney failures are due to diabetes.
- B. Kidney failure can be detected by the presence of protein in urine.
- C. People with end-stage renal disease (ESRD) have kidneys that perform at less than 10% of normal.
  1. People with ESRD need a kidney transplant.
  2. People with ESRD undergo regular dialysis to remain alive.
     - a) A dialysis machine consists of membranous tubes in fluid.
     - b) The patient's blood is circulated through the machine.
     - c) Metabolic wastes move from the blood, through the membranes, and into the fluid.
     - d) Substances added to the fluid can also diffuse into the blood.
     - e) Dialysis removes more urea than do the kidneys, so it only needs to be done three to four times per week.

Practice Questions

Matching

1. elimination
2. aldosterone
3. dialysis
4. ureter
5. excretion
6. glomerulus
7. nephron
8. renal pelvis
9. acidosis
10. renin
11. renal cortex
12. hyponatremia
13. glomerular capsule
14. urethra
15. urea

a. mass of capillaries
b. caused by low blood pH
c. removal of solid wastes
d. releases urine from the body
e. causes swelling of tissues
f. metabolic waste removal by machine
g. hormone that increases active transport of sodium
h. layer of the kidney that contains nephrons
i. hollow center of the kidney that passes urine from nephrons to the ureter
j. gathers plasma from capillaries
k. filtration of metabolic wastes
l. structure in the kidney that filters waste
m. moves urine to the bladder
n. modified waste product from protein metabolism
o. hormone that activates angiotensinogen

Fill-in-the-Blank

16. During the ____________ reflex, the ____________ urethral sphincter relaxes and the detrusor muscle ____________.

17. ____________ in the walls of capillaries in the ____________ allow water to flow into nephron tubules.

18. Unless it gets too low, blood pH is maintained by ____________ ions and ____________.

19. ____________ are composed of calcium, phosphorus, protein crystals, and ____________ and can block the flow of urine through the ____________.

20. ANP is a ____________ that increases excretion by inhibiting the release of ____________ and aldosterone.

21. The flow of urine into the ____________ triggers the urge to ____________.

22. ____________ is caused by the buildup and ____________ of uric acid in the joints.

23. ____________ is produced when proteins are broken down during cellular ____________.

24. The concentration of salts in the blood is too ____________ in someone who is suffering from hyponatremia.

25. Patients with ESRD undergo ____________ while waiting for a ____________ transplant.

Labeling

Use the terms below to label Figure 11.1.

renal capsule
renal cortex
renal medulla
renal pelvis
renal pyramid
ureter

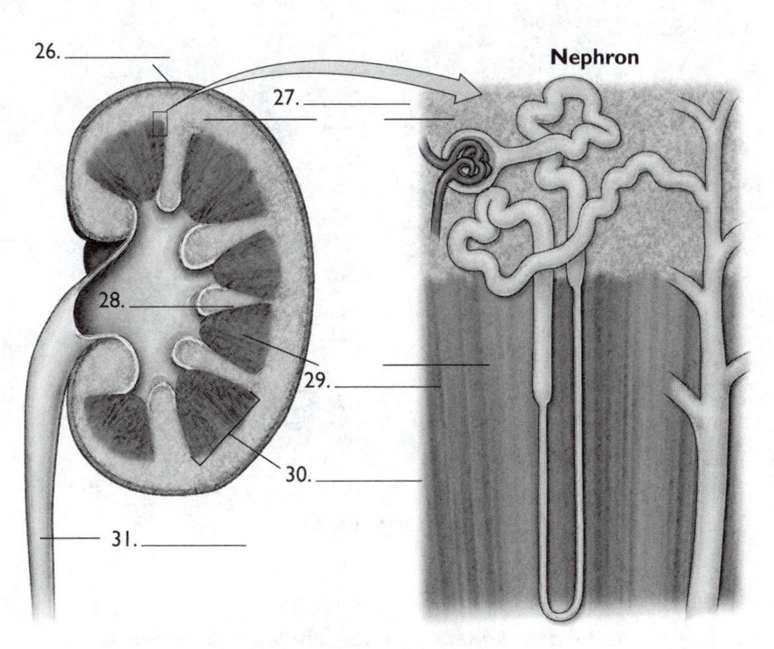

Figure 11.1

Use the terms below to label Figure 11.2. Any term listed twice will be used twice.

Cl^-
glucose or amino acids
H_2O
Na^+
Na^+
sodium cotransporter

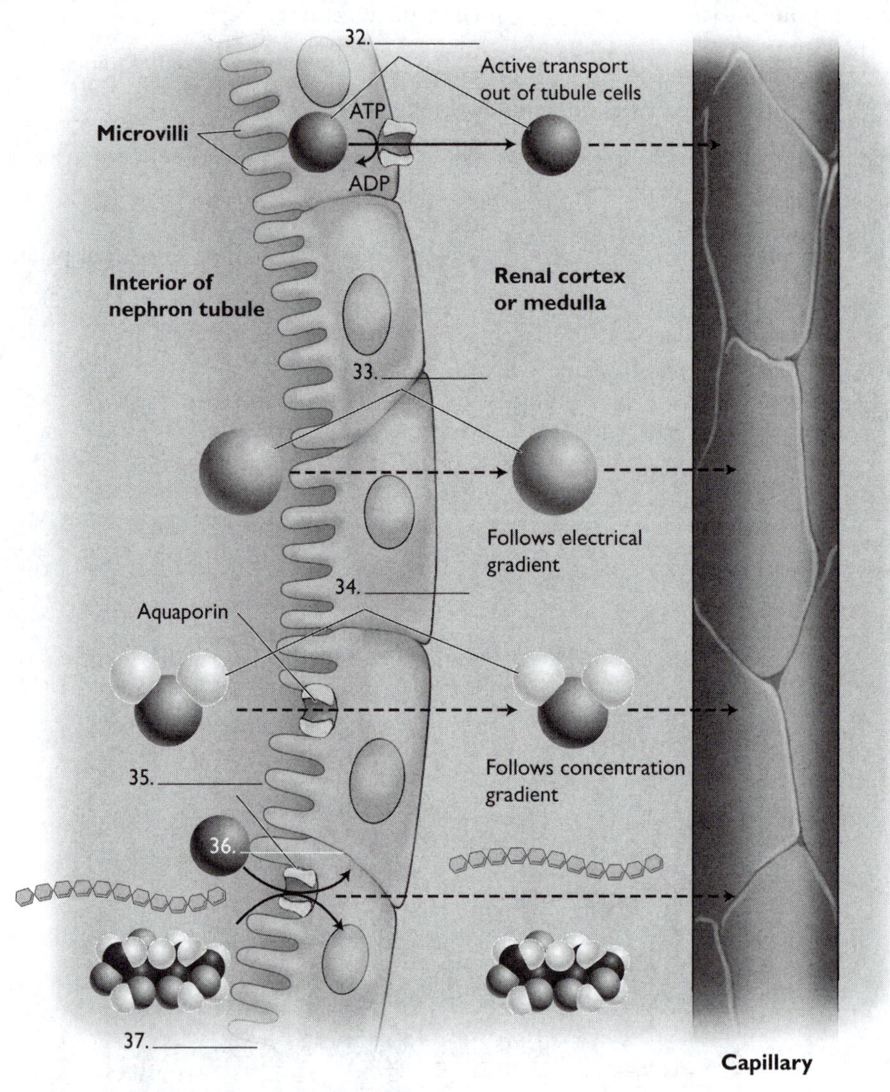

Figure 11.2

Roots to Remember

Use your knowledge of the root words presented in this chapter to answer the following questions.

38. Nephritis is inflammation of what organ?
39. When a patient experiences renal failure, what has happened?
40. What could urinalysis show?

Word Choice

Circle the word or phrase that correctly completes each sentence.

41. The (external/internal) urethral sphincter controls the movement of urine out of the body.
42. The outermost layer of tissue in the kidney is called the renal (cortex/capsule).
43. Control of the detrusor muscle is (voluntary/involuntary).

44. (Bilirubin/Urobilin) is the substance that makes urine yellow.

45. After sodium ions are actively transported out of epithelial cells in a nephron tubule, chloride ions diffuse (into/out of) the tubule.

46. Red blood cells (are/are not) filtered by nephrons.

Crossword Puzzle

Use the clues below to fill in the puzzle.

Across

47. DNA metabolite
48. have the longest urethra
49. arteriole leaving the glomerulus
50. produced by muscle activity
51. involuntary action
52. nephron shape

Down

53. to move via active transport
54. collecting structure
55. toxic to cells
56. a function of nephrons
57. pelvis, cortex, or medulla

Sequencing

Put the following events in order by numbering each event, from 1 to 8.

58. ____________ Aldosterone is produced by the adrenal glands.

59. ____________ Renin is released.

60. ____________ Angiotensin II is produced.

61. ____________ Blood pressure declines.

62. ____________ Angiotensinogen produced by the liver is activated.

63. ____________ Water flows from nephrons into kidney tissue.

64. ____________ Juxtaglomerular cells expand.

65. ____________ Active transport of sodium from the distal tubule is increased.

Critical Thinking

66. Why are you less likely to feel the urge to urinate during a workout?
67. What would happen if the bladder operated under a negative-feedback loop?
68. People who have high blood pressure are often given diuretics. How might this affect their salt balance?

Practice Test

1. Which of the following do not diffuse from nephrons back into the bloodstream?
   A. water molecules
   B. glucose molecules
   C. sodium ions
   D. inorganic chemicals

2. The interior of a nephron tubule is also called the
   A. lumen.
   B. capsule.
   C. cortex.
   D. collecting duct.

3. Where is detrusor muscle found?
   A. in the kidney
   B. in the bladder
   C. in the urethra
   D. in the urethral sphincters

4. Which of the following best describes urine?
   A. Urine is a combination of liquid and solid wastes.
   B. Urine contains all wastes produced by body cells.
   C. Urine is composed solely of excess water and urea.
   D. Urine is a combination of water, metabolic wastes, and ions.

5. During the process of urination, which of the following is controlled by a voluntary response?
   A. the internal urethral sphincter
   B. the stretch receptors
   C. the external urethral sphincter
   D. the micturition reflex

6. All of the molecules below help to buffer blood *except*
   A. ammonia.
   B. bicarbonate.
   C. carbon dioxide.
   D. hydrogen phosphate.

7. A long-distance runner drinks too much water, and her blood becomes diluted. What is the body's response?
   A. The body increases the excretion of hydrogen ions.
   B. The body inhibits the production of ADH.
   C. The body increases the production of angiotensin II.
   D. The body inhibits the production of ANP.

8. Which part of the nephron can extend into the renal medulla?
   A. the nephron loop
   B. the glomerular capsule

C. the proximal tubule
D. the distal tubule

9. A patient complains of swollen, painful toes. Which substance may be the cause of the patient's symptoms?
A. uric acid
B. creatinine
C. urea
D. bilirubin

10. Nephrons are found in the
A. renal cortex and renal medulla only.
B. renal medulla and renal pelvis only.
C. renal capsule, renal cortex, and renal medulla only.
D. renal cortex, renal medulla, and renal pelvis only.

11. As sodium ion concentration increases in the renal cortex,
A. water and chloride ion concentrations decrease.
B. water concentration decreases and chloride ion concentration increases.
C. water and chloride ion concentrations increase.
D. water concentration increases and chloride ion concentration decreases.

12. Which of the following is controlled by positive feedback?
A. pH balance in the blood
B. water balance in tissues
C. blood pressure
D. urination

13. The ascending arm in the nephron loop is
A. impermeable to sodium ions.
B. impermeable to water.
C. impermeable to chloride ions.
D. impermeable to all ions.

14. When solutes such as glucose increase in blood,
A. water flows out of blood and into tissues.
B. blood pressure decreases.
C. water volume in blood also increases.
D. the kidneys can be damaged.

15. What can you do to avoid hyponatremia during a long-distance race?
A. Eat small, salty snacks.
B. Reduce consumption of water.
C. Seek help if you feel disoriented or nauseated.
D. All of the above.

Answer Key

Matching

1. c; **2.** g; **3.** f; **4.** m; **5.** k; **6.** a; **7.** l; **8.** i; **9.** b; **10.** o; **11.** h; **12.** e; **13.** j; **14.** d; **15.** n

Fill-in-the-Blank

16. micturition; internal; contracts; **17.** Pores; glomerulus; **18.** bicarbonate; hemoglobin; **19.** Kidney stones; uric acid; ureter; **20.** diuretic; rennin; **21.** urethra; urinate; **22.** Gout; crystallization; **23.** Ammonia; respiration; **24.** low; **25.** dialysis; kidney

Labeling

26. renal capsule; **27.** renal cortex; **28.** renal pelvis; **29.** renal medulla; **30.** renal pyramid; **31.** ureter; **32.** Na^+; **33.** Cl^-; **34.** H_2O; **35.** sodium cotransporter; **36.** Na^+; **37.** glucose or amino acids

Roots to Remember

38. kidney; **39.** The kidneys have shut down or ceased to work; **40.** whether there are substances in urine that could indicate disease

Word Choice

41. external; **42.** capsule; **43.** involuntary; **44.** Urobilin; **45.** out of; **46.** are not

Crossword Puzzle

Across

47. uric acid; **48.** males; **49.** efferent; **50.** creatinine; **51.** reflex; **52.** loop

Down

53. secrete; **54.** duct; **55.** ammonia; **56.** filter; **57.** renal

Sequencing

58. 6; **59.** 3; **60.** 5; **61.** 1; **62.** 4; **63.** 8; **64.** 2; **65.** 7

Critical Thinking

66. During a workout, water is being excreted by sweat glands, so there is less water to be excreted by the kidneys.

67. The flow of urine through the urethra would inhibit detrusor muscle contraction, so the bladder would never fully empty.

68. Diuretics cause the excretion of water and sodium chloride, so salt concentration in the blood would decrease.

Practice Test

1. D; **2.** A; **3.** B; **4.** D; **5.** C; **6.** C; **7.** B; **8.** A; **9.** A; **10.** A; **11.** C; **12.** D; **13.** B; **14.** C; **15.** D

CHAPTER 12

THE IMMUNE SYSTEM: WILL MAD COW DISEASE BECOME AN EPIDEMIC?

Learning Goals

1. Describe how genetic and infectious diseases differ.
2. Describe the structure of a typical bacterium and a typical virus.
3. List the means by which infectious diseases are transmitted.
4. Describe the physical and chemical barriers that make up the first line of defense.
5. State the actions of white blood cells, proteins, inflammation, and fever in the second line of defense.
6. Compare and contrast the roles that B cells and T cells play in the immune response.
7. Describe how the immune system differentiates between self and nonself.
8. Explain why it is necessary to have a flu shot every year but only one inoculation against some other diseases.
9. Explain why your immune system is usually more effective at fighting off infection after a previous exposure.
10. Outline a genetic reason why one person might die of an infectious disease whereas another person recovers.

Chapter Outline

I. Infectious Agents
  A. Overview
    1. Pathogens are organisms that cause disease.
      a) Contagious pathogens can be spread between organisms.
      b) Infectious pathogens are able to grow inside the body.
        (1) Most infectious pathogens are parasites.
        (2) Parasites obtain nutrients and shelter from a host organism but do not benefit the host.
    2. Pathogens are often microscopic organisms, or microbes.
  B. Bacteria
    1. Bacteria are diverse, numerous, and microscopic.
      a) Common bacteria are rod shaped, spherical, or spiral.
      b) Bacteria are microscopic prokaryotes.
        (1) Prokaryotes are single-celled organisms without nuclei.
        (2) Prokaryotes do not contain membrane-bounded organelles.
    2. Bacterial DNA is coiled inside a nucleoid region.
      a) A bacterial chromosome is double stranded and circular.
      b) Extrachromosomal DNA is found in small, circular structures called plasmids.
    3. Most bacterial cells are surrounded by a rigid cell wall.
      a) The cell wall is composed of carbohydrates and proteins.
      b) A gelatinous capsule usually surrounds the cell wall and helps bacteria attach to cells.
    4. Bacteria may have flagella to aid in movement.
    5. Bacteria may have pili to help them attach to other bacteria and share genes.
    6. Bacteria reproduce by binary fission.
      a) Each cell gives rise to two identical cells.
        (1) The single chromosome within the cell is copied.
        (2) Each copy attaches to the plasma membrane.
        (3) Growth separates the two copies.

(4) The plasma membrane pinches into two cells.
(5) The two cells separate.
b) Bacterial populations grow exponentially.
7. Most symptoms of bacterial disease are caused by toxins secreted by bacteria.
8. Many bacteria are beneficial to humans.
a) Bacterial flora in the human intestinal tract secrete vitamins.
b) Beneficial flora prevent harmful flora from colonizing the body.
c) Newborns develop bacterial flora from exposure to their mother and the outside world.

C. Viruses
1. Viruses are not living organisms.
a) They cannot replicate without a host cell.
b) They are not composed of cells.
c) They cannot make their own proteins.
d) They lack cytoplasm and membrane-bounded organelles.
2. Viruses are packets of RNA or DNA surrounded by a protein coat.
a) A protein coat called a capsid surrounds the virus.
(1) Many viruses have a viral envelope outside the capsid.
(a) The envelope is derived from the cell membrane of the host cell.
(b) The virus infects the host cell by fusing its envelope with the host cell's membrane.
(2) Viruses without envelopes infect host cells when capsid proteins bind with receptor proteins in the host cell plasma membrane.
b) The genome of a virus varies.
(1) It can be single stranded or double stranded.
(2) It can be linear or circular.
3. The virus replicates within the host cell.
a) The virus genome is copied.
b) The virus uses host cell ribosomes and amino acids to make viral proteins.
(1) These proteins build new capsids.
(2) These proteins synthesize some envelope proteins.
c) Daughter virus cells exit the host cell.
d) Daughter virus cells infect other host cells.
4. Viruses can remain latent, or inactive, within the body.

D. Eukaryotic Pathogens
1. Eukaryotic pathogens include single-celled protozoans, worms, and fungi.
2. Protozoans can be spread by food or water that has been contaminated by animal feces.
3. Worms can damage internal organs and tissues.
4. Fungi can secrete digestive enzymes that damage tissue.

E. Prions
1. Prions are proteins produced by the brain that cause spongiform encephalopathy when misfolded.
a) A rare human form of spongiform encephalopathy, called Creutzfeldt-Jakob disease (CJD), generally causes memory loss and loss of coordination in the elderly.
b) A new form of the disease (nvCJD) has surfaced.
(1) This form affects younger people.
(2) In this new form, misfolded proteins refold properly folded proteins, so that they become misfolded.
(a) The brain becomes clogged with misfolded proteins.
(b) Clogged cells cease to function.
(c) Clogged cells burst and release prions that can damage other cells.
(d) The brain becomes spongy with holes.
2. Prions have no DNA or RNA.
3. Prions do not degrade easily.

II. Transmission of Infectious Agents
A. Direct Contact
1. Exposure to body fluids can lead to infection.
2. Pathogens can be transmitted through touch.
3. Some pathogens can be transmitted through the air when infected people cough or sneeze.

B. Indirect Contact
   1. Some pathogens can survive on inanimate objects.
   2. The pathogen is transmitted when you touch the object.
C. Vector-Borne Transmission
   1. Vectors are intermediate organisms that transmit pathogens.
   2. Ticks, mosquitoes, raccoons, bats, and dogs are examples of vectors.
D. Ingestion
   1. Stomach acids can kill many pathogens.
   2. Some pathogens, such as prions, can survive an acidic environment.
   3. Healthy cattle can become infected with mad cow disease if they are fed meal made from cattle with misfolded prions.

III. The Body's Response to Infection: The Immune System
A. First Line of Defense: Skin and Mucous Membranes
   1. Nonspecific defenses do not specifically target pathogens.
   2. Nonspecific defenses can be physical or chemical.
   3. The skin provides nonspecific defenses from pathogens.
      a) The skin is a physical barrier.
      b) The skin has a low pH to repel pathogens.
      c) Glands in the skin secrete chemicals to slow bacterial growth or trap or break down microbes.
   4. Mucous membranes provide nonspecific defenses from pathogens.
      a) Mucous membranes secrete mucus to trap pathogens.
         (1) Mucous membranes are found in the respiratory, digestive, urinary, and reproductive tracts.
         (2) Trapped pathogens can be removed from the body through excretion, coughing, or sneezing.
      b) Digestive secretions kill microbes in the stomach.
      c) Vomiting can also remove pathogens.
B. Second Line of Defense: White Blood Cells, Inflammation, Defensive Proteins, and Fever
   1. White Blood Cells
      a) Phagocytes are white blood cells that engulf and digest any invaders.
         (1) Neutrophils are phagocytes that destroy bacterial cells and some fungi.
         (2) Macrophages are phagocytes that grab and engulf invaders using pseudopodia.
            (a) Macrophages travel in lymphatic fluid.
            (b) Macrophages clean up old, dead body cells.
            (c) Macrophages release chemicals to stimulate white blood cell production.
      b) Eosinophils are white blood cells that do not engulf pathogens.
         (1) Eosinophils surround invaders and secrete digestive enzymes that irritate or destroy the pathogens.
         (2) Eosinophils attack protozoans and worms.
      c) Natural cell killers are another type of white blood cell that attacks pathogens.
         (1) They attack tumor cells and body cells infected by viruses.
         (2) Natural killer cells release chemicals that cause plasma membranes of target cells to disintegrate.
   2. Inflammation
      a) When tissue is damaged, the body triggers an inflammatory response.
         (1) Histamine is produced by mast cells and white blood cells called basophils.
         (2) Histamine promotes vasodilation at the injury site.
            (a) Vasodilation allows capillary walls to pull apart and make room for phagocytes.
            (b) Increased cellular activity at the injury site causes warmth, redness, pain, and swelling.
      b) After the inflammatory response, pus is drained from the injury site.
         (1) Pus consists of dead cells, dead microbes, and tissue fluid.
         (2) If pus cannot be drained, it may collect in an abscess.
   3. Defensive Proteins
      a) Interferon is a protein that helps to protect uninfected cells.
         (1) Infected cells release interferon when they die.
         (2) Interferons bind to receptors on uninfected cells and stimulate those cells to produce proteins that inhibit viral reproduction.

b) Complement proteins circulate in the blood and complement other body defenses.
   (1) Complement proteins can coat microbial surfaces, making it easier for these microbes to be engulfed by macrophages.
   (2) Complement proteins can make holes in microbial membranes.
   (3) Complement proteins can increase the inflammatory response.

4. Fever
   a) Body temperature increases as macrophages release chemicals called pyrogens.
   b) Higher body temperature inhibits bacterial growth and stimulates healthy cells.

C. Third Line of Defense: Lymphocytes

1. Antigens
   a) Molecules that cause the host to react are called antigens.
      (1) Antigens are proteins or carbohydrates on the surface of pathogens or infected cells.
      (2) Each individual has a set of characteristic proteins on his or her own body cells.
         (a) These proteins are called major histocompatibility complex (MHC) proteins.
         (b) Another person's MHC proteins may be an antigen for you.
   b) Lymphocytes are white blood cells that recognize and attack specific antigens.
      (1) There are B lymphocytes and T lymphocytes.
         (a) B cells recognize and attack small microorganisms and their toxins.
         (b) T cells recognize and attack larger cells.
            (i) T cells respond to infected body cells or cancer cells.
            (ii) T cells respond to transplanted tissues.
            (iii) T cells respond to parasitic worms and fungi.
      (2) Lymphocytes use antigen receptor proteins to recognize antigens.
         (a) Antigen receptor proteins are either attached to the lymphocyte surface or secreted by the lymphocyte.
         (b) Antigen receptor proteins bind to the antigen.

2. Antibodies
   a) B cells secrete antibodies.
   b) Antibodies are special proteins that locate and mark antigens.
      (1) Each antibody has a four-chain structure.
         (a) Two heavy polypeptide chains form a Y.
         (b) Two light polypeptide chains are attached on the shorter sides of the Y.
      (2) Each antibody has constant and variable regions.
         (a) A constant region at one end of each of the four chains contains a fixed sequence of amino acids that reflects the antibody's class.
         (b) A variable region at the other end of each chain has a variable amino acid sequence that is specific for an antigen.
   c) Each antibody belongs to one of five immunoglobulin classes.
      (1) Classes are IgG, IgM, IgA, IgD, and IgE.
      (2) Each immunoglobulin has a specific function and location in the body.
   d) A nursing baby receives antibodies from its mother.
      (1) These antibodies confer passive immunity to the baby.
      (2) Passive immunity lasts as long as the antibody remains in the baby's blood.
      (3) Active immunity must be gained through exposure to antigens.

3. Allergy
   a) Allergy is an immune response that occurs when no pathogen is present.
   b) A nonharmful substance is called an allergen when it triggers an allergic reaction.
      (1) An allergen triggers B cells to produce IgE antibodies.
      (2) IgE antibodies bind to mast cells and basophils.
      (3) When the allergen enters the body again, it binds to the IgE antibodies.
      (4) Binding causes the cells to release histamine, which causes an allergic reaction.

D. Anticipating Infection

1. Lymphocyte Production
   a) Bone marrow stem cells produce lymphocytes.
      (1) Lymphocytes that continue development in the bone marrow become B cells.
      (2) Lymphocytes that develop in the thymus gland become T cells.
   b) Lymphocytes recognize trillions of different antigens.

2. Antigen Diversity
   a) B and T cells have evolved a mechanism to generate trillions of antigen receptors from a few genes.
   b) As B and T cells develop, DNA segments that code for antibodies rearrange themselves.
      (1) Each unique DNA arrangement codes for a different receptor protein.
      (2) These receptor proteins move to the surface of lymphocytes and become unique antigen receptors.
3. Self Versus Nonself
   a) The body generally keeps the immune system in homeostasis.
      (1) Developing lymphocytes with receptors that bind to self-proteins are usually destroyed.
      (2) Developing lymphocytes with receptors that do not bind to self-proteins are allowed to mature.
   b) Autoimmune diseases occur when a person's immune system attacks its own body.
      (1) In people with multiple sclerosis, T cells with receptors for proteins on nerve cells attack nerve cells in the brain.
      (2) In people with insulin-dependent diabetes, T and B cells attack cells in the pancreas that produce insulin.
      (3) In people with lupus, self-antibodies attack the nuclei of all cells.
      (4) In people with rheumatoid arthritis, the immune system attacks the synovial membranes that line some joints.

E. Humoral and Cell-Mediated Immunity
1. Humoral Immunity
   a) The protection provided by B cells is called humoral immunity.
   b) The B cell that responds to an antigen creates copies of itself called plasma cells.
      (1) Plasma cells fight infection.
      (2) The population of plasma cells is called a clonal population.
      (3) Cells in the clonal population are identical, and all secrete identical antibodies.
      (4) Cells in the clonal population called memory cells "remember" the infection and help to fight it if it ever enters the body again.
         (a) Vaccinations introduce inactive viruses or bacteria, or pieces of the pathogen, to stimulate the immune system to produce memory cells.
         (b) These cells can fight the active version of the pathogen if it ever infects the body.
         (c) Viruses with quickly changing strains, such as influenza, require development and administration of new vaccines each year.
   c) Antibodies produced by B cells mark the pathogen for destruction.
      (1) Antibody binds to antigen to form the antibody–antigen complex.
         (a) This complex signals phagocytes to attack the pathogen.
         (b) Complement proteins activate and break down the pathogen when they encounter an antibody–antigen complex.
      (2) Some antibodies cause pathogens to agglutinate, which makes them unable to infect cells.
2. Cell-Mediated Immunity
   a) The protection provided by T cells is called cell-mediated immunity.
   b) When they encounter a pathogen, T cells divide and produce memory cells, cytotoxic T cells, and helper T cells.
      (1) Cytotoxic T cells attack and kill body cells that are infected with virus.
         (a) Viral proteins are found on the surface of infected body cells.
         (b) Cytotoxic T cells bind to those proteins and destroy the cells.
      (2) Helper T (T4) cells detect invaders and alert B and T cells.
         (a) Helper T cells secrete a substance that boosts cytotoxic T cell response.
         (b) HIV destroys helper T cells, which shuts down much of the body's immune system response.
   c) Macrophages alert T cells to foreign antigens.
      (1) A macrophage engulfs and digests a foreign particle.
      (2) The macrophage displays parts of the pathogen's antigen on its own surface.
      (3) The macrophage, now called an antigen-presenting cell (APC), presents the antigen to a T cell with a matching receptor.

(4) The T cell produces specific memory cells, cytotoxic T cells, and helper T cells to fight the pathogen.

F. No Immune Response to Prions
1. Prions are refolded versions of normal prions, so the body does not recognize them as foreign.
2. Currently, there are no effective treatments for prion diseases.

IV. Preventing an Epidemic of Prion Diseases
A. An epidemic is a contagious disease that spreads quickly.
B. Epidemiologists work to understand and prevent epidemics.
C. Preventative measures are in place to prevent the spread of mad cow disease.
1. Laws prohibit the feeding of mammal products to cattle.
2. Laws prohibit use of certain cow parts in products meant for human use.
D. Cases of mad cow disease still occur in the U.S. and Canada.

Practice Questions

Matching

1. basophil
2. pathogen
3. lupus
4. humoral immunity
5. histamine
6. antigen
7. eosinophil
8. macrophage
9. antibody
10. prion
11. capsid
12. B lymphocyte
13. prokaryote
14. interferon
15. neutrophil

a. white blood cells that release histamine
b. white blood cell that battles protozoans
c. phagocyte that destroys bacteria and some fungi
d. protein produced by infected cells that protects uninfected cells
e. protection provided by B cells
f. protein surrounding a virus
g. foreign molecule
h. autoimmune disease
i. organism that causes disease
j. protein that binds to a specific antigen
k. single-celled organism without a nucleus
l. stimulates the inflammatory response
m. phagocyte that destroys bacteria and stimulates other immune cells
n. secretes antibodies to specific antigens
o. protein produced by brain cells

Fill-in-the-Blank

16. A protozoan is a ____________ that is attacked by white blood cells called ____________ when it invades the body.
17. A breast-feeding infant gains ____________ immunity to ____________ to which its mother was exposed.
18. Rheumatoid arthritis is an ____________ disease in which the body's own immune system attacks the ____________ membranes of joints in the fingers and toes.
19. The extrachromosomal DNA in a bacterium, called a ____________, is located inside the ____________ of the bacterial cell.
20. Some bacteria are able to move around using ____________.
21. Mutant prions that are ____________ can cause properly folded prions to ____________.
22. The ____________ in the stomach is part of the body's ____________ line of immune defense.
23. Macrophages have ____________ that allow them to move and to engulf ____________.

24. A bacterium that has entered the body may be attacked by ____________ proteins that coat its surface or destroy its ____________.

25. A ____________ consists of inactivated pathogen or portions of a pathogen that spur the immune system to produce ____________ cells.

Labeling

Use the terms below to label Figure 12.1. Any term listed twice will be used twice.

antibody
antigen
antigen
B cell
B-cell receptor
T cell
T-cell receptor

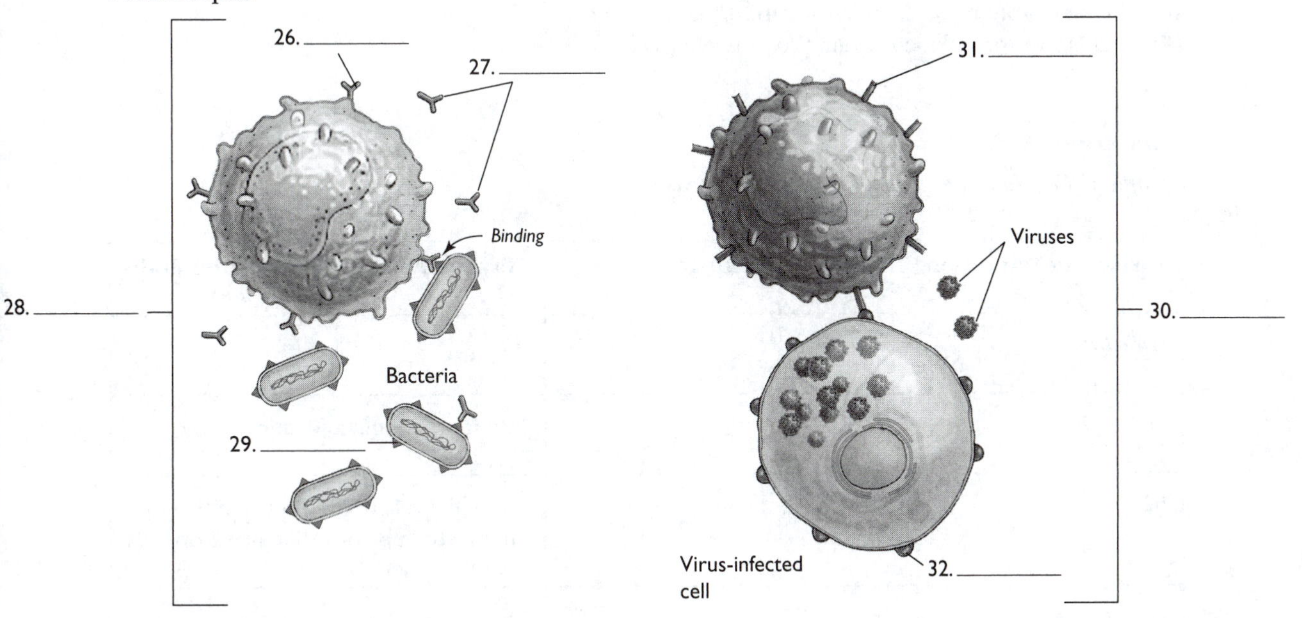

Figure 12.1

Use the terms below to label Figure 12.2.

antigen-binding site
constant regions
heavy chain
light chain
variable regions

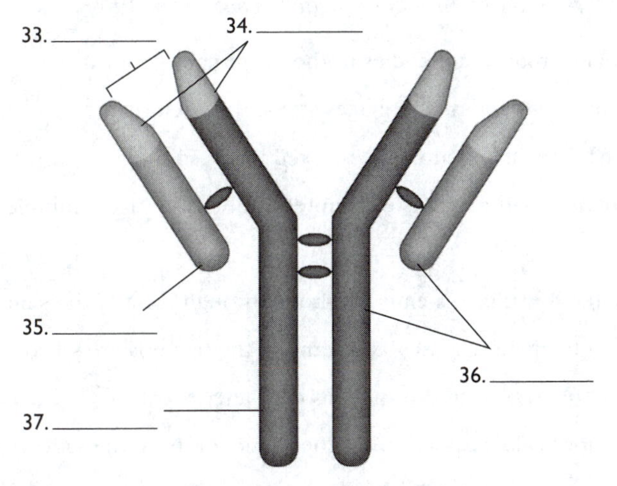

Figure 12.2

Roots to Remember

Use your knowledge of the root words presented in this chapter to answer the following questions.

38. What does the term *false feet* describe?
39. Encephalitis causes inflammation in what organ?
40. What kind of tissue is likely to be affected by rheumatic fever?

Word Choice

Circle the word or phrase that correctly completes each sentence.

41. A virus is a (living/nonliving) particle.
42. A (neutrophil/eosinophil) cell is not a phagocyte.
43. (B cells/T cells) produce a clonal population of plasma cells.
44. (Tetanus/Hepatitis) is caused by a bacterium.
45. T cells are part of the (specific/nonspecific) immune system defense.
46. The viral protein coat is called a (capsid/plasmid).
47. Viral DNA is transcribed within the (host/viral) cell.

Table Completion

Complete the following table on infectious diseases.

Disease or Condition	Type of Pathogen	Way the Pathogen Enters the Body
Giardiasis	48.	49.
50.	51.	Bite from an infected animal
52.	53.	Through broken skin after stepping on an infected nail or other sharp object
Hepatitis C	54.	55.
Tapeworm	56.	57.

Sequencing

Put the following events in order by numbering each event, from 1 to 8.

58. ____________ B cells produce antibodies to the bacteria.
59. ____________ Neutrophils and macrophages attack the bacteria.
60. ____________ The bacterial infection is destroyed.
61. ____________ Complement proteins encounter antibody–antigen complexes and destroy bacterial cells.
62. ____________ Bacterial pathogens enter the body through a cut in the skin.
63. ____________ Body temperature rises, and some of the bacteria are killed.
64. ____________ Antibodies bind to the antigens on bacterial cells.
65. ____________ Memory cells respond to another infection from the same type of bacteria.

Critical Thinking

66. Are people likely to have a very similar mix of digestive flora or a distinct mix of digestive flora? Why?
67. Why is it possible for people with type 2 diabetes mellitus to recover from the disease with a change in lifestyle, while it is not possible for people with type 1 (insulin-dependent) diabetes to recover?
68. Why do dead viruses used in vaccinations still trigger an immune response?

Practice Test

1. What viral disease can be spread by mosquitoes?
   A. hepatitis
   B. mononucleosis
   C. West Nile virus
   D. anthrax

2. Which of the following white blood cells could recognize and attack cells that have been infected with the influenza virus?
   A. natural killer cells
   B. cytotoxic T cells
   C. basophils
   D. B cells

3. The antibody-antigen complex is most like
   A. a lock and key.
   B. predator and prey.
   C. two bones and a joint.
   D. peanut butter and jelly.

4. Which of the following statements is true?
   A. Lymphocytes are produced in the bone marrow, and they also mature there.
   B. Lymphocytes are produced in bone marrow, but only B cells mature there.
   C. Lymphocytes are produced in bone marrow, but only T cells mature there.
   D. Lymphocytes are produced in bone marrow and mature in the thymus gland.

5. Lysosomes that digest invading pathogens are found in
   A. basophils.
   B. B cells.
   C. T cells.
   D. macrophages.

6. You receive a flu shot, but you get the flu anyway. What is likely to have happened?
   A. Your body did not produce any memory cells, so it could not fight off the flu.
   B. The shot you got contained a different strain of the flu from the strain that you contracted.
   C. The flu shot interfered with the natural immunity that you had to the flu.
   D. The shot you received had not been refrigerated properly, so all the flu particles in the shot died.

7. Which type of antibody responds to an allergen?
   A. IgA
   B. IgB
   C. IgD
   D. IgE

8. Why are helper T cells essential to a successful immune response?
   A. They produce antibodies that can mark and destroy invading pathogens.
   B. They attack and destroy cells that have been infected by a pathogen.

C. They secrete a substance that strengthens the response of B and T cells to a pathogen.
D. They ingest pathogens and display their antigens to trigger a response from cytotoxic T cells.

9. Prions are dangerous when they
A. become misfolded.
B. take a helical shape.
C. invade the brain.
D. are ingested.

10. Which of the following is an example of transmission of an infectious agent by indirect contact?
A. A boy who has a cold sneezes in a crowd of people.
B. A flu virus on a telephone receiver is passed to a person making a call.
C. A tick carrying the Lyme disease bacterium bites a camper.
D. A cow ingests feed made with infectious prions.

11. Mucus traps and removes pathogens in each of the following body structures *except*
A. the spleen.
B. the esophagus.
C. the nose.
D. the trachea.

12. Which type of food poisoning could result from creamy potato salad that is left out at a picnic?
A. botulism
B. salmonellosis
C. *E. coli* enteritis
D. staphylococcal infection

13. What is a capsid?
A. the protein coat around a virus
B. the gelatinous coating around a bacterium
C. the nucleic acid within a virus
D. the extrachromosomal DNA molecule in a bacterium

14. While breastfeeding, a baby
A. obtains active immunity from its mother.
B. can get antigens from its mother.
C. obtains bacterial flora from its mother.
D. gains memory cells from its mother.

15. The bacteria on a cutting board are able to divide once every 30 minutes. If you start with one bacterium, how many bacteria would be on the cutting board after 5 hours?
A. 512
B. 1,024
C. 2,048
D. 4,096

Answer Key

Matching

1. a; **2.** i; **3.** h; **4.** e; **5.** l; **6.** g; **7.** b; **8.** m; **9.** j; **10.** o; **11.** f; **12.** n; **13.** k; **14.** d; **15.** c

Fill-in-the-Blank

16. pathogen; eosinophils; **17.** passive; antigens; **18.** autoimmune; synovial; **19.** plasmid; nucleoid region; **20.** flagella; **21.** misfolded; misfold; **22.** mucous membrane; first; **23.** pseudopodia; pathogens; **24.** complement; membrane; **25.** vaccination; memory

Labeling

26. B-cell receptor; **27.** antibody; **28.** B cell; **29.** antigen; **30.** T cell; **31.** T-cell receptor; **32.** antigen; **33.** antigen-binding site; **34.** variable regions; **35.** light chain; **36.** constant regions; **37.** heavy chain

Roots to Remember

38. pseudopodia; **39.** brain; **40.** connective

Word Choice

41. nonliving; **42.** eosinophil; **43.** B cells; **44.** Tetanus; **45.** specific; **46.** capsid; **47.** host

Table Completion

48. Protozoan; **49.** Cysts in contaminated water or food are ingested; **50.** *Rabies*; **51.** Virus; **52.** *Tetanus*; **53.** Bacterium; **54.** Virus; **55.** Intravenous drug use, body piercing, tattoos; **56.** Worm; **57.** Ingestion of raw or undercooked pork or beef

Sequencing

58. 4; **59.** 2; **60.** 7; **61.** 6; **62.** 1; **63.** 3; **64.** 5; **65.** 8

Critical Thinking

66. People are likely to have very distinct digestive flora, because everyone grows up in a different environment with different microorganisms.

67. In insulin-dependent diabetes, the immune system attacks and destroys the pancreatic cells that make insulin. Therefore, the pancreas loses the ability to produce insulin. These cells are not necessarily destroyed in type 2 diabetes, so a change in lifestyle may allow remission of the disease.

68. These dead viruses still have antigens on their surface that trigger an immune response and allow the body to produce memory cells that can fight an invasion from the live virus.

Practice Test

1. C; **2.** B; **3.** A; **4.** B; **5.** D; **6.** B; **7.** D; **8.** C; **9.** A; **10.** B; **11.** A; **12.** D; **13.** A; **14.** C; **15.** C

CHAPTER 13

SEXUALLY TRANSMITTED INFECTIONS: THE CERVICAL CANCER VACCINE

Learning Goals

1. Define *sexually transmitted infection*.
2. List the common sexually transmitted infections, describe their causative agents, and explain how (if at all) they may be cured or prevented.
3. Describe the link between HPV, genital warts, and cervical cancer.
4. Explain how HIV causes the symptoms of AIDS.
5. Describe why HIV eventually wins its battle with the immune system and how combination drug therapy can lengthen survival.
6. List the recommendations for the individual behaviors that best prevent HIV and other sexually transmitted infections.

Chapter Outline

I. The Old Epidemics
   A. Overview
      1. Sexually transmitted infections (STIs) are infections that are transferred via sexual contact.
      2. STIs can lead to STDs, or sexually transmitted diseases.
      3. STDs have a long history in human populations.
   B. The Eukaryotes: Pubic Lice and Trichomoniasis
      1. Pubic lice are insects of the species *Phthirus pubis*.
         a) Pubic lice are also called "crabs."
         b) Pubic lice cause itching and discomfort.
            (1) Pubic lice bite the skin to take a blood meal.
            (2) Females lay eggs at the base of pubic hair.
         c) Treatments include washing with insecticidal soap and cleaning bedding.
      2. Trichomoniasis is caused by the protozoan *Trichomonas vaginalis*.
         a) Trichomonas mainly causes symptoms in women.
            (1) Trichomonas colonizes the vagina.
            (2) It causes yellow-green discharge and itching.
         b) Trichomonas can be treated with antibiotics.
      3. Candida is a fungus that causes yeast infection.
         a) Yeast is a normal inhabitant of the vagina.
         b) Stress, illness, and antibiotic treatment may cause overgrowth of yeast.
         c) Symptoms of yeast infection include itching and discharge.
         d) Yeast infection can be treated with antifungal agents.
         e) A yeast infection may be a sign of an underlying STI.
   C. The Bacteria: Chlamydia, Gonorrhea, and Syphilis
      1. Chlamydia and gonorrhea are the most common STIs, and a patient who has one of these diseases often also has the other.
         a) Chlamydia is caused by the *Chlamydia trachomatis* bacterium.
         b) Gonorrhea ("the clap") is caused by the *Neisseria gonorrhoeae* bacterium.
         c) Both diseases are spread by vaginal discharge or semen.
            (1) Both commonly infect the urethra, anus, cervix, endometrium, and oviducts.
            (2) Both can (rarely) infect mucous membranes of the eye, mouth, and throat.

- d) Symptoms are different for males and females.
  - (1) Infected males may experience painful urination or discharge.
  - (2) Infected females may be asymptomatic.
- e) Screening for chlamydia and gonorrhea is common and beneficial.
  - (1) Infection can be cured with antibiotics.
  - (2) Untreated infection can cause pelvic inflammatory disease (PID) in women.
    - (a) PID can cause debilitating pain.
    - (b) PID can damage the reproductive tract and cause infertility.
    - (c) PID can cause fatal blood infection.
  - (3) Women with untreated chlamydia or gonorrhea are at greater risk for HIV infection.
- f) Some strains of gonorrhea and chlamydia have developed resistance to certain antibiotics.

2. Syphilis is caused by the *Treponema pallidum* bacterium.
   - a) Syphilis is less common than chlamydia and gonorrhea.
   - b) Syphilis has three phases.
     - (1) During the primary phase, a painless sore appears on genitals or the anus.
       - (a) The sore is highly infectious.
       - (b) After it disappears, bacteria remain in the infected person's body.
     - (2) During the secondary phase, the infected person may experience flulike symptoms, skin rash, hair loss, fever, and sore throat.
     - (3) During the third phase, the *Treponema* bacteria cause damage to the heart, brain, liver, and other internal organs.
       - (a) In this phase, the infected person may experience paralysis, blindness, or dementia.
       - (b) The final phase may not appear for 5–20 years after the initial infection.
   - c) Untreated syphilis in a pregnant woman can cause infection in the fetus.
     - (1) This condition is called congenital syphilis.
     - (2) Congenital syphilis can cause bone deformation, blindness, or stillbirth.
   - d) Syphilis can be treated with penicillin.

D. The Viruses: Herpes, Hepatitis, and Genital Warts

1. Herpes Simplex
   - a) Herpes simplex is caused by one of the herpes simplex viruses.
     - (1) Herpes simplex, type 1 (HSV-1) commonly causes cold sores.
     - (2) Herpes simplex, type 2 (HSV-2) commonly causes genital herpes.
   - b) Herpes viruses produce blisters that burst and leave sores or ulcers.
     - (1) These viruses are transmitted through contact with sores or ulcers.
     - (2) Herpes viruses can also be transmitted through skin-to-skin contact when no sores are present.
   - c) Active HSV in a pregnant woman can infect her baby during birth.
     - (1) HSV in infants can cause brain damage, blindness, or death.
     - (2) Pregnant women with HSV often deliver through cesarean section.
   - d) Herpes viruses remain in the body for life.
     - (1) People with herpes experience random outbreaks.
     - (2) Acyclovir and other medications can reduce outbreaks by disrupting viral reproduction.
2. Hepatitis
   - a) Hepatitis viruses damage the liver.
     - (1) As the liver is damaged, an infected person shows symptoms from fatigue to mental confusion.
     - (2) Bilirubin builds up and causes the skin to become jaundiced or yellow.
   - b) Hepatitis B virus (HBV) is most commonly transmitted sexually.
     - (1) The body can generally fight and conquer HBV.
     - (2) In some cases, HBV can lead to chronic infection.
       - (a) Chronic HBV causes cirrhosis of the liver.
       - (b) Cirrhosis can lead to liver cancer, liver failure, or death.
     - (3) Some antiviral drugs can treat HBV.
   - c) Hepatitis can be transmitted from mother to child during childbirth.
   - d) Hepatitis B can be transmitted from blood-to-blood contact.
   - e) There is a vaccine for hepatitis B.
     - (1) Immunization has dramatically reduced the number of HBV cases in the U.S.
     - (2) The HBV vaccine has a number of health effects.
       - (a) It prevents chronic liver disease.
       - (b) It prevents liver cancer that is caused by chronic liver disease.

3. Human papillomavirus (HPV) causes warts.
   a) There are many strains of HPV.
      (1) HPV-1 causes plantar warts.
      (2) HPV-2 causes warts on the hands.
      (3) Four other strains cause genital warts.
   b) All strains of HPV are spread by skin-to-skin contact.
   c) Some strains of HPV are associated with the development of cervical cancer.
   d) The Gardasil vaccine prevents infection from two strains that are associated with most cervical cancers: HPV-16 and HPV-18.
      (1) With widespread use, the Gardasil vaccine could prevent thousands of cases of cervical cancer.
      (2) Widespread use would inhibit the spread of the disease.
   e) The Gardasil vaccine also prevents infection from two strains of HPV that cause 90% of the cases of genital warts: HPV-6 and HPV-11.
      (1) Vaccination against these strains will benefit men.
      (2) If men receive the Gardasil vaccine, the infection rate from HPV will go down dramatically.

II. The New Epidemic—AIDS

A. A Disease of the Immune System

1. Acquired immune deficiency syndrome, or AIDS, is a worldwide epidemic.
2. The human immunodeficiency virus (HIV) causes AIDS.
3. HIV kills or disables helper T (T4) cells.
   a) Disabling these cells causes immune deficiency.
   b) Diseases that the body could normally fight can cause death.
      (1) *Pneumocystis jirovecii* is found in the body at low levels.
      (2) When the immune system is weak, *Pneumocystis* can cause pneumonia and lung damage.
      (3) Other opportunistic infections can also attack the body, leading to death.
4. Most HIV is spread through unprotected sexual intercourse.
5. HIV reproduces inside body cells.
   a) HIV binds to the protein receptor CD4 and a co-receptor on the T4 cell membrane.
   b) It releases its RNA and proteins into the T4 cell.
   c) Viral RNA is converted to viral DNA in the cell by the action of reverse transcriptase.
      (1) Reverse transcription is the converse of transcription.
      (2) Because it uses reverse transcription, HIV is called a retrovirus.
   d) The viral DNA inserts itself into the cell's genome.
   e) In many cases, the viral DNA takes control of the cell.
      (1) The cell makes new copies of viral RNA.
      (2) It translates its genes into virus proteins.
      (3) It assembles new viruses.
      (4) The virus copies bud off the cell membrane and infect other cells.
   f) In some cases, the viral DNA remains dormant in the cell.

B. The Course of HIV Infection

1. Early symptoms of HIV infection are flulike.
2. After 6–12 weeks, the immune system develops a specific response to HIV.
   a) Most infected individuals produce HIV antibodies.
   b) Tests for HIV indicate whether a person is HIV positive (has antibodies) or HIV negative (does not have antibodies).
3. After the specific immune response, the level of T4 cells rebounds.
   a) An infected individual becomes asymptomatic.
   b) HIV can remain dormant in cells for years.
   c) The body continues to fight—and win—against active HIV cells.
4. The immune system and virus take part in an arms race.
   a) New HIV variants appear.
   b) The body develops antibodies against these variants.
5. Eventually, a variant arises that the body cannot fight.
   a) No antibodies are produced.
   b) The variant replicates unchecked.

c) Large numbers of T4 cells are killed or damaged.
d) The immune system weakens, signaling the onset of AIDS.

C. Treating HIV Infection
1. Early drug therapies failed because HIV quickly became resistant.
2. Combination drug therapy is now used.
a) This therapy is called highly active antiretroviral therapy, or HAART.
b) It uses three or more drugs to attack the virus.
c) Combination drug therapy has lengthened the life span of HIV-positive individuals in the U.S. from 7 years to 24 years.
3. Combination drug therapy is very expensive.
4. This therapy is difficult to follow and can have unpleasant side effects.
5. Prevention efforts are diminished as combination drug therapy makes HIV/AIDS appear to be less of a threat.

D. Preventing HIV/AIDS
1. HIV is transmitted only through direct contact with body fluids.
a) Blood, semen, and vaginal fluid can carry HIV.
b) Unprotected sex is the main mode of transmission.
c) Needle sharing can spread HIV.
d) Deep kissing is not likely to spread HIV, unless there are cuts in the mouth.
2. HIV is not spread by tears, sweat, coughing, or sneezing.
3. HIV cannot survive outside the body.
4. HIV is not transmitted by insect bites.
5. Active STDs can increase the risk of HIV infection.
a) Ulcers or sores allow blood-to-blood transmission.
b) T4 are likely to gather around lesions and become infected by HIV.
c) Some STI infections can increase the number of HIV particles in body fluid.
6. The best treatment for HIV/AIDS is prevention.
a) Abstain from sexual activity.
b) If you are sexually active, be monogamous.
c) Use a condom during sexual intercourse.
d) Promptly treat any STI that appears.
e) Don't use illicit drugs or share needles.
7. There is evidence that explaining the risks associated with sexual activity leads to reduced activity or safer sex practices.

Practice Questions

Matching

1. retrovirus
2. chlamydia
3. syphilis
4. HPV
5. congenital syphilis
6. trichomoniasis
7. asymptomatic
8. HIV
9. AIDS
10. gonorrhea
11. PID
12. herpes simplex virus
13. STD
14. hepatitis
15. monogamous

a. specific virus that attacks T4 cells
b. when no symptoms are present
c. one of these can cause cold sores
d. often occurs with gonorrhea
e. having only one sexual partner
f. any virus that reproduces using reverse transcription
g. can cause blindness or bone deformations in a fetus
h. caused by the human immunodeficiency virus
i. sores appear at the infection site
j. causes inflammation of the liver
k. caused by a protozoan
l. known as "the clap"
m. virus that causes warts
n. disease that is transmitted sexually
o. can result from untreated chlamydia

Fill-in-the-Blank

16. After DNA is produced from HIV RNA, using ____________, the viral DNA inserts into the host cell's ____________.

17. Syphilis is caused by a ____________ called *Treponema pallidum*.

18. Infection from HPV, or the ____________, can be prevented with a ____________ or by using a ____________ during intercourse.

19. It is likely that a ____________ who has trichomoniasis would show no symptoms.

20. Gonorrhea can be treated with an ____________ and herpes can be treated with an ____________ medication.

21. ____________ infection is not necessarily spread sexually, but it can be a sign of an ____________.

22. Partners who only kiss have a ____________ risk of spreading STIs.

23. Pubic lice are ____________ organisms that can be treated with ____________ soap.

24. As ____________ weakens the immune system during the course of many years, ____________ infections can attack the body.

25. Within a human T4 cell, viral DNA is ____________ by the cell to make many copies of viral ____________.

Labeling

Use the terms below to label Figure 13.1. Any term listed more than once should be used more than once.

HIV particle
human DNA
reverse transcriptase
viral enzyme
viral DNA
viral DNA
viral protein
viral RNA
viral RNA
viral RNA
viral RNA
viral surface proteins

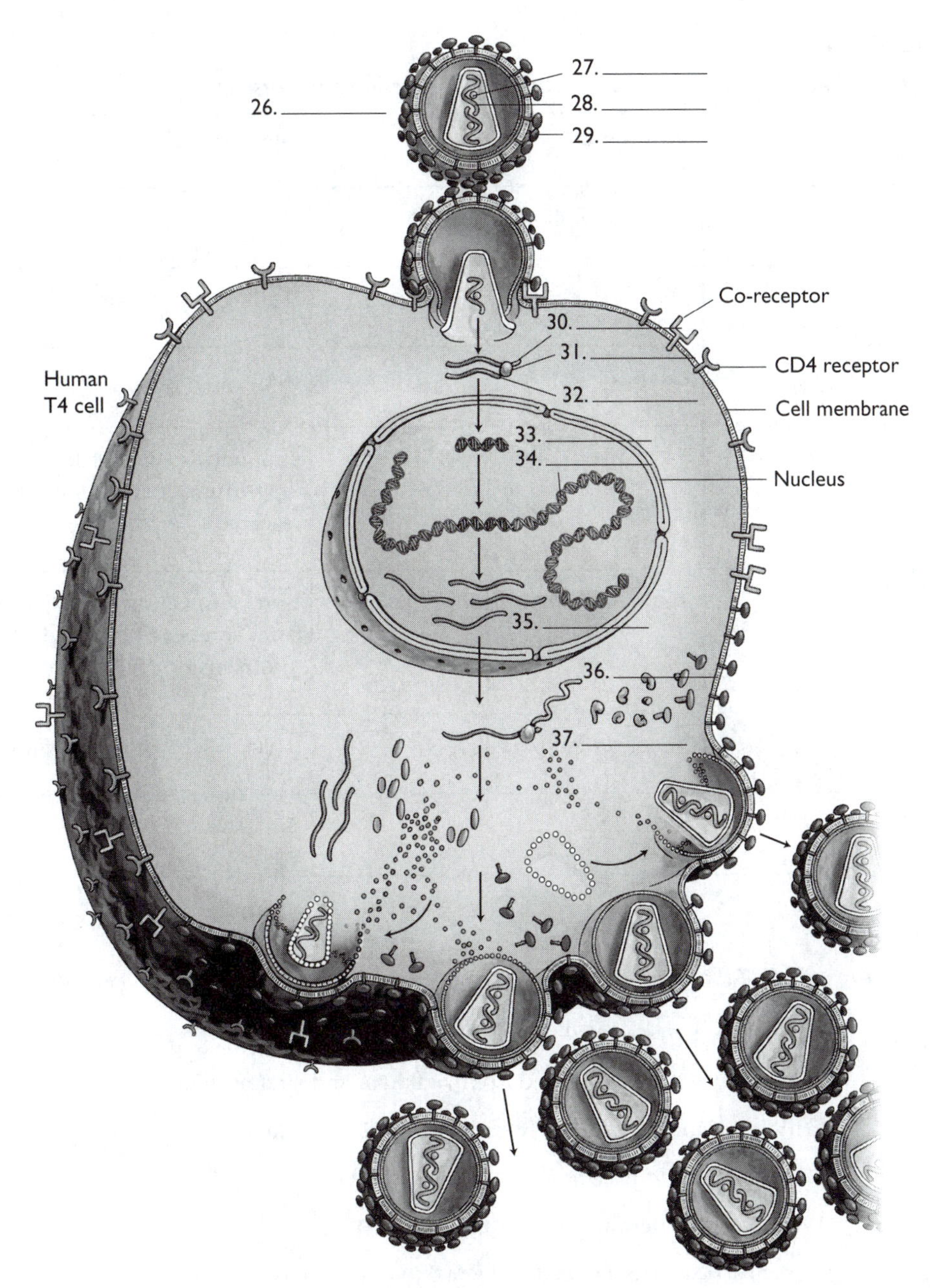

Figure 13.1

Roots to Remember

Use your knowledge of the root words presented in this chapter to answer the following questions.

38. *Hepatica* is a small woodland plant. Its name comes from the shape of its leaves. What human organ do the leaves resemble?
39. The medical term for gum disease is *gingivitis.* What is a likely symptom of gum disease?
40. A cancer cell may look atypical. What does *atypical* mean?

Word Choice

Circle the word or phrase that correctly completes each sentence.

41. When you practice abstinence, you engage in (no/limited) sexual activity.
42. (Entry/Protease) inhibitors block the process that converts viral proteins to active enzymes.
43. Chlamydia (can/cannot) be easily cured with antibiotics.
44. (Trichomoniasis/Gonorrhea) causes yellow-green vaginal discharge in infected women.
45. The most common bacterial STD in the United States is (chlamydia/gonorrhea).
46. The final stage of syphilis can cause (blindness/PID).
47. Immune system cells (can/cannot) make antibodies to HIV when it first enters the body.

Table Completion

Complete the following table on sexually transmitted infections and diseases.

Disease	Pathogen	Symptoms
48.	Human papillomavirus	49.
Pubic lice	50.	Itching in pubic area
51.	*Candida*	52.
53.	54.	Immune system weakens and opportunistic infections cause damage
55.	*Treponema pallidum*	Sore at infection site; rash, fever, joint soreness; neurological problems or death if untreated
56.	*Neisseria gonorrhoeae*	Thick discharge or no symptoms
Herpes simplex	HSV-2	57.

Sequencing

Put the following events of the HIV reproductive cycle in order by numbering each event, from 1 to 8.

58. ____________ Viral DNA is transcribed to produce copies of viral RNA.
59. ____________ Viral particles are assembled from viral RNA and viral proteins.
60. ____________ HIV binds to a T4 cell and releases viral RNA into the cell.
61. ____________ Viral mRNA is translated into viral proteins.
62. ____________ Viral enzymes perform reverse transcription.
63. ____________ Virus particles are released and attack other T4 cells.
64. ____________ Part of the host cell membrane encloses each viral particle.
65. ____________ Viral DNA inserts itself into DNA in the host cell.

Critical Thinking

66. Why isn't the HIV virus considered dormant during the asymptomatic phase of untreated HIV infection?
67. Can kissing cause transmission of HSV-2?
68. What is the likely cause of the increase in the number of syphilis infections since 2000?

Practice Test

1. What parts of the host cell are present in an HIV virus particle that has just been created?
   A. host-cell membrane only
   B. host-cell DNA only
   C. host-cell RNA only
   D. host-cell RNA and host-cell membrane

2. Which of the following STDs cannot be cured?
   A. herpes simplex
   B. syphilis
   C. gonorrhea
   D. trichomoniasis

3. The risk of transmitting an STI during protected anal sex is
   A. low.
   B. moderate.
   C. high.
   D. very high.

4. The life cycle of pubic lice is similar to all of the following *except*
   A. fleas.
   B. mosquitoes.
   C. house flies.
   D. ticks.

5. Chlamydia is caused by a
   A. fungus.
   B. bacterium.
   C. virus.
   D. protozoan.

6. A woman complains to her gynecologist about vaginal discharge. Which of the following diseases would *not* cause the woman's symptoms?
   A. yeast infection
   B. trichomoniasis
   C. chlamydia
   D. herpes

7. Someone who has gonorrhea is also likely to have
   A. chlamydia.
   B. AIDS.
   C. syphilis.
   D. pubic lice.

8. Syphilis is least damaging to the body
   A. when it is congenital.
   B. during its earliest stage.
   C. during its secondary stage.
   D. during its third stage.

9. A male patient has a painless sore on his penis that disappears within a few weeks. Which STD is the patient likely to have?
   A. syphilis
   B. gonorrhea
   C. genital warts
   D. hepatitis B

10. Who would benefit from the HPV vaccine?
    A. uninfected females only
    B. uninfected males only
    C. uninfected males and females
    D. infected males and females

11. Your friend says that she doesn't use exercise equipment at the gym because she is afraid of getting HIV. You tell your friend that
    A. she is smart. HIV is transmitted by body fluids, so it could be transmitted through sweat.
    B. she might be right. HIV can be passed along in body fluids, and sweat is a body fluid.
    C. she shouldn't worry so much. HIV transmission through sweat is rare.
    D. she shouldn't worry at all. HIV isn't transmitted by sweat and it cannot survive outside the body.

12. What is happening during the time that HIV is asymptomatic in the body?
    A. HIV is dormant within body cells.
    B. HIV is reproducing but hiding from the immune system.
    C. New HIV variants are evolving, and the immune system is fighting them.
    D. Scientists have not yet determined what is happening during this time.

13. Hepatitis B is *not* likely to be transmitted
    A. through sexual contact.
    B. by contaminated needles.
    C. during childbirth.
    D. through insect bite.

14. Which class of the combination drugs used to fight HIV infection blocks HIV from binding to T4 cells?
    A. entry inhibitors
    B. non-nucleoside reverse transcriptase inhibitors
    C. nucleoside analogs
    D. protease inhibitors

15. Which sexually transmitted disease could lead to cancer?
    A. genital warts
    B. hepatitis B
    C. both A and B
    D. neither A nor B

Answer Key

Matching

1. f; **2.** d; **3.** i; **4.** m; **5.** g; **6.** k; **7.** b; **8.** a; **9.** h; **10.** l; **11.** o; **12.** c; **13.** n; **14.** j; **15.** e

Fill-in-the-Blank

16. reverse transcription; genome; **17.** bacterium; **18.** human papillomavirus; vaccine; condom; **19.** man; **20.** antibiotic; antiviral; **21.** Yeast; STI; **22.** low; **23.** eukaryotic; insecticidal; **24.** HIV; opportunistic; **25.** transcribed; RNA

Labeling

26. HIV particle; **27.** viral enzyme; **28.** viral RNA; **29.** viral surface proteins; **30.** viral RNA; **31.** reverse transcriptase; **32.** viral DNA; **33.** viral DNA; **34.** human DNA; **35.** viral RNA; **36.** viral protein; **37.** viral RNA

Roots to Remember

38. the liver; **39.** inflammation of the gums; **40.** not typical or not normal

Word Choice

41. no; **42.** Protease; **43.** can; **44.** Trichomoniasis; **45.** chlamydia; **46.** blindness; **47.** can

Table Completion

48. Genital warts; **49.** Bumps on genital area; abnormal cell changes on cervix; **50.** *Phthirus pubis*; **51.** Yeast infection; **52.** White vaginal discharge and vaginal itching; **53.** AIDS; **54.** HIV; **55.** Syphilis; **56.** Gonorrhea; **57.** Blisters in the genital area

Sequencing

58. 4; **59.** 6; **60.** 1; **61.** 5; **62.** 2; **63.** 8; **64.** 7; **65.** 3

Critical Thinking

66. An active battle between the virus and the immune system is actually taking place during the asymptomatic phase of infection. The virus is constantly producing new variants that are then attacked by the immune system.

67. In most cases, oral herpes is HSV-1. HSV-1 can be spread by kissing. However, if HSV-2 has been transmitted to the mouth by genital contact, it can be transmitted by kissing.

68. An increase in the frequency of unprotected sex.

Practice Test

1. A; **2.** A; **3.** B; **4.** C; **5.** B; **6.** D; **7.** A; **8.** B; **9.** A; **10.** C; **11.** D; **12.** C; **13.** D; **14.** A; **15.** C

CHAPTER 14

BRAIN STRUCTURE AND FUNCTION: ATTENTION DEFICIT DISORDER

Learning Goals

1. List the three types of neurons and their functions.
2. Describe the structure of a neuron.
3. Describe the events that occur during an action potential.
4. Describe how a nerve impulse is transmitted from one neuron to the next.
5. Explain what happens during synaptic integration.
6. Explain how blocking neurotransmitter breakdown in the synapse affects neural function.
7. Identify the major parts of the brain and their functions.
8. Explain the role of the limbic system and identify two of its major structures.
9. List the two types of nerves found in the PNS and explain how they differ from a mixed nerve.
10. Describe the functions of the somatic and autonomic divisions of the peripheral nervous system.

Chapter Outline

I. Nervous System Tissues
   A. Overview
      1. The nervous system receives, interprets, and responds to messages.
      2. Nervous tissue consists of two types of cells.
         a) Neurons can carry electrical and chemical messages.
         b) Neuroglia support and nourish neurons.
      3. Signals travel through neurons, between neurons, and from neurons to effectors.
         a) Effectors are muscles and glands that are responsive to nervous system signals.
         b) Effectors allow the body to respond to internal and external stimuli.
      4. Sensory receptors detect sensory input from the environment.
         a) Sensory receptors are neurons or other cells.
         b) Sensory receptors detect internal or external changes.
         c) There are sensory receptors for general senses.
            (1) General senses include temperature, pain, touch, pressure, and body position (proprioception).
            (2) General sensory receptors are located throughout the body.
         d) There are sensory receptors for special senses.
            (1) Special senses include smell, taste, equilibrium, hearing, and vision.
            (2) Sensory receptors for special senses are located in complex organs.
      5. The nervous system is divided into the central nervous system (CNS) and peripheral nervous system (PNS).
         a) The CNS includes the brain and spinal cord.
         b) The PNS includes the nerves extending through the rest of the body.
   B. Neuron Structure
      1. Neurons have a branching structure.
         a) Signals are received into the neuron through dendrites.
         b) Dendrites extend from a cell body that contains a nucleus and organelles.

c) Signals travel from dendrites to a long axon.
d) The axon terminates in axon terminals, or terminal boutons.

2. There are three categories of neurons.
   a) Sensory neurons carry signals to the CNS.
   b) Motor neurons carry signals from the CNS toward effectors.
   c) Interneurons link sensory and motor neurons.
3. A myelin sheath coats and protects many neurons.
   a) Neuroglial cells called Schwann cells create a myelin sheath around axons in the PNS.
      (1) Schwann cells contain myelin in their plasma membranes.
      (2) Myelin is a white lipid.
   b) Myelin determines the type of matter in CNS tissue.
      (1) CNS tissue with myelinated axons is called white matter.
      (2) CNS tissue without myelinated axons is called gray matter.
   c) Myelin prevents sideways message transmission.
   d) Unmyelinated patches between Schwann cells, called the nodes of Ranvier, increase signal conduction.
      (1) Impulses jump from node to node.
      (2) Nerve impulses can travel 100 times faster on a myelinated axon than on one with no myelin.
   e) Myelinated tissue is often bundled together.
      (1) Nerves are bundles of myelinated axons in the PNS.
      (2) Nerve tracts are bundles of white matter in the CNS.

C. The Creation of Nerve Impulses

1. Overview
   a) Neurons transmit electrical charges called nerve impulses.
   b) Stimuli (such as light or sound) can excite sensory neurons.
      (1) Excited neurons transmit nerve impulses to the CNS.
      (2) The CNS sends information through motor nerves.
      (3) Transmitted information elicits a response from muscles or glands.
2. Resting Potential
   a) Electrical potential is the difference in electric charge or voltage between two regions.
   b) Resting potential is the net negative charge inside an axon compared to its outside.
      (1) Resting potential results from a difference in ion concentrations.
         (a) The concentration of sodium (Na^+) ions is greater outside the axon than inside.
         (b) The concentration of potassium (K^+) ions is greater inside the axon than outside.
      (2) Sodium-potassium pumps in the cell membrane maintain the difference in ion concentrations.
         (a) These pumps actively transport sodium out of the cell.
         (b) These pumps actively transport potassium into the cell.
         (c) Each pump transports three sodium molecules for every two potassium molecules.
         (d) More positive ions are found outside the cell than inside.
3. Action Potential
   a) Action potential is the brief reversal of an axon's electrical charge.
      (1) Action potential is a wave of current that travels down the axon.
      (2) Action potential occurs above a certain stimulus level.
      (3) An intense stimulus causes the axon to increase the frequency of action potentials.
   b) An action potential relies on depolarization and repolarization.
      (1) Depolarization
         (a) When a neuron is stimulated, some sodium channels open to let sodium ions into the cell.
         (b) That portion of the cell's interior becomes less negative.
         (c) As more sodium channels open, that portion of the cell loses its charge difference and becomes depolarized.
         (d) Sodium channels are activated in a wave, causing depolarization down the length of the neuron.
      (2) Repolarization
         (a) As each sodium channel closes, a potassium channel opens.

(b) Potassium ions leave the cell, and the cell interior regains a net negative charge.
(c) This process is called repolarization.

c) An axon undergoes a refractory period after an action potential.
(1) Sodium gates remain closed during the refractory period.
(2) Another action potential cannot occur during the refractory period.

D. Neurotransmitters Carry Signals Between Neurons

1. Signals are transmitted between two neurons across the synapse.
a) The synapse contains
(1) the terminal bouton of the presynaptic neuron;
(2) the space between neurons, also called the synaptic cleft; and
(3) the plasma membrane of the postsynaptic neuron.

2. Neurotransmitters carry signals across the synapse.
a) Vesicles in the terminal bouton contain neurotransmitters.
b) The electrical impulse traveling down the axon causes neurotransmitters to be released.
c) Neurotransmitters diffuse across the synapse.
d) Neurotransmitters bind to receptors on the postsynaptic cell.
e) This binding excites the postsynaptic cell and generates an action potential.

3. Neurotransmitters are removed after evoking a response.
a) Enzymes in the synapse break down some neurotransmitters.
b) The presynaptic neuron may reuptake the neurotransmitters.

4. The removal of neurotransmitters prevents overstimulation of neurons.

E. Neurotransmitters and Disease

1. Each type of neuron secretes one type of neurotransmitter.

2. Defects in neurotransmission can cause disease.
a) Alzheimer's disease may stem from impaired function of acetylcholine.
(1) Acetylcholine is a neurotransmitter at nerve-muscle synapses.
(2) Alzheimer's patients experience deterioration in mental function.
(3) The enzyme acetylcholinesterase, which breaks down acetylcholine, can improve mental function.
b) Parkinson's disease is likely caused by malfunction in neurons that produce dopamine.
(1) Without dopamine, nerve cells fire without regulation.
(2) Parkinson's patients have trouble controlling movements, as nerve cells fire randomly.

F. Synaptic Integration

1. Different axons can synapse with one neuron, sending many signals.
a) Excitatory signals move the neuron toward action potential.
b) Inhibitory signals move the neuron away from action potential.

2. Each neuron combines signals through integration.
a) A neuron will fire if excitatory signals outweigh inhibitory signals.
b) A neuron will not fire if inhibitory signals outweigh excitatory signals.

G. Neurotransmission, ADD, and Ritalin

1. People with ADD may have lower dopamine levels.
a) Dopamine suppresses the responsiveness of neurons to stimuli.
b) People with low dopamine may become over-responsive.

2. Lower dopamine levels can result from an overabundance of dopamine receptors on the presynaptic cell.
a) Dopamine receptors remove dopamine from the synapse during reuptake.
b) An overabundance of receptors may remove too much dopamine.

3. Stimulants block the actions of the dopamine reuptake receptor.
a) Ritalin is a stimulant that is used to control ADD.
b) Ritalin and other stimulants can temporarily increase focus and energy.
c) Ritalin and other stimulants have side effects.
(1) They may cause depression, fatigue, poor concentration, and irritability when they wear off.
(2) Some stimulants may cause personality changes, nausea, and headaches.
d) Students may abuse Ritalin.
(1) The side effects of Ritalin may make students feel like the drug is helping them learn when it isn't.

(2) There is no evidence that Ritalin helps people who do not have low dopamine levels.

(3) Abuse of stimulants such as Ritalin may lead to psychotic episodes, seizures, delusions, and death.

4. There are other hypotheses about the biological cause of ADD.
5. ADD can also be caused by environmental factors.

II. The Central Nervous System

A. Overview

1. The central nervous system (CNS) interprets and acts on sensory information.
2. The CNS controls intelligence, learning, memory, and emotion.
3. The CNS consists of the brain and spinal cord, which are protected by tissues and fluid.
   a) The brain is protected by the bony skull.
   b) The spinal cord is protected by the bony vertebrae of the spine.
   c) Three layers of connective tissue, called the meninges, protect both the brain and spinal cord.
   d) Cerebrospinal fluid surrounds and cushions the CNS.
4. Damage to the CNS may be permanent.
   a) CNS neurons usually do not divide, so they cannot be repaired by cell division.
   b) Damage to spinal cord neurons can cause permanent paralysis.
   c) Brain injury can cause permanent brain damage.

B. Spinal Cord

1. The spinal cord is located down the back of the body.
   a) It extends from the base of the brain to the lowermost rib.
   b) It is housed within the spinal column.
2. The spinal cord is composed of nerve fibers.
   a) White matter is arranged around gray matter.
      (1) White matter conducts signals.
      (2) Gray matter contains cell bodies, dendrites, and synapses.
   b) Outer white matter is divided into dorsal and ventral columns.
   c) A central canal in the spinal cord contains cerebrospinal fluid.
3. Nerve tracts carry information.
   a) Information is carried between the spinal cord and brain.
   b) Information is carried between different parts of the spinal cord.
   c) Ascending tracts carry sensory information.
   d) Descending tracts carry nerve impulses that induce motor responses.
4. Spinal nerves carry information from sensory receptors in the skin to the spinal cord.
   a) The dorsal root of each spinal nerve contains sensory fibers that enter gray matter.
   b) The ventral root of each spinal nerve contains motor fibers that exit gray matter.
   c) The roots join to form a spinal nerve that leaves the vertebral column.

C. The Brain

1. Overview
   a) The brain directs and coordinates body activities.
      (1) The brain receives information from nerves and the spinal cord.
      (2) The brain integrates that information.
      (3) The brain generates a response to that information.
   b) The brain contains billions of neurons and neuroglial cells.
   c) The brain has four interconnected ventricles or chambers.
2. Cerebrum
   a) The cerebrum fills the upper part of the skull.
   b) The cerebrum controls language, memory, sensations, and logic.
   c) The two hemispheres of the cerebrum are divided into four lobes.
      (1) The temporal lobe processes auditory information and some visual information, and is involved with memory and emotion.
      (2) The occipital lobe processes visual information.
      (3) The parietal lobe processes touch information.
      (4) The frontal lobe processes voluntary muscle movements and organizes future behavior.

d) The cerebral cortex is the folded surface of the cerebrum.
  (1) Folding of the cortex increases surface area.
  (2) The cortex contains areas that process sensory information.
e) A fissure divides the right and left cerebral hemispheres.
  (1) The corpus callosum at the base of the fissure is a network of nerve fibers that connects the two hemispheres.
  (2) The caudate nuclei deep within each hemisphere coordinate movement.
f) The primary somatosensory area in the parietal lobe receives sensory input from the skin.
  (1) Specific regions within the parietal lobe receive input from specific areas of the body.
  (2) Sensitive areas of the body take up more space in the parietal lobe.
g) The primary motor area in the frontal lobe initiates motor activity.
  (1) Specific regions within the frontal lobe control movements in specific areas of the body.
  (2) Areas of the body that perform complex movements take up more space in the frontal lobe.

3. Thalamus and Hypothalamus
  a) The thalamus and hypothalamus are located between the two cerebral hemispheres.
  b) The thalamus is a message relay center.
    (1) The thalamus receives messages related to pain, pressure, and temperature.
    (2) The thalamus represses some of these messages.
    (3) The thalamus relays some of these messages to the cerebrum.
  c) The hypothalamus controls body functions.
    (1) The hypothalamus controls blood pressure and body temperature.
    (2) The hypothalamus plays a role in controlling sex drive, pleasure, pain, hunger, and thirst.
    (3) The hypothalamus releases hormones to regulate gamete production and the menstrual cycle.
4. Cerebellum
  a) The cerebellum is located beneath the cerebral hemispheres.
  b) The cerebellum has two hemispheres that are connected by a band of neurons.
  c) The cerebellum controls balance and coordination.
5. Brain Stem
  a) The brain stem is located beneath the thalamus and hypothalamus.
  b) The brain stem controls survival functions of the body.
    (1) The brain stem controls heartbeat, respiration, swallowing, and coughing.
    (2) The brain stem controls reflexes.
  c) The brain stem is composed of three regions.
    (1) The midbrain is highest on the brain stem and controls sensitivity to light and sound.
    (2) The pons is located below the midbrain and allows messages to move between the brain and spinal cord.
    (3) The medulla oblongata is a continuation of the spinal cord that sends messages between the brain and spinal cord.
    (4) The pons and medulla oblongata also regulate breathing rate.
  d) Brain functions are divided into left and right hemispheres.
    (1) Each hemisphere controls the opposite side of the body.
    (2) The left hemisphere controls reading, speech, and mathematical function.
    (3) The right hemisphere controls spatial perception and artistic function.
  e) The reticular formation extends through the brain stem and into the cerebral cortex.
    (1) The reticular formation is a network of neurons.
    (2) The reticular formation analyzes and filters sensory input.
    (3) The reticular formation also keeps the cerebral cortex alert.
6. Brain Activity During Sleep
  a) The reticular activating system (RAS) is located in the reticular formation.
  b) Some RAS neurons transmit continuous action potentials to the cerebrum, which keep you awake and alert.

c) When RAS neurons release the neurotransmitter serotonin, neurons in the brain are inhibited and you can sleep.

D. ADD and the Structure and Function of the Brain

1. Sleep deprivation is associated with ADD diagnosis.
2. Structural brain differences may cause ADD symptoms.
   a) One study found that the corpus callosum in people with ADD was smaller than in people without ADD.
   b) One study found that the caudate nucleus in the right hemisphere of people with ADD was smaller than the nucleus in those who did not have ADD.
   c) Other studies have found that the cortex and cerebellum were smaller in people with ADD.
3. Functional brain differences may cause ADD symptoms.
   a) The reticular formation may send too much information to the cerebral cortex in people with ADD.
   b) One study found that people with ADD had decreased blood flow through the right caudate nucleus.
   c) One study found that the cortex in people with ADD was less active, and so it may not be as effective at predicting the consequences of actions.
4. Structural and functional differences in the brain could be caused by genetics or the environment.

III. The Limbic System and Memory

A. Limbic System Structures

1. The hypothalamus controls self-gratifying behaviors, such as hunger, thirst, and sexual desire.
2. The hippocampus (in the temporal lobe) keeps the prefrontal area aware of past experience.
3. The amygdala (in the temporal lobe) processes emotions and memories and gives experiences emotional overtones.

B. Memory

1. Short-term or working memory stores information for a few hours.
   a) New sensory information is stored in short-term memory.
   b) Short-term memory storage is stimulated by action potentials in the limbic system.
   c) Neurons that fired when information was first stored are primed to fire again in the short term.
2. Long-term memory stores information that can be retrieved much later in the future.
   a) Repetition may help move information into long-term memory.
   b) Long-term memory centers are located in the cerebral cortex.
   c) As memory is stored, neurons create additional synapses that allow future memory retrieval.

IV. The Peripheral Nervous System

A. Overview

1. The peripheral nervous system (PNS) is the center that relays sensory and motor neurons between the CNS and the rest of the body.
2. Nerves and ganglia make up the PNS.
   a) Nerves are bundles of axons.
   b) Ganglia are clusters of cell bodies of peripheral neurons.

B. The Nerves of the Peripheral Nervous System

1. Cranial Nerves
   a) Cranial nerves carry impulses to and from the brain.
   b) There are 12 pairs of cranial nerves, identified by Roman numerals.
   c) Cranial nerves differ in structure.
      (1) Some contain only sensory fibers that carry information to the CNS.
      (2) Some contain only motor fibers that carry information away from the CNS.
      (3) Mixed nerves contain both sensory and motor fibers.
   d) Cranial nerves control mainly the head, neck, and face.
2. Spinal Nerves
   a) Spinal nerves carry impulses to and from the spinal cord.
   b) There are 31 pairs of spinal nerves.
   c) Each spinal nerve is a mixed nerve that originates from two roots.
      (1) The dorsal root contains sensory fibers that carry impulses toward the spinal cord.

(2) The ventral root contains motor fibers that carry impulses away from the spinal cord.

d) Each spinal nerve controls the region where it is located.

C. Somatic System

1. The somatic system controls body functions related to sense and response.
   a) The somatic system controls skeletal muscle movement.
   b) The somatic system allows you to sense your environment.
2. Nerves in this system carry
   a) impulses from sensory receptors to the CNS, and
   b) motor commands to skeletal muscle.
3. Some somatic system actions are due to reflexes.
   a) Reflexes are involuntary movements of skeletal muscle.
   b) Reflexes are controlled by a reflex arc that does not involve the brain.
      (1) Sensory neurons receive a signal.
      (2) Interneurons pass the signal along.
      (3) Motor neurons carry the signal to a skeletal muscle.
      (4) The skeletal muscle responds.
   c) Reflexes help the body to quickly avoid danger.

D. The Autonomic Nervous System

1. The autonomic system controls subconscious body functions.
   a) It controls smooth and cardiac muscle movement.
   b) It controls release of substances from glands.
   c) It regulates metabolism, digestion, thermoregulation, and cardiovascular functions.
2. This system can be divided into two divisions based on location.
   a) The sympathetic division contains autonomic fibers that exit thoracic and lumbar segments on the spinal cord.
   b) The parasympathetic division contains autonomic fibers that exit the brain stem or the sacral segments of the spinal cord.
3. The sympathetic and parasympathetic systems work antagonistically.
   a) Parasympathetic nerves dominate when the body is not receiving external signals.
      (1) Energy is used for digestion and thermoregulation.
      (2) The body is not on alert.
   b) Sympathetic nerves dominate when the body perceives outside stimuli.
      (1) Emergency situations trigger a fight-or-flight response.
      (2) Energy is diverted from body maintenance tasks.

V. What Causes ADD?

A. ADD has no definitive biological cause.

B. Environmental factors may aggravate ADD.

1. The number of cases of ADD in the U.S. has risen dramatically in the last decade.
2. Evolution could not effect biological change this quickly.
3. Societal changes may be negatively affecting children.

C. Diagnostic changes may account for increased ADD cases.

1. Doctors may be better at diagnosing ADD.
2. Doctors may be misdiagnosing ADD.

D. Treatment options depend on whether ADD is caused by biology or environment.

1. If it is simply caused by biology, using medication alone may make sense.
   a) Medication may have side effects, however.
      (1) Ritalin is a stimulant that may cause insomnia, loss of appetite, nervousness, and decreased growth.
      (2) Use of Ritalin may lead to increased use of stimulants when children become adults.
   b) Investigation of alternative treatments is important.
2. Modifying the environment may help patients with ADD.
   a) Parents may be counseled about parenting strategies.
   b) Behavior modification may be used.
   c) Teachers can minimize distractions in the classroom.
3. Effective treatments address issues related to natural tendencies in the brain as well as imbalances in the person's environment.

Practice Questions

Matching

1. neuroglia
2. fissure
3. cerebellum
4. thalamus
5. effector
6. Schwann cell
7. hippocampus
8. axon
9. midbrain
10. synapse
11. dopamine
12. amygdala
13. spinal nerve
14. refractory period
15. pons

a. bridge between brain and spinal cord
b. delivers electrical signals to another cell
c. controls balance and coordination
d. cells that nourish neurons
e. contain dorsal and ventral roots
f. part of the limbic system that communicates with the prefrontal area
g. deep groove between hemispheres
h. neurotransmitter that suppresses neuron response to stimuli
i. contains the lipid myelin
j. can cause you to feel sad or happy during an experience
k. part of the brain that first receives sensations of pain
l. time during which no action potential may start
m. muscles or glands that respond to nervous signals
n. can make you sensitive to light
o. region between two neurons

Fill-in-the-Blank

16. ____________ is a ____________ sense of the body's position.
17. The brain and ____________ are part of the ____________ nervous system.
18. The pons, midbrain, and ____________ are found in the ____________.
19. Someone with a damaged ____________ would be unable to filter out ____________ sensory information, such as constant traffic noise.
20. The ninth pair of ____________ nerves are ____________ nerves that carry information from the pharynx and to the pharyngeal muscles.
21. When you touch a hot surface, a signal travels from touch receptors in your hand, to your ____________, to interneurons, to ____________, and then to the muscles in your arm.
22. Ritalin blocks dopamine ____________ so that dopamine stays in the synapse for a ____________ time.
23. When ____________ channels in an axon open, the ions that flood in cause the ____________ of the nerve cell.
24. The ____________ bouton of a presynaptic neuron sends ____________ across the ____________ to a dendrite of the postsynaptic neuron.
25. When you go bird watching, your ____________ lobe helps you to identify the bird by its song and your ____________ lobe helps you to identify the bird by the patterns on its feathers.

Labeling

Use the terms below to label Figure 14.1. Any term listed twice will be used twice.

axon
cell body
cell body
dendrites
effector
interneuron
motor neuron
node of Ranvier
Schwann cell nucleus
sensory neuron
sensory receptor
terminal bouton

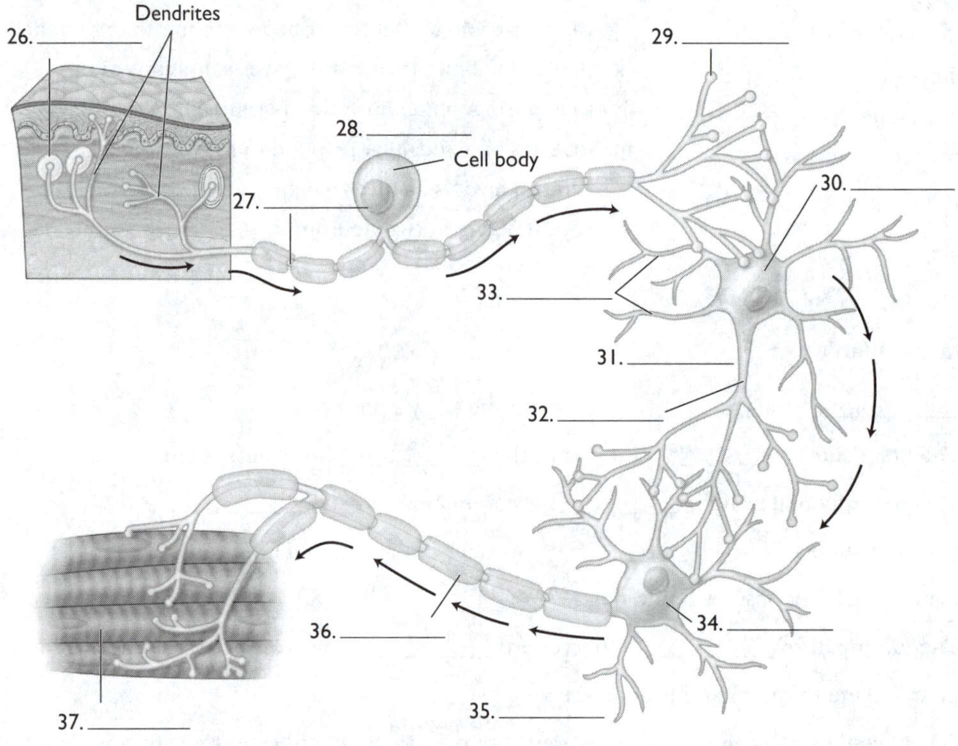

Figure 14.1

Roots to Remember

Use your knowledge of the root words presented in this chapter to answer the following questions.

38. When someone calls a person "cerebral," what do they mean?
39. Why is the meaning of the word *cortex* a good description for the cerebral cortex?
40. When you feel sympathy for someone, what are you doing (according to the word's roots)?
41. According to its word root, where is the hypothalamus located?
42. How is the reticular formation like a television channel?

Word Choice

Circle the word or phrase that correctly completes each sentence.

43. The (midbrain/pons) is the highest structure on the brain stem.
44. The cerebellum and (caudate nuclei/cerebral cortex) help a hurdler to leap and run.
45. A neuron will achieve an action potential if it receives a majority of (inhibitory/excitatory) signals.
46. The structures that release neurotransmitters into the synaptic cleft are called (vesicles/gated channels).
47. Electrical impulses jump from (node to node/dendrite to dendrite) on a neuron.

Crossword Puzzle

Use the clues below to fill in the puzzle.

Across

48. neurons that carry information to effectors
49. I am I, according to this lobe
50. lipid protection
51. matter filled with Schwann cells
52. potential that depolarizes
53. some cell bodies are located in these

Down

54. a system missing the brain
55. direction of diffusing potassium ions during repolarization
56. cord or nerves
57. encloses the brain
58. bundles in the CNS

Table Completion

Fill in the table below using your knowledge of the structures in the brain.

Structure	Location	Function
Cerebellum	Below the cerebrum	59.
60.	Base of the fissure	61.
62.	Within the cerebrum	Processes voluntary muscle movements
63.	64.	Controls hunger and thirst
65.	Below the pons	Facilitates communication between the brain and spinal cord

Critical Thinking

66. Multiple sclerosis is an autoimmune disease in which the immune system attacks neuroglial cells that produce myelin. What is a likely effect of this disease?
67. Why might you be able to sleep while a loud air conditioner is running but wake up when your roommate quietly enters the room and closes the door?
68. Why couldn't Ritalin be used to treat the symptoms of Parkinson's disease?

Practice Test

1. Signals are carried from one neuron to the next by
   A. chemical transmitters only.
   B. electrical impulses only.
   C. both chemical transmitters and electrical impulses.
   D. neither chemical transmitters nor electrical impulses.

2. Structures that remove waste from a nerve cell are found in the
   A. dendrites.
   B. cell body.
   C. axon.
   D. terminal boutons.

3. Which of the following drugs does *not* block reuptake of neurotransmitters?
   A. caffeine
   B. amphetamines
   C. MDMA
   D. cocaine

4. When you are quietly reading a book, which of the following parts of the brain is likely to be the most active?
   A. cerebellum
   B. temporal lobe
   C. thalamus
   D. occipital lobe

5. Which part of the brain controls breathing rate and other automatic body functions?
   A. cerebrum
   B. cerebellum
   C. brain stem
   D. thalamus

6. What is an effector?
   A. a nerve cell that triggers body movement
   B. a muscle or gland that responds to a nerve signal
   C. a receptor that senses outside stimuli
   D. a nerve cell that links sensory and motor neurons

7. During resting potential,
   A. the concentration of Na^+ ions is greater inside the axon than outside.
   B. the concentration of K^+ ions is greater inside the axon than outside.
   C. the concentration of Na^+ ions inside and outside the axon is the same.
   D. the concentration of K^+ ions inside and outside the axon is the same.

8. As you are studying this material, you should be hoping for increased activity in your
   A. hippocampus.
   B. cerebral cortex.

C. medulla oblongata.
D. caudate nuclei.

9. What is a vesicle?
A. It is one of the four chambers of the brain.
B. It is a pocket of tissue that cushions the brain.
C. It is the root that contains motor fibers.
D. It is a saclike structure that releases neurotransmitters.

10. Dopamine stimulates a nerve cell when it is taken up by receptors
A. on the terminal bouton.
B. on the dendrites.
C. in the synaptic cleft.
D. in the axon.

11. When you stub your toe, the pain impulse travels from your toe to the CNS through
A. sensory neurons.
B. motor neurons.
C. nerve tracts.
D. interneurons.

12. Which portion of the brain is involved with consciousness?
A. medulla oblongata
B. corpus callosum
C. cerebral cortex
D. hypothalamus

13. Where are myelinated axons found in the spinal cord?
A. in the dorsal column only
B. in the ventral column only
C. in both dorsal and ventral columns
D. in the central canal only

14. Which of the following is *not* a function of neuroglial cells in the brain?
A. Neuroglial cells provide nutrients to very active brain cells.
B. Neuroglial cells protect brain cells from pathogens.
C. Neuroglial cells produce substances that cushion and protect brain cells.
D. Neuroglial cells improve signal transmission between brain cells.

15. A neuron in a refractory period
A. cannot undergo an action potential.
B. is in the process of being depolarized.
C. has a net positive charge inside its cytoplasm.
D. has negative electrical potential.

Answer Key

Matching

1. d; **2.** g; **3.** c; **4.** k; **5.** m; **6.** i; **7.** f; **8.** b; **9.** n; **10.** o; **11.** h; **12.** j; **13.** e; **14.** l; **15.** a

Fill-in-the-Blank

16. Proprioception; general; **17.** spinal cord; central; **18.** medulla oblongata; brain stem; **19.** reticular formation; repetitive; **20.** cranial; mixed; **21.** spinal cord; motor neurons; **22.** receptors; longer; **23.** sodium; depolarization; **24.** terminal; neurotransmitters; synaptic cleft; **25.** temporal; occipital

Labeling

26. sensory receptor; **27.** node of Ranvier; **28.** sensory neuron; **29.** terminal bouton; **30.** cell body; **31.** interneuron; **32.** axon; **33.** dendrites; **34.** cell body; **35.** motor neuron; **36.** Schwann cell nucleus; **37.** effector

Roots to Remember

38. that the person is brainy or an intellectual; **39.** The cerebral cortex is a covering, like the bark of a tree; **40.** You are joining with them in their pain; **41.** under or below the thalamus; **42.** Both are networks that broadcast information.

Word Choice

43. midbrain; **44.** caudate nuclei; **45.** excitatory; **46.** vesicles; **47.** node to node

Crossword Puzzle

Across

48. motor; **49.** parietal; **50.** myelin; **51.** white; **52.** action; **53.** ganglia

Down

54. peripheral; **55.** out; **56.** spinal; **57.** meninges; **58.** tracts

Table Completion

59. Controls balance, muscle movement, and coordination; **60.** Corpus callosum; **61.** Allows communication between right and left hemispheres; **62.** Frontal lobe; **63.** Hypothalamus; **64.** Below the thalamus; **65.** Medulla oblongata

Critical Thinking

66. This disease would likely cause unsuccessful transmission of nerve signals, as nervous impulses jumped sideways or slowed down without the insulation of the myelin sheath.

67. Your reticular formation prevents the brain from reacting to a repetitive sound, but it may arouse the cerebrum if a new, nonrepetitive sound is heard.

68. Parkinson's disease is thought to be caused by a malfunctioning of neurons that produce dopamine. It is not caused by a too-rapid uptake of dopamine. Ritalin only increases the amount of time dopamine spends in the synapse. It cannot increase the release of dopamine, or fix malfunctioning neurons that produce dopamine.

Practice Test

1. A; **2.** B; **3.** A; **4.** D; **5.** C; **6.** B; **7.** B; **8.** B; **9.** D; **10.** B; **11.** A; **12.** C; **13.** C; **14.** D; **15.** A

CHAPTER 15

THE SENSES: DO HUMANS HAVE A SIXTH SENSE?

Learning Goals

1. Compare and contrast the five types of sensory receptors and the type of stimuli to which they respond.
2. Compare and contrast the general and special senses.
3. Describe the steps involved in conscious perception of a stimulus.
4. Describe the two proprioceptors and how they function to maintain body sense and muscle tone.
5. List the stimuli to which touch receptors in the skin respond.
6. Describe the structure of a taste bud and explain how taste is sensed.
7. Compare and contrast the olfactory sense with that of taste.
8. Explain how sound is sensed in the inner ear.
9. Summarize the structure and function of the vestibular apparatus.
10. Compare and contrast rod and cone cells in the eye and summarize the function of each in vision.
11. Describe how visual stimuli are processed in the brain.

Chapter Outline

I. Sensing and Perceiving
   A. Sensory Receptors
      1. The senses encompass the structures and processes that receive and interpret environmental stimuli.
      2. Humans have five types of sensory receptors.
         a) Mechanoreceptors respond to changes in receptor shape or in the shape of adjacent cells.
         b) Thermoreceptors respond to temperature changes.
         c) Photoreceptors respond to light.
         d) Chemoreceptors respond to chemical stimuli.
         e) Nociceptors, or pain receptors, respond to tissue damage or extreme temperature.
      3. Sensory receptors have two focal points.
         a) Externoreceptors sense conditions outside the body.
         b) Internoreceptors sense conditions within the body.
      4. Sensory receptors are part of the general or special senses.
         a) Receptors for touch, pressure, vibration, temperature, body position, and pain are receptors for general or somatic senses.
         b) Receptors for sight, hearing, taste, smell, and equilibrium are receptors for the special senses.
            (1) Receptors for special senses are located in specialized structures.
            (2) These structures are called sense organs.
      5. Sensory receptors have evolved to sense aspects of our environment that aid in survival.
   B. Reading and Understanding the Environment
      1. Stimuli lead to changes in the sensory receptor.
         a) A stimulus causes the membrane of the sensory receptor to depolarize.
         b) Depolarization triggers a cascade of events that may allow us to perceive and respond to the stimulus.
         c) The stimulus must trigger a large enough change to stimulate a sensory neuron.
         d) If stimulated, the sensory neuron carries information to the central nervous system.

2. A sensed stimulus is called a sensation.
3. Our body's interpretation of a sensation is called perception.
4. Sensory adaptation allows the body to turn down perception.
   a) Neural adaptation occurs when the receptor stops sending a signal after it receives constant stimulation.
   b) Habituation occurs when the reticular formation stops sending signals to the conscious brain.

II. The General Senses
  A. Proprioception
    1. Proprioceptors sense our body position in space.
    2. Proprioceptors are always active.
    3. Muscle spindles are one type of proprioceptors.
      a) Muscle spindles are specialized muscle fibers wrapped with sensory nerve endings.
      b) These sensory receptors sense muscle stretching and maintain muscle tone.
      c) Muscle spindles stimulate enough muscle mass to maintain body position.
    4. Golgi tendon organs are another type of proprioceptor.
      a) Golgi tendon organs measure the tension within tendons.
      b) These organs provide feedback to maintain balance and posture.
      c) If Golgi tendon organs sense high tension, they trigger rapid relaxation in the attached muscle.
  B. The Sense of Touch
    1. Mechanoreceptors allow you to sense touch by transmitting signals when they are deformed.
    2. Mechanoreceptors are unevenly distributed throughout the body.
      a) Higher densities are found in the hands, near the mouth, and around the genitals.
      b) Lower densities are found on the abdomen and back.
      c) The receptor field, which is the area monitored by a particular sense receptor, ranges from less than 1 mm up to 7 cm for touch.
    3. Touch receptors are at the ends of individual nerve cells.
      a) Receptors without an outer covering are highly sensitive.
        (1) Receptors that wrap around hairs are highly sensitive.
        (2) Merkel disks, receptors at the base of the epidermis, can sense light pressure on the skin.
      b) Encapsulated receptors, with a sheath of connective tissue covering the nerve ending, are less sensitive or can adapt to constant stimuli.
        (1) Meissner's corpuscles, at the base of the epidermis in hairless skin, detect light pressure and vibration, but adapt to the stimuli.
        (2) Ruffini corpuscles, deep in the dermis, respond to harder pressure.
        (3) Pacinian corpuscles, deep in the dermis, respond and adapt to harder pressure.
  C. Temperature and Pain
    1. Thermoreceptors
      a) Thermoreceptors in the skin sense moderate changes in external temperature.
        (1) Many of these receptors respond to cool temperatures.
        (2) Fewer of these receptors respond to warm temperatures.
      b) Chemicals in plants, such as peppermint and garlic, can activate thermoreceptors.
      c) Thermoreceptors adapt quickly to stable temperatures.
    2. Pain Receptors
      a) Pain receptors sense temperature extremes.
      b) Pain receptors are free nerve endings in the skin, joints, bone, and blood vessels.
      c) Pain receptors can be triggered by mechanical or chemical disturbance.
      d) Internal organs have few pain receptors.
        (1) Internal organ damage may cause referred pain in a different part of the body.
        (2) Arm pain in a person experiencing a heart attack is referred pain.
      e) Pain is a mechanism of survival.
        (1) Fast pain can trigger withdrawal reflex.
        (2) Slow pain can signal damage.
      f) Chronic pain is usually slow pain that persists.
        (1) Aspirin or other anti-inflammatory drugs can reduce pain by interfering with the release of prostaglandins.

(2) Morphine and codeine mimic pain-relieving chemicals produced in the brain, such as endorphins.

III. The Chemical Senses

A. Taste

1. Gustatory (taste) receptors sense chemicals in foods.
2. Taste receptors (taste buds) are densely distributed on the tongue and are also found in small numbers in the mouth and throat.
   a) Taste buds are located in large numbers on papillae.
   b) Each taste bud is made of taste cells.
   c) Each taste cell is covered with taste hairs.
3. Receptors on the taste hairs of taste buds bind to chemicals in food called tastants.
   a) Binding of a tastant releases neurotransmitters from the taste cell.
   b) These neurotransmitters send a signal to the associated sensory neuron.
4. A tastant may have a sweet, salty, sour, bitter, or meaty quality.
   a) The quality of the tastant activates a particular receptor.
   b) Food items may have more than one tastant.
5. Different receptors are concentrated in different areas of the tongue.
   a) Sweet receptors are concentrated on the tip of the tongue.
   b) Bitter receptors are concentrated at the back of the tongue.
6. Different tastants trigger different reactions.
   a) Sweet, salty, and meaty tastes trigger a swallowing reflex.
   b) Bitter tastes trigger a gag reflex.
      (1) A bitter taste may signal poisonous or spoiled food.
      (2) Bitter taste receptors are highly sensitive.

B. Smell

1. Olfactory (smell) receptors sense chemical odorants.
2. Olfactory receptors are concentrated on the roof of the nasal passage.
   a) The receptive surface of the olfactory receptor is a hair-covered knob that contains the dendrite of a sensory neuron.
   b) The hairs project into the nasal cavity and detect chemical odorants dissolved in surrounding mucus.
3. Olfactory receptors are replaced every 60 days.
4. Cell receptors on olfactory neurons are highly sensitive and specific.
   a) Olfactory receptors allow us to distinguish thousands of smells.
   b) Olfactory receptors allow us to detect as little as four molecules of odorant.
5. Information from olfactory receptors travels to the olfactory bulb in the brain.
   a) This information is partially processed and sent to the cerebral cortex.
   b) Olfactory information passes through the limbic system, where it can trigger emotion.
6. Reaction to certain odors is instinctual.
   a) Sweet or meaty scents trigger pleasure.
   b) Odors from human waste or spoiled food trigger disgust.
7. Humans are much less sensitive to odors than are other mammals.
   a) Pheromones are important nonverbal signals in other mammals.
   b) Humans have limited ability to sense pheromones.

IV. Senses of the Ear

A. Hearing

1. Sounds are compression waves with different frequencies and amplitudes.
   a) Mechanoreceptors in the ears sense sound waves.
   b) The nervous system translates these waves into the perception of sound.
2. The ear is a complicated receiver for sound waves.
   a) The disk-shaped outer ear (pinna) funnels waves into the auditory canal.
   b) Waves travel down the auditory canal to the eardrum, or tympanic membrane.
   c) The eardrum vibrates and sends the vibrations to ossicles in the middle ear.
   d) The ossicles translate the vibrations onto a membrane in the inner ear called the oval window.
      (1) The oval window is much smaller than the eardrum.
      (2) Transferring wave energy from a large membrane to a smaller one amplifies the vibration.

e) The cochlea in the inner ear receives sound waves.
   (1) Vibrations in the oval window are transferred to the fluid in the outer cochlea tube.
   (2) Fluid passes over the top of the interior tube, or cochlear duct, and moves around the tube to a membrane called the round window.
   (3) The round window allows sound energy to move into the middle ear and out the Eustachian tube.

f) The cochlear duct contains auditory receptors, called hair cells, which are covered with hairlike structures.
   (1) The lower surface of hair cells rests on the basilar membrane.
   (2) The tectorial membrane rests on the tips of the hair cells.

g) A wave of fluid causes part of the basilar membrane to vibrate.
   (1) This vibration pushes hair cells into the tectorial membrane.
   (2) As the hairs bend, they trigger release of a neurotransmitter.
   (3) The neurotransmitter stimulates a sensory neuron.
   (4) A signal is sent to the brain through the auditory nerve.

h) Hair cells on the basilar membrane range in length and tension.
   (1) Hairs at one end are narrow and stiff.
   (2) Hairs at the other end are longer and looser.

i) Different-frequency sounds cause vibrations in different parts of the basilar membrane.
   (1) Low-frequency sounds cause vibrations in the looser sections of the membrane.
   (2) High-frequency sounds cause vibrations in the stiffer sections of the membrane.

j) Louder sounds activate more hair cells and cause them to bend more dramatically.

3. Damaged cochlea can be replaced by a cochlear implant.
   a) The implant transmits sound waves from a receiver to the cochlea, which has been fitted with electrodes.
   b) Different sounds stimulate different electrodes.
   c) Electrodes send signals through the auditory nerve to the brain.
4. The location of the ears allows us to triangulate and identify the source of sounds.
5. The ears are greatly attuned to human speech.
   a) The ear can capture sound waves that correspond to frequencies of human speech.
   b) The basilar membrane responds best to the frequencies of human speech.
6. Extremely low-frequency sounds, called infrasounds, are too low for the ear to detect.
   a) Infrasounds often accompany earthquakes or volcanic eruptions.
   b) Infrasounds can be sensed by other mammals.
   c) Infrasounds may cause a feeling of dread in humans, even though they cannot hear the sounds.

B. A Sense of Balance
1. Mechanoreceptors in the inner ear sense head movement and the head's position in space.
2. The vestibular apparatus is the sensory organ that senses equilibrium.
   a) The vestibular apparatus is attached to the cochlea.
   b) The apparatus contains a basal sac and three semicircular canals.
      (1) Each canal projects into a different three-dimensional direction.
      (2) When the head moves in one of these directions, the fluid in the corresponding canal moves in the opposite direction.
      (3) Moving fluid impacts a cupula at the base of the canal.
      (4) Hairs on sensory cells that project into the cupula are deformed and send signals to the brain.
   c) The vestibule of the vestibular apparatus is divided into two sections.
      (1) The utricle is the larger section.
      (2) The saccule is the smaller section.
   d) Both sections of the vestibule contain otolith organs.
      (1) The organ is composed of hair cells embedded in gelatin.
      (2) Calcium carbonate "ear stones" float in the gelatin in response to gravity.
      (3) The ear stones deform the hair cells, sending signals about head position to the brain.
3. Signals from the vestibular apparatus are integrated with other sensory signals.
   a) These integrated signals help you to keep your balance.
   b) Discrepancies between signals from the vestibular apparatus and other sensory organs can cause motion sickness or vertigo.

V. Vision
  A. Focusing Light
    1. Visual receptors in the eyes respond to visible light.
      a) Visible light is light energy between the wavelengths of 400 nm and 700 nm.
      b) Colors of the rainbow make up the visible spectrum, and each has a different wavelength.
    2. Eyes are the organs of sight.
      a) The eyeball is a sphere-shaped cavity surrounded by connective tissue.
        (1) The opaque sclera covers the eyeball.
        (2) The cornea allows light to pass into the eye.
      b) Aqueous humor is found in a chamber behind the cornea.
        (1) Aqueous humor is liquid that is continually recycled.
        (2) Blockage in the channel that drains aqueous humor can lead to glaucoma, or high pressure in the eye.
      c) Light entering the eye is processed by a variety of structures.
        (1) The iris is a disk-shaped muscle that controls the amount of light entering.
        (2) The pupil is the opening adjusted by the iris that allows light to travel to the back of the eye.
        (3) The lens is a structure made of layered proteins that focuses the light.
          (a) A ciliary body holds the lens in place.
          (b) The muscles in the ciliary body contract or relax to change the curvature of the lens, which focuses the viewed image.
        (4) Light passes from the lens through vitreous humor.
        (5) Light shines on the retina at the back of the eye.
          (a) The retina contains photoreceptors.
          (b) The coroid behind the retina contains capillaries that provide the photoreceptors with oxygen and nutrients.
      d) Eyes can be affected by conditions that impair vision.
        (1) In myopia, the relaxed lens focuses the image in front of the retina.
          (a) Myopia is also called nearsightedness.
          (b) In myopia, the eyeball is elongated.
          (c) Myopia can occur during development, if a child focuses on too many close objects.
          (d) Myopia can be corrected with a concave lens.
        (2) In hyperopia, the focal point falls behind the retina.
          (a) Hyperopia is also called farsightedness.
          (b) In hyperopia, the eyeball is shortened.
          (c) Hyperopia develops as we age and our lenses do not relax completely during close focusing.
          (d) Hyperopia can be corrected with a convex lens.
        (3) In astigmatism, the cornea is misshapened and light scatters before reaching the retina.
          (a) In astigmatism, the light does not focus on the retina.
          (b) Astigmatism can be corrected with an irregularly shaped lens.
        (4) Cataracts can develop when lens proteins become malformed and cloudy.
          (a) Cataracts can result from diabetes or other disease, or aging.
          (b) Artificial lenses can be surgically implanted to replace lenses affected by cataracts.
        (5) When retinal detachment occurs, the retina is torn and vitreous humor leaks in between the retina and choroid.
          (a) Symptoms of retinal detachment include blurred vision and flashes of light.
          (b) Laser treatment can repair retinal detachment.
          (c) Untreated retinal detachment can cause blindness.
  B. Photoreceptors
    1. Photochemicals in the photoreceptors change shape when absorbing light energy.
    2. Photochemicals are located in rods and cones.
      a) Rods are very sensitive and help with night vision.
      b) Cones are sensitive to color and detail.
      c) Rods and cones cannot be replaced.
        (1) Macular degeneration breaks down the retina.
        (2) During macular degeneration, gaps appear in the visual field.

(3) Macular degeneration may be age-related or it may be caused by high blood pressure or diabetes.

d) Rods and cones are distributed unevenly.

(1) Cones are concentrated at the eye's focal point, called the fovea.

(2) Rods are absent from the fovea but abundant around it.

e) Rods and cones contain rhodopsin.

(1) Rhodopsin is composed of retinal and the protein opsin.

(2) Variation in opsin structure allows absorbance of different light wavelengths.

(a) Rhodopsin in rods is sensitive to middle wavelengths.

(b) Different cones respond mainly to red, green, or blue wavelengths.

(3) Rhodopsin disassembles when light strikes it.

(a) Bright light causes rhodopsin levels to decline.

(b) When entering a darker room from a brighter one, rhodopsin levels need to recover.

f) When rods and cones are stimulated, they send signals to associated sensory neurons called bipolar cells.

g) Bipolar cells pass the signal to ganglion cells.

(1) The axons of ganglion cells make up the optic nerve.

(2) Ganglion cells integrate information from many rods and cones.

(a) Signals from hundreds of rods can be integrated.

(b) Signals from a few dozen cones can be integrated.

(c) Cones provide a clearer picture than rods do.

(d) Rods are good at detecting movement.

C. Vision and Perception

1. Colors we see do not perfectly represent the light that is being reflected or emitted.

a) Green photoreceptors outnumber red and blue receptors.

b) Objects may reflect or emit a range of wavelengths, but our eye may pick up mainly one.

2. Some people have colorblindness.

a) Colorblindness is caused by the lack of one or more types of photoreceptor.

b) Red-green colorblindness is most common.

3. We cannot see many of wavelengths of electromagnetic radiation.

4. The brain processes images as they enter the eye.

a) Information is integrated by ganglion cells.

b) Signals are passed through the optic nerve to the optic chiasm.

(1) At the optic chiasm, information from each eye is shared.

(2) Visual information travels to both hemispheres of the brain.

(3) Because humans process information from both eyes, the brain can perceive relative distances between objects.

c) Signals travel from the optic chiasm to the lateral geniculate nucleus.

(1) Information is broken down into components.

(2) Packets of information are sent to the visual cortex for reassembly.

(3) The brain fills in missing information.

(a) We have a blind spot where the optic nerve exits the retina.

(b) The blind spot lacks photoreceptors.

(c) Our visual cortex fills in information from this blind spot.

VI. Predicting the Future

A. Understanding Premonitions

1. Premonitions may be the result of recall bias.

a) People are more likely to recall premonitions that came true than the many others that did not.

b) There may have been danger signals that you do not remember that caused you to act.

2. Premonitions may be a result of shared experience.

B. Expanding the Receptive Field

1. Modern humans can use tools to expand our receptive fields.

a) Tools such as surveillance cameras and infrared receivers enhance sight.

b) Tools such as weather satellites and seismographs expand our ability to sense environmental conditions.

2. Modern humans can use the scientific method to model and understand natural processes.

3. Expanding the receptive field can help us to predict events.

Practice Questions

Matching

1. cupula
2. myopia
3. Merkel disks
4. hyperopia
5. hair cell
6. photoreceptor
7. rhodopsin
8. pinna
9. proprioceptor
10. nociceptor
11. glaucoma
12. odorant
13. retina
14. vertigo
15. cochlea

a. fluid-filled organ that receives sound
b. receptor that senses body position
c. structure at the back of the eye
d. touch receptors at the base of the epidermis
e. receptor that senses light
f. structure at the base of a semicircular canal
g. high pressure in the eye
h. auditory receptors
i. nearsightedness
j. receptor that senses pain
k. farsightedness
l. chemical sensed by olfactory receptors
m. outer ear
n. inappropriate sense of motion
o. pigment in rods and cones

Fill-in-the-Blank

16. When sound energy reaches the ____________ membrane, it causes the membrane to vibrate, which causes ____________ to bend.
17. Hundreds of taste buds are found on each ____________ on the ____________.
18. Humans use ____________ vision to ____________, which allows them to perceive the relative locations of objects.
19. The ear is a ____________ organ that contains auditory receptors that interpret sound waves and the ____________ that helps to sense body position and maintain balance.
20. In the most common form of ____________, the affected person lacks either the red or ____________ photoreceptor.
21. ____________ receptors, found in the nasal passage, bind with chemical ____________.
22. Chemicals called ____________ can cause inflammation of tissues, which can cause ____________.
23. ____________ corpuscles adapt to continuous hard pressure, unlike ____________ corpuscles, which do not adapt.
24. The ____________ are proprioceptors that prevent tearing in muscles.
25. Receptors can become ____________ to stimuli that are constantly sending signals.

Labeling

Use the terms below to label Figure 15.1.

cornea
iris
lens
optic nerve
pupil
retina

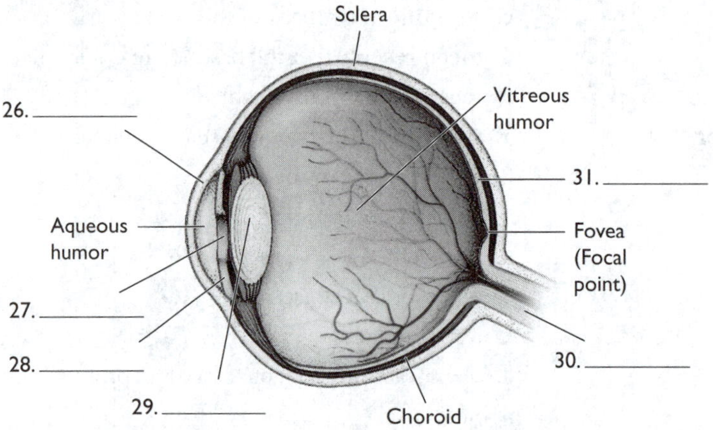

Sclera
26. ________
Vitreous humor
31. ________
Aqueous humor
Fovea (Focal point)
27. ________
28. ________
30. ________
29. ________
Choroid

Figure 15.1

Use the terms below to label Figure 15.2.

auditory canal
cochlea
eardrum
Eustachian tube
ossicles
pinna

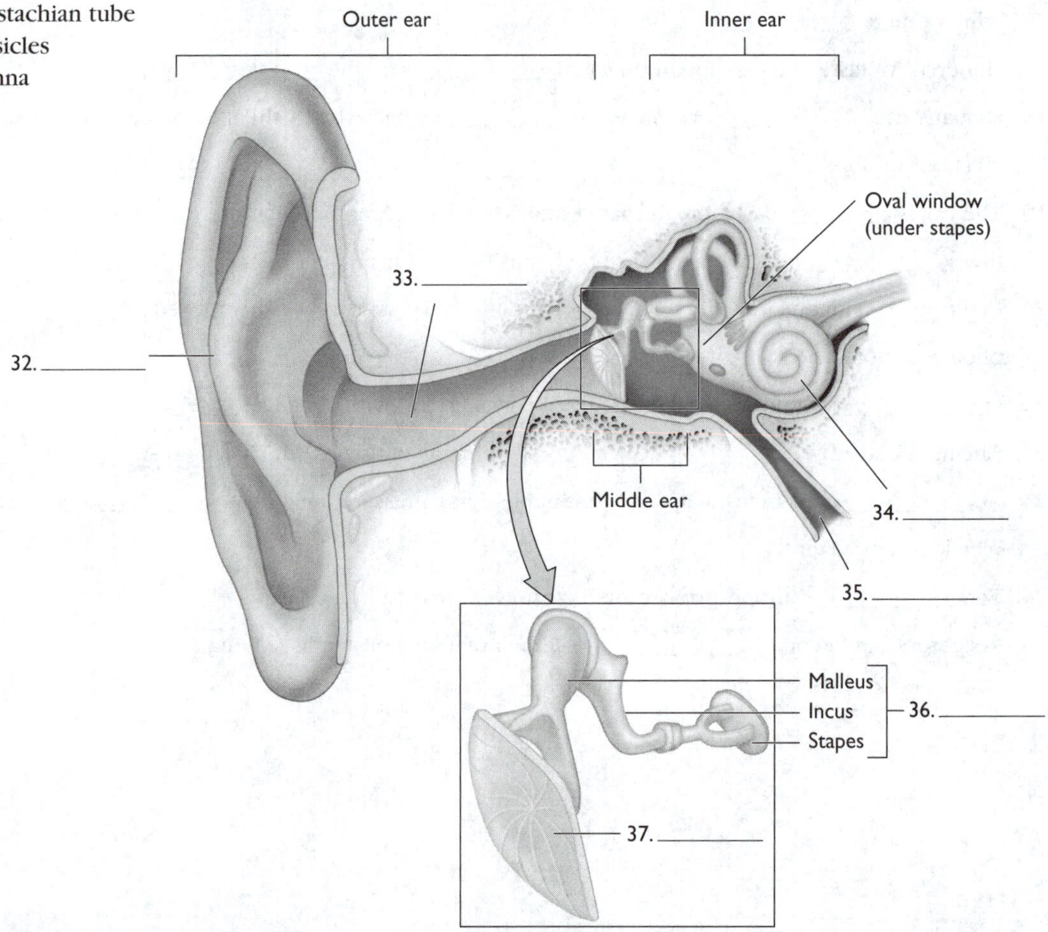

Outer ear
Inner ear
Oval window (under stapes)
33. ________
32. ________
Middle ear
34. ________
35. ________
Malleus
Incus
Stapes
36. ________
37. ________

Figure 15.2

Roots to Remember

Use your knowledge of the root words presented in this chapter to answer the following questions.

38. *Lithos* is a Greek word root that refers to stone. What is the literal meaning of *otolith*?
39. According to its word roots, what is an ossicle?
40. How do word roots help you to remember the order of some colors in the color spectrum?

Word Choice

Circle the word or phrase that correctly completes each sentence.

41. Each odor receptor (is associated with/is not associated with) an intermediate sensory neuron.
42. A green wavelength of visible light is (shorter/longer) than a blue wavelength.
43. Meissner's corpuscles are located at the base of the (dermis/epidermis).
44. Most temperature receptors respond to (cool/warm) temperatures.
45. Damage to the esophagus can result in referred pain in the (abdomen/chest).
46. A high concentration of bitter taste receptors is found (at the back/on the tip) of the tongue.
47. (Hormones/Pheromones) transmit chemical messages that many nonhuman animals can sense with the vomeronasal organ.
48. The (cornea/lens) is located behind the iris and pupil of the eye.

Table Completion

Complete the following table listing conditions that affect vision.

Condition	Description of Condition	Correction Used
49.	Cornea is misshapened, so light scatters before hitting the retina	50.
Myopia	51.	52.
53.	54.	Convex lens

Sequencing

Put the following events related to hearing in order by numbering each event, from 1 to 11.

55. ____________ Sound impulses are transmitted by the auditory nerve.
56. ____________ Sound waves hit the eardrum and cause it to vibrate.
57. ____________ Sound waves cause portions of the basilar membrane to vibrate.
58. ____________ Sound waves travel up the vestibular canal.
59. ____________ The round window absorbs vibrations from sound waves.
60. ____________ Hair cells bend in response to the vibrating basilar membrane.
61. ____________ Vibrations are transferred to the oval window and become amplified.
62. ____________ Sound waves move down the tympanic canal.
63. ____________ Ossicles in the middle ear vibrate.
64. ____________ Energy from sound waves travels down the Eustachian tube.
65. ____________ Sound waves are transmitted along the cochlear duct.

Critical Thinking

66. Why might you become unable to detect odors when you have a bad cold?
67. An orange slice tastes sweet and sour when you eat it. Right after you eat a sweet cookie, an orange slice tastes mostly sour. Why might this occur?
68. Music at a rock concert can be louder than a jackhammer. Describe how this music could cause damage to your hearing.

Practice Test

1. Cataracts affect which part of the eye?
   A. cornea
   B. choroid
   C. retina
   D. lens

2. Which structure separates the outer ear from the middle ear?
   A. eardrum
   B. oval window
   C. round window
   D. cochlear duct

3. When you are resting, muscles in your body
   A. are also completely at rest.
   B. take turns being activated.
   C. activate only when you toss and turn.
   D. are in a state of constant alertness.

4. Which of the following types of radiation has the shortest wavelength?
   A. radio waves
   B. X-rays
   C. ultraviolet light
   D. microwaves

5. Which receptors are *not* likely to be triggered by very spicy Mexican food?
   A. mechanoreceptors
   B. thermoreceptors
   C. pain receptors
   D. gustatory receptors.

6. You hear a high-pitched car alarm that sounds very far away. What happens in your ear when you hear the alarm?
   A. The stiff part of the basilar membrane vibrates strongly.
   B. The stiff part of the basilar membrane vibrates weakly.
   C. The loose part of the basilar membrane vibrates strongly.
   D. The loose part of the basilar membrane vibrates weakly.

7. What is the function of the round window in the ear?
   A. It transfers vibrations to the ossicles.
   B. It amplifies sound energy.
   C. It allows sound energy to dissipate and leave the ear.
   D. It helps to monitor the position of the head.

8. The fruit bat is a nocturnal animal, so it is active at night. It is likely that the eyes of a fruit bat have
   A. many more rods than cones.
   B. many more cones than rods.

C. an equal number of rods and cones.
D. no rods and very few cones.

9. The majority of tastants in a glass of lemonade are likely to be
   A. bitter.
   B. sweet.
   C. sour.
   D. salty.

10. The area in the eye where the optic nerve exits the retina
   A. has no photoreceptors.
   B. is rich in rhodopsin.
   C. is called the optic chiasm.
   D. can be destroyed by macular degeneration.

11. Where is the utricle located?
   A. in the middle ear
   B. in the nasal passages
   C. in the eye
   D. in the vestibular apparatus

12. Of the following body parts, which would be most sensitive to touch?
   A. the forehead
   B. the back
   C. the hand
   D. the knee

13. Damage to the liver is likely to result in
   A. fast pain at the organ site.
   B. slow pain at the organ site.
   C. chronic pain at the organ site.
   D. referred pain in another part of the body.

14. Eyeglasses with convex lenses are used to correct
   A. myopia.
   B. hyperopia.
   C. astigmatism.
   D. glaucoma.

15. After a visual signal leaves the ganglion cells, it travels to the
   A. optic chiasm.
   B. optic nerve.
   C. lateral geniculate nucleus.
   D. visual cortex.

Answer Key

Matching

1. f; **2.** i; **3.** d; **4.** k; **5.** h; **6.** e; **7.** o; **8.** m; **9.** b; **10.** j; **11.** g; **12.** l; **13.** c; **14.** n; **15.** a

Fill-in-the-Blank

16. basilar; hair cells; **17.** papilla; tongue; **18.** binocular; triangulate; **19.** sense; vestibular apparatus; **20.** color blindness; green; **21.** Olfactory or Smell; odorants; **22.** rostaglandins; pain; **23.** Ruffini; Pacinian; **24.** Golgi tendon organs; **25.** habituated

Labeling

26. cornea; **27.** pupil; **28.** iris; **29.** lens; **30.** optic nerve; **31.** retina; **32.** pinna; **33.** auditory canal; **34.** cochlea; **35.** Eustachian tube; **36.** ossicles; **37.** eardrum

Roots to Remember

38. bone stone or stone bone; **39.** a little bone; **40.** *Infra-* means "beneath" or "below," so you know that it is below the red wavelength. That helps you to remember that red is at one end of the visible light spectrum. *Ultra-* means "beyond," so you know that it is above violet. That helps you to remember that violet is at the other end of the visible light spectrum.

Word Choice

41. is not associated with; **42.** longer; **43.** epidermis; **44.** cool; **45.** chest; **46.** at the back; **47.** Pheromones; **48.** lens

Table Completion

49. Astigmatism; **50.** Irregularly shaped lens; **51.** Eye is elongated, so the focal point for far away objects falls in front of the retina; **52.** Concave lens; **53.** Hyperopia; **54.** Eye is shortened, so the focal point for close objects falls behind the retina

Sequencing

55. 8; **56.** 1; **57.** 6; **58.** 4; **59.** 10; **60.** 7; **61.** 3; **62.** 9; **63.** 2; **64.** 11; **65.** 5

Critical Thinking

66. Excess mucus builds up when you have a bad cold. This mucus may block the passage of scent molecules to olfactory receptors.

67. The sweet tastants from the cookie bind to and stimulate sweet taste receptors. These receptors need a little time to recover before they can be stimulated again. If you eat an orange immediately after eating the cookie, sweet tastants cannot trigger many sweet taste receptors. However, sour tastants in the orange can trigger sour taste receptors. Therefore, the orange tastes sour.

68. Intense sound waves could cause the hair cells in the inner ear to vibrate with such force that the cells sustain damage. If hair cells break, you may lose the ability to hear certain frequencies.

Practice Test

1. D; **2.** A; **3.** B; **4.** B; **5.** A; **6.** B; **7.** C; **8.** A; **9.** C; **10.** A; **11.** D; **12.** C; **13.** D; **14.** B; **15.** A

CHAPTER 16

THE ENDOCRINE SYSTEM: WORRIED SICK

Learning Goals

1. Summarize the basic structure and function of the endocrine system.
2. Contrast the mechanisms by which protein and steroid hormones elicit a response from cells.
3. State the difference between an exocrine gland and an endocrine gland.
4. Describe how the hypothalamus and pituitary gland interact to maintain homeostasis in different body systems.
5. List the nine primary endocrine organs, describe where they are found in the body, and summarize the function of the major hormones that each organ produces.
6. Explain how the thymus serves as a junction between endocrine and immune systems.
7. Provide two examples of locally acting hormones and describe their effects.
8. Describe the endocrine dysfunctions that cause Cushing's syndrome and Graves' disease.
9. Explain how blood-glucose and calcium homeostasis are maintained.
10. Describe how the actions of prostaglandins differ from those of other hormones.

Chapter Outline

I. An Overview of the Endocrine System
  A. Overview
    1. The endocrine system consists of endocrine glands and the hormones they secrete.
      a) Endocrine glands secrete hormones into the bloodstream.
      b) Exocrine glands secrete products through ducts.
  B. Hormones: Chemical Messengers
    1. Hormones are chemicals that move through the blood and stimulate specific tissues or cells.
      a) A hormone's target tissues or target cells are the tissues or cells that are stimulated by the hormone.
      b) Target tissues and cells contain protein receptors for the specific hormone.
      c) The hormone-receptor system is like a lock and key.
    2. Hormones elicit a response through two mechanisms.
      a) Hormones can bind to surface receptors and trigger an internal change in the cell.
      b) Hormones can diffuse across the cell membrane and bind to receptors inside the cell to trigger a response.
      c) Type of response depends on the chemistry of the hormone.
        (1) Lipid-soluble hormones can cross a cell membrane.
        (2) Most protein hormones cannot cross a cell membrane.
    3. Protein hormones use a signal transduction pathway to trigger a response.
      a) A signal transduction pathway is a chain reaction that relays a message from the outside of the cell to the inside.
        (1) A first messenger protein hormone binds to a receptor on the target cell's surface.
        (2) The receptor changes shape.
        (3) The modified receptor signals an intracellular enzyme.
          (a) Molecules of ATP within the cell are converted to cyclic AMP (cAMP).
          (b) Cyclic AMP (a secondary messenger) stimulates a cascade of enzyme actions.
      b) A signal transduction pathway leads to the conversion of glycogen to glucose.
    4. Steroid hormones are synthesized from cholesterol.
      a) Steroid hormones are nonpolar molecules that can diffuse across the cell membrane.
        (1) Inside the cell, these hormones bind to receptors within the cytoplasm.

(2) The hormone-receptor complex travels to the cell nucleus.
(3) The complex triggers the transcription of genes that code for enzymes and other proteins.

b) In smooth muscle, a hormone-receptor complex triggers the production of angiotensin II receptors.
(1) Angiotensin II, a protein hormone, binds to these receptors.
(2) Binding causes vessels to narrow, increasing blood pressure.

C. Endocrine Glands
1. Endocrine system organs include nine glands.
2. Endocrine tissue is also extraglandular and spread throughout the body.
3. Endocrine tissues and glands produce more than 50 hormones.

D. Stress and the Endocrine System
1. Physical stress and psychological stress have similar effects on the body.
2. The stress response prepares the body for fight or flight.
a) Heart rate, breathing rate, and blood pressure increase.
b) Glucose enters the bloodstream.
c) Digestion stops.
d) Perception of pain is reduced.
e) Alertness and memory improve.
3. The stress response results from actions in both the nervous and endocrine systems.
a) A stressor causes activation of the neurons in the sympathetic nervous system.
(1) Sympathetic neurons stimulate the heart, blood vessels, and lungs to increase oxygen delivery to muscles.
(2) Other neurons stimulate endocrine glands to release hormones.
b) The adrenal glands are endocrine organs responsible for most physiological effects of the stress response.
(1) There are two adrenal glands in the body.
(2) An adrenal gland is located on top of each kidney.
(3) The inner layer of the adrenal gland is called the medulla.
(a) The cells of the medulla are neurons at the end of the nerve pathway that extends to the brain.
(b) When medulla neurons fire, epinephrine and norepinephrine are released from medulla cells and secreted into adrenal capillaries.
(i) Epinephrine is also known as adrenaline.
(ii) Norepinephrine is also known as noradrenaline.
(iii) These chemicals increase heart rate, redirect blood flow from digestive organs to muscles, and dilate the pupils.
(4) The outer layer of the adrenal gland is called the cortex.
(a) The cortex is also involved in the stress response.
(b) The cortex releases two types of steroid hormones.
(i) Mineralocorticoids retain minerals.
(a) Aldosterone is the most common human mineralocorticoid.
(b) Aldosterone regulates blood sodium and potassium.
(ii) Glucocorticoids bind to cortisol receptors.
(a) Cortisone is a glucocorticoid that is produced in response to stress.
(b) Cortisol is the active form of cortisone, and it can cause a long-term stress response.
(c) Cortisol has an anti-inflammatory effect.

II. The Endocrine System and Homeostasis
A. The Control Center: The Hypothalamus
1. The hypothalamus controls and coordinates reactions to stress.
a) The hypothalamus regulates blood pressure, heart rate, and breathing rate.
b) The hypothalamus enhances alertness and memory.
c) The hypothalamus links the nervous and endocrine systems.
2. The hypothalamus directs activity in the pituitary gland.
a) The pituitary gland is called the "master gland" because its hormones affect a wide range of tissues.
b) The pituitary gland is made up of the anterior and posterior pituitary.

3. The hypothalamus produces hormones.
    a) Antidiuretic hormone (ADH) and oxytocin are produced by neuroendocrine cells that extend into the posterior pituitary.
        (1) These two hormones are stored in the posterior pituitary until they are triggered to release.
        (2) ADH and oxytocin affect a variety of tissues.
    b) Regulatory hormones target cells in the anterior pituitary.
        (1) These hormones are produced in tiny amounts.
        (2) They travel through a portal of interconnected capillary beds between the two organs.
        (3) Releasing hormones trigger pituitary hormone production.
        (4) Inhibiting hormones inhibit pituitary hormone production.
4. The hypothalamus goes into action when stimulated by a stressor.
    a) The hypothalamus sends nerve impulses to the adrenal medulla.
    b) The adrenal medulla releases epinephrine and norepinephrine.
    c) Neurons storing ADH and oxytocin are triggered to release these hormones into the bloodstream.
        (1) ADH constricts blood vessels.
            (a) Blood pressure increases.
            (b) Sugar and oxygen delivery to large muscles increases.
        (2) Oxytocin promotes affiliative behavior.
        (3) Oxytocin acts on the uterus and mammary glands.
            (a) During labor, oxytocin is part of the positive feedback loop that increases contractions.
            (b) After birth, oxytocin is part of the positive feedback loop that allows milk release.

B. Turning Down Hormone Release Through Negative Feedback Loops
1. Endocrine responses are usually part of a negative feedback loop.
2. A negative feedback loop along the hypothalamus-pituitary-adrenal axis occurs during the stress response.
    a) The hypothalamus releases corticotrophin-releasing hormone (CRH).
    b) CRH promotes the release of adrenocorticotropic hormone (ACTH).
    c) ACTH travels to the adrenals and triggers glucocorticoid release.
    d) The increase in glucocorticoids triggers a decrease in the release of ACTH and CRH.
    e) Glucocorticoid levels then drop and the stress response ends.
3. Individuals who experience repeated stress responses retain high glucocorticoid levels.
    a) Glucocorticoid levels are a measure of chronic stress.
    b) Chronic stress can have many negative effects.

III. Other Endocrine Glands

A. The Pituitary: Regulation of Growth
1. The pituitary gland releases growth hormone (GH).
    a) GH promotes cell division and growth of muscle and bone.
        (1) GH promotes the uptake of amino acids.
        (2) Increased uptake of amino acids leads to increased protein synthesis.
    b) Effects of growth from GH take place before puberty.
    c) In adults, GH regulates metabolism by promoting the use of fats as an energy source.
    d) People with pituitary dwarfism have a genetic mutation that leads to insufficient GH release.
    e) Children with stress dwarfism have experienced abuse or neglect that inhibits the release of GH.
2. The pituitary gland releases prolactin, which is associated with milk production in nursing mothers.

B. The Gonads: Sex-Specific Characteristics
1. Activities of the gonads are influenced by the hypothalamus and pituitary gland.
    a) The hypothalamus produces gonadotropin-releasing hormone (GnRH).
        (1) GnRH regulates the release of follicle-stimulating hormone (FSH) and luteinizing hormone (LH).
            (a) Both FSH and LH promote the production of sex cells and sex hormones.
                (i) In women, FSH promotes egg development and LH promotes ovulation.
                (ii) In men, FSH promotes sperm development and LH stimulates the testes to produce testosterone.
        (2) GnRH secretion does not begin until puberty.

b) Stressful situations can reduce fertility.
(1) Hormone production is reduced.
(a) The release of GnRH is suppressed.
(b) The responsiveness of the pituitary gland to GnRH is reduced.
(c) Less FSH and LH are produced.
(2) In children, suppression of hormones can result in delayed puberty.
(3) In women, suppression of hormones can result in suppression of menstruation, ovulation, and sex drive.
(4) In men, suppression of hormones can result in reduction in sperm count and sex drive.
2. Sex hormones are steroid hormones produced by the gonads.
a) Testosterone is more abundant in men.
b) Estrogen and progesterone are more abundant in women.
c) Sex hormones promote the development of secondary sexual characteristics during puberty.
(1) Pubic hair is produced in both sexes.
(2) The pelvis widens and fat deposition increases in girls.
(3) The voice deepens and muscles develop in boys.
d) Anabolic steroids are synthetic testosterone derivatives that are used to enhance athletic performance.
(1) Anabolic steroid use can lead to many physical problems.
(2) Use of anabolic steroids is banned by organized sports.
C. The Pancreas: Regulation of Blood-Glucose Levels
1. Pancreatic islets are endocrine cells in the pancreas.
a) Alpha cells produce glucagon, which raises blood-sugar levels by increasing the breakdown of glycogen in the liver.
b) Beta cells produce insulin, which lowers blood-sugar levels by increasing uptake of glucose.
c) Delta cells produce somatostatin, which inhibits secretion of glucagon and insulin.
2. Stress can cause an increase in blood sugar.
a) The sympathetic nervous system triggers the release of glucagon.
b) Glucocorticoids maintain higher blood sugar levels.
(1) They make fat cells temporarily insulin resistant.
(2) They block the action of leptin.
(a) Appetite increases when leptin is blocked.
(b) The person eats more food, which leads to higher blood sugar.
c) Beta cells release insulin to return blood-sugar levels to normal when glucocorticoids decrease.
d) The body expends metabolic energy to control blood-sugar levels.
e) Insulin resistance and overactive beta cells can lead to type 2 diabetes.
3. People with Cushing's syndrome have abnormally high glucocorticoid levels.
a) There are a number of causes for high glucocorticoid levels.
(1) High levels may be caused by adrenal or pituitary gland diseases.
(2) High levels may be caused by glucocorticoid drugs.
b) People with Cushing's syndrome store excess fat in the midsection and face.
c) People with Cushing's syndrome have a high risk of type 2 diabetes.
4. Stress on the pancreas can cause an increase in LDL cholesterol.
D. The Thyroid and Parathyroid: Metabolism and Development
1. The thyroid is located on the anterior surface of the trachea.
a) The thyroid is shaped like a bow tie.
b) The thyroid has four parathyroid glands on its back surface.
2. The thyroid and parathyroid work together to maintain calcium homeostasis.
a) Calcitonin produced by the thyroid decreases blood-calcium levels by promoting calcium deposition on bones.
b) Parathyroid hormone produced by the parathyroid increases blood-calcium levels.
(1) It promotes calcium release from bones.
(2) It promotes increased absorption of calcium from the digestive tract.
(3) It prevents excretion of calcium in urine.
3. The thyroid produces T_3 and T_4 hormones.
a) These are protein hormones that can cross the cell membrane.
b) These hormones activate genes within cells to boost ATP production.

c) T_3 and T_4 hormones control the body's basal metabolism rate.
d) Stressors that increase ATP demand can increase production of T_3 and T_4 hormones.
(1) The stressor stimulates the hypothalamus to secrete TRH (TSH-releasing hormone).
(2) TRH stimulates the release of TSH (thyroid-stimulating hormone) from the anterior pituitary.
(3) TSH stimulates the secretion of T_3 and T_4 by the thyroid.
e) Diseases of the thyroid can lead to under- and overproduction of thyroid hormones.
(1) Hypothyroidism, or low thyroid hormone levels, can produce weight gain and lethargy.
(2) Hyperthyroidism, or oversecretion of thyroid hormones, can cause insomnia and hyperactivity.
(a) Graves' disease causes hyperthyroidism.
(b) Symptoms of Graves' disease can include protruding eyes and goiter.
(c) Graves' disease is associated with severe stress.
(d) Graves' disease can be cured by removal of part of the thyroid.

E. The Pineal Gland: Hormonal Effects of Light and Darkness
1. The pineal gland is located deep within the brain.
a) The pineal gland is attached to the optic nerve.
b) The pineal gland responds to light cues.
2. The pineal gland produces melatonin.
a) Melatonin induces sleep.
b) This hormone ebbs and flows daily.
(1) Production rises in the evening up to the middle of the night.
(2) Production declines from the middle of the night until morning.
c) Glucocorticoid levels lower as melatonin rises.
(1) Disrupted sleep can keep glucocorticoid levels high.
(2) High glucocorticoid levels keep the brain awake.
d) In regions where winter days are short, melatonin levels may remain elevated.
(1) This may result in seasonal affective disorder (SAD).
(2) SAD causes lethargy and depression.
e) Melatonin is a powerful antioxidant.
(1) Low melatonin levels are associated with high cancer rates.
(2) Night workers may experience low melatonin levels and higher risk for disease.

F. The Thymus: Junction of the Endocrine and Immune Systems
1. The thymus is involved in the development and maturation of T cells.
2. This gland produces the hormones thymosin and thymopoietin.
3. The thymus is most active during childhood.
4. The thymus shrinks and becomes less active as we age.
a) Glucocorticoids accelerate the process of shrinking.
b) It is not clear whether a shrinking thymus causes a decline in immune response.

G. Other Tissues That Produce Hormones
1. Many organs that have main functions that are not endocrine produce hormones.
a) The kidneys
(1) produce erythropoietin, which stimulates red blood cell production;
(2) modify vitamin D into calcitriol, which stimulates absorption of calcium from the intestines; and
(3) release angiotensin to regulate blood pressure.
b) The heart releases atrial natriuretic peptide (ANP) to regulate blood pressure.
c) The digestive tract produces gastrin during food intake.
(1) Gastrin stimulates the stomach to release hydrochloric acid.
(2) Hydrochloric acid and digestive enzymes break down proteins.
2. Endocrine tissues throughout the body produce prostaglandins.
a) Prostaglandins have a local effect.
b) Prostaglandins promote inflammation at damaged sites.
(1) Inflammation increases blood flow to the area.
(2) Inflammation signals immune system cells to travel to the site.
3. A variety of body tissues release growth factors.
a) Growth factors stimulate cell growth and division.
b) Most growth factors have a local effect.

c) Insulin-type growth factor (IGF) is the most prominent growth factor.
  (1) IGF is produced by the liver.
  (2) IGF promotes damaged tissue repair.
  (3) High levels of IGF in experimental animals correlate to increased life span and longer good health.
  (4) Chronic stress can lower IGF levels.

IV. Combating Stress
  A. Chronic stress can trigger many diseases.
  B. The stress response can have value, by increasing alertness and energy level.
  C. Individuals with Addison's disease have an inadequate stress response.
    1. Glucocorticoid levels are abnormally low.
      a) Low glucocorticoid levels may result from nonfunctional adrenal glands.
        (1) When adrenal glands do not function, bronzing of the skin may occur as excess ACTH stimulates melanin production.
        (2) A bronzed look makes the patient appear sun-tanned and healthy.
      b) Low glucocorticoid levels may also result when the pituitary fails to produce ACTH.
    2. With low glucocorticoids, the body cannot replenish blood glucose when a stressor occurs.
    3. Lack of blood glucose can cause fatigue, weakness, and low blood pressure.
    4. Too much stress can be fatal for a patient with Addison's disease.
  D. The following behaviors can help to manage the stress response.
    1. Maintain a regular sleep schedule.
    2. Establish a regular exercise routine.
    3. Practice meditation.
    4. Modify emotional responses to stress.
    5. Focus on controllable factors and minimize factors that are beyond your control.
    6. Seek professional help if stress remains a problem.

Practice Questions

Matching

1. thymus	a. acts on bone to lower blood-calcium levels
2. glucagon	b. one of the primary sex hormones in women
3. norepinephrine	c. associated with milk production in nursing mothers
4. calcitonin	d. produces T_3 and T_4 hormones
5. cortisol	e. involved in the development and maturation of T cells
6. melatonin	f. "master gland" producing hormones that regulate many tissues
7. pituitary gland	g. also known as noradrenaline
8. cortisone	h. produces hormones associated with the stress response
9. gastrin	i. active form of a stress hormone that gives rise to a long-term stress response
10. progesterone	j. stimulates the production of stomach acid
11. leptin	k. raises blood-glucose levels
12. thyroid gland	l. helps to reduce appetite
13. adrenal gland	m. helps to regulate sleep
14. parathyroid hormone	n. acts on bone and other tissues to raise blood-calcium levels
15. prolactin	o. produced in response to stress; converts to cortisol

Fill-in-the-Blank

16. After a protein hormone binds to the surface of a target cell, the modified receptor signals an enzyme to convert ____________ within the cell to ____________.

17. The adrenal ____________ secretes ____________ and norepinephrine.

18. The ____________ produces the hormone ____________, which acts on the uterus and mammary glands in women.

19. When adrenocorticotropic hormone (ACTH) is released into the bloodstream, it travels to the ____________ and triggers the release of ____________.

20. In the pancreas, ____________ cells produce glucagon and ____________ cells produce insulin.

21. When blood-calcium levels are too low, the ____________ gland releases hormones that signal ____________ to break down bone.

22. People with seasonal affective disorder (SAD) have ____________ melatonin levels that can lead to lethargy and ____________.

23. The ____________ is largest and most active during childhood but shrinks during adulthood due to the action of ____________.

24. Many endocrine tissues produce ____________, which are hormones that have a localized effect.

25. Hormones called ____________ stimulate cell growth and division; the most prominent of these hormones, ____________, promotes tissue repair.

Labeling

Use the terms below to label Figure 16.1.

adrenal
gonads
hypothalamus
ovary
pancreas
parathyroid
pineal
pituitary
testis
thymus
thyroid

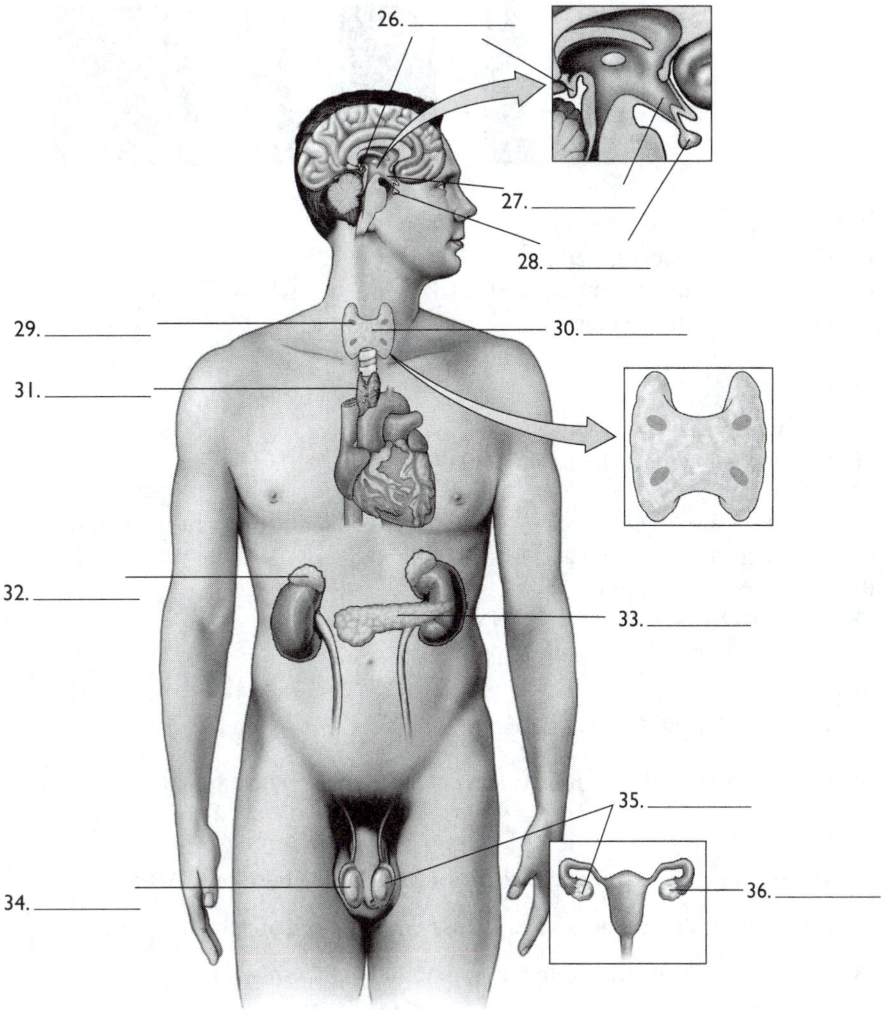

26. ____________
27. ____________
28. ____________
29. ____________
30. ____________
31. ____________
32. ____________
33. ____________
34. ____________
35. ____________
36. ____________

Figure 16.1

Roots to Remember

Use your knowledge of the root words presented in this chapter to answer the following questions.

37. Bright's disease is also known as chronic nephritis. What organ does this disease affect?
38. Both the thalamus and hypothalamus are structures in the brain. Is the hypothalamus likely to be above or below the thalamus?
39. The root *-emia* relates to blood. The Greek word *glykys* means "sweet." What condition does someone with hyperglycemia have?

Crossword Puzzle

Across

40. gland that produces melatonin
41. part of the pituitary that secretes ADH
42. endocrine control center

Down

43. hormone released by fat cells
44. structure containing alpha, beta, and delta cells
45. sex hormone
46. To reduce stress, you should ____________ on factors that you can control.
47. type of thyroid disease that can cause goiter
48. stimulated by oxytocin
49. group of glands working together to regulate stress

Word Choice

Circle the word or phrase that correctly completes each sentence.

50. Vitamin D is modified into (calcitriol/calcitonin), which stimulates calcium absorption from the intestines.
51. The pituitary gland (produces/secretes) antidiuretic hormone (ADH).
52. (Protein/Steroid) hormones can diffuse cross the cell membrane.
53. The adrenal glands are located on top of the (kidneys/thyroid).
54. Corticotropin-releasing hormone (CRH) and adrenocorticotropic hormone (ACTH) levels are maintained by a (positive/negative) feedback loop.

Table Completion

Complete the following table on hormones.

Hormone	Secreted By	Action
55.	Pituitary	Water reabsorption by kidneys
Insulin	56.	57.
58.	59.	Regulate sodium and potassium levels in blood
Luteinizing hormone (LH)	60.	61.
Follicle-stimulating hormone (FSH)	Pituitary	62.
Androgens	63.	Promotes development of sperm and male-specific sex characteristics
Thymosin	64.	65.

Critical Thinking

66. How might chronic stress cause chronic high blood pressure?
67. Working at a computer before going to sleep may cause sleeplessness. Why?
68. Both the testes and ovaries respond to the FSH and LH hormones. What does this suggest about the developmental origins of these reproductive organs?

Practice Test

1. Which hormone stimulates the release of T_3 and T_4 hormones?
   A. ADH
   B. GH
   C. ACTH
   D. TSH

2. A student is startled by the backfiring of a car. In response to the loud noise, the neurons in the student's posterior pituitary release
   A. epinephrine.
   B. ADH.
   C. norepinephrine.
   D. ACTH.

3. How does oxytocin affect men?
   A. It promotes affiliative behavior.
   B. It constricts blood vessels.
   C. It reduces the stress response.
   D. It does not have any effect on men.

4. How does a steroid hormone act on a cell?
   A. It binds to receptors in the cell membrane.
   B. It triggers a signal transduction pathway.
   C. It binds to receptors in the cell cytoplasm.
   D. It triggers binding of RNA polymerase to angiotensin II receptors.

5. Which of the following endocrine glands is located in the throat?
   A. thyroid
   B. pituitary
   C. thymus
   D. adrenal

6. A runner drinks too much water during a race, and the sodium concentration in her blood becomes too low. Which hormone is likely to be released?
   A. aldosterone
   B. antidiuretic hormone
   C. progesterone
   D. prolactin

7. What effect do T_3 and T_4 hormones have on the body?
   A. They maintain the body's normal basal metabolic rate.
   B. They maintain blood-calcium homeostasis in the body.
   C. They maintain blood-glucose homeostasis in the body.
   D. They promote the development and maturation of T cells.

8. Low melatonin levels are likely to cause
   A. lethargy.
   B. rapid heart beat.
   C. insomnia.
   D. diabetes.

9. It is late afternoon, and you are starting to feel hungry. What hormone is your body likely to be releasing as your blood-glucose levels drop?
   A. insulin
   B. glucagon
   C. calcitonin
   D. somatostatin

10. Which ailment results from abnormally high glucocorticoid levels?
    A. Graves' disease
    B. SAD
    C. Cushing's syndrome
    D. Addison's disease

11. Which hormone causes localized swelling when you stub your toe?
    A. cortisol
    B. adrenocorticotropic hormone
    C. cortisone
    D. prostaglandin

12. Where is gonadotropin-releasing hormone (GnRH) produced?
    A. in the gonads
    B. in the pituitary gland
    C. in the hypothalamus
    D. in the thyroid

13. Which of the following is *not* secreted by the pituitary gland?
    A. adrenocorticotropic hormone
    B. epinephrine

C. prolactin
D. growth hormone

14. Meditation can reduce stress on the body by reducing
A. glucocorticoid levels.
B. melatonin levels.
C. the activity of brain neurons.
D. blood-glucose levels.

15. A large meal would trigger the release of
A. erythropoietin.
B. leptin.
C. angiotensin.
D. gastrin.

Answer Key

Matching

1. e; **2.** k; **3.** g; **4.** a; **5.** i; **6.** m; **7.** f; **8.** o; **9.** j; **10.** b; **11.** l; **12.** d; **13.** h; **14.** n; **15.** c

Fill-in-the-Blank

16. ATP; cyclic AMP (cAMP); **17.** medulla; epinephrine; **18.** hypothalamus; oxytocin; **19.** adrenals; glucocorticoids; **20.** alpha; beta; **21.** parathyroid; osteoclasts; **22.** elevated; depression; **23.** thymus; glucocorticoids; **24.** prostaglandins; **25.** growth factors; insulin-type growth factor (IGF)

Labeling

26. pineal; **27.** hypothalamus; **28.** pituitary; **29.** parathyroid; **30.** thyroid; **31.** thymus; **32.** adrenal; **33.** pancreas; **34.** testis; **35.** gonads; **36.** ovary

Roots to Remember

37. the kidney; **38.** below; **39.** high blood glucose

Crossword Puzzle

Across

40. pineal; **41.** posterior; **42.** hypothalamus

Down

43. leptin; **44.** islet; **45.** estrogen; **46.** focus; **47.** hyper; **48.** labor; **49.** axis

Word Choice

50. calcitriol; **51.** secretes; **52.** Steroid; **53.** kidneys; **54.** negative

Table Completion

55. Antidiuretic hormone (ADH); **56.** Pancreas; **57.** Lowers blood-glucose levels; **58.** Mineralocorticoids; **59.** Adrenals; **60.** Pituitary; **61.** Ovulation and testosterone production; **62.** Egg and sperm production; **63.** Testes; **64.** Thymus; **65.** Helps T cells to mature

Critical Thinking

66. People under chronic stress may have high cortisol levels in their blood, resulting from the release of cortisone. Cortisol increases blood pressure. People under chronic stress may also experience more frequent release of epinephrine and norepinephrine. These hormones trigger the release of ADH, which constricts blood vessels, thereby increasing blood pressure. High glucocorticoid levels due to chronic stress can also increase LDL cholesterol in the blood. This type of cholesterol can narrow arteries and increase blood pressure.

67. The light from the computer screen might interfere with the buildup of melatonin levels. Increased alertness, caused by working before going to bed, can also blunt the effects of melatonin and interfere with sleep.

68. It suggests that testes or ovaries can develop from the same structure.

Practice Test

1. D; **2.** B; **3.** A; **4.** C; **5.** A; **6.** A; **7.** A; **8.** C; **9.** B; **10.** C; **11.** D; **12.** C; **13.** B; **14.** A; **15.** D

CHAPTER 17

DNA SYNTHESIS, MITOSIS, AND MEIOSIS: CANCER

Learning Goals

1. Describe the differences between a benign and a malignant tumor.
2. Describe the process of DNA synthesis.
3. Name the enzyme that facilitates DNA synthesis.
4. Define the term *complementary* as it applies to DNA nucleotides.
5. Draw the metaphase chromosomes of a cell with four chromosomes.
6. Describe the function that microtubules of the spindle apparatus play in cell division.
7. List the stages of mitosis, and describe what occurs during each stage.
8. Describe the events of cytokinesis.
9. Compare and contrast the events of meiosis I and meiosis II.
10. Describe what crossing over accomplishes.

Chapter Outline

I. What is Cancer?
  A. Overview
    1. Cancer begins when a single cell replicates without regulation.
    2. Unregulated cell division forms a tumor.
      a) Slow-growing tumors that do not invade surrounding structures are benign.
      b) Tumors that invade surrounding tissues are malignant.
        (1) Cells from a malignant tumor may break away and start new cancers in other parts of the body through metastasis.
        (2) Cancer cells can travel via the lymphatic or circulatory systems.
    3. Cancer cells are different from normal cells.
      a) They divide without regulation.
      b) They invade surrounding tissue.
      c) They move to other locations.
  B. Risk Factors for Cancer
    1. Tobacco Use
      a) Smoking is the leading cause of 90% of all lung cancers.
      b) Tobacco use is the cause of one-third of all cancer deaths.
        (1) Cigar smokers have increased rates of lung, larynx, esophagus, and mouth cancers.
        (2) People who chew tobacco have increased rates of mouth, gum, and cheek cancers.
        (3) People exposed to secondhand smoke have increased lung cancer rates.
      c) Tobacco smoke contains more than 20 known carcinogens.
      d) Chemicals in tobacco smoke can affect normal cell function.
        (1) They may increase cell division.
        (2) They may inhibit the cell's ability to repair damaged DNA.
        (3) They may prevent cells from dying normally.
        (4) They may disrupt the transport of substances across cell membranes.
        (5) They may inhibit enzyme reactions.
        (6) They may increase the generation of free radicals.
          (a) Free radicals remove electrons from other molecules.

(b) Removal of electrons can damage DNA and other molecules.
(c) Damage to DNA and other molecules can lead to cancer.

2. A High-Fat, Low-Fiber Diet
   a) A diet high in fat and low in fiber can increase the risk of cancer.
   b) A diet rich in fruits, vegetables, grains, beans, and other plant products can reduce cancer risk.
      (1) Fruits and vegetables are rich in antioxidants, which can neutralize the electrical charge on free radicals.
      (2) Antioxidants may also minimize the number of free radicals in cells.
3. Lack of Exercise
   a) Exercise keeps the immune system functioning effectively.
   b) The immune system can help recognize and destroy cancer cells.
4. Obesity
   a) Obesity can increase the risk for breast, uterine, ovarian, colon, gallbladder, and prostate cancers.
   b) Abundant fatty tissue may increase the odds of hormone-sensitive cancers.
5. Excess Alcohol Consumption
   a) Having more than one or two alcoholic drinks per day can increase cancer risk.
   b) People who drink and smoke increase their odds of cancer in a multiplicative way.
6. Increasing Age
   a) An aging immune system becomes weaker and cannot eliminate cancer cells as effectively.
   b) Cumulative damage can also explain the increased cancer risk associated with aging.

II. Overview of Cell Division

A. Overview

1. Cell division allows tissues to heal and organisms to reproduce.
2. Humans all start from a single cell produced during sexual reproduction and end up with trillions of body cells.
3. A cell must first copy its DNA before dividing.
   a) DNA is located in the nucleus.
   b) DNA is organized in structures called chromosomes.
      (1) DNA molecules are in an uncondensed, stringy form, called chromatin, before cell division.
      (2) Right before cell division, the chromatin becomes tightly wrapped, forming a condensed chromosome.
   c) When a chromosome replicates, it produces two sister chromatids.
      (1) Sister chromatids are identical copies of DNA.
      (2) Sister chromatids are attached at a centromere.

B. DNA Replication

1. During replication, the DNA molecule splits in half lengthwise up the double helix.
   a) New nucleotides are added to each side of the parent molecule.
   b) This process produces two identical daughter molecules, each composed of a strand of nucleotides from the parent molecule and a newly synthesized strand.
2. Free nucleotides in the nucleus form the new strands.
   a) Free nucleotides form hydrogen bonds with complementary nucleotides on each parent strand.
   b) DNA polymerase catalyzes covalent bonds between nucleotides as they are added.

III. The Cell Cycle and Mitosis

A. Overview

1. The cell cycle includes all the events that occur as a parent cell produces two daughter cells through cell division.
2. The cell cycle is constantly occurring in the cells of many body tissues.

B. Interphase: Normal Functioning and Preparations

1. During interphase, a cell performs normal functions and produces proteins.
2. Different cells spend different amounts of time in interphase.
   a) Skin cells and other cells that divide frequently spend a less-than-average amount of time in interphase.

b) Nerve cells and other cells that seldom divide spend a greater-than-average amount of time in interphase.

3. Interphase is separated into three phases.
   a) During the G_1 phase (first gap or growth), the machinery in the cell duplicates and the cell grows larger.
   b) During S phase (synthesis), DNA in the chromosomes replicates.
   c) During G_2 phase (second gap), proteins that drive mitosis are synthesized and the cell continues to grow and prepare for mitosis.

C. Mitosis: The Nucleus Divides

1. Mitosis is the division of the cell nucleus.
2. Mitosis is divided into four stages.
   a) Prophase is the first stage of mitosis.
      (1) Replicated chromosomes condense.
      (2) Microtubules (spindle fibers) form and grow from opposite poles of the cell.
      (3) The nuclear envelope breaks down.
      (4) Centrioles at the poles anchor microtubules as they form.
   b) Metaphase is the second stage of mitosis.
      (1) Replicated chromosomes align along the equator of the cell.
      (2) Microtubules help the chromosomes to line up in single file.
   c) Anaphase is the third stage of mitosis.
      (1) The centromere splits and microtubules shorten.
      (2) Shortening of microtubules pulls each sister chromatid to opposite poles of the cell.
   d) Telophase is the final stage of mitosis.
      (1) Nuclear envelopes re-form around daughter nuclei.
      (2) Chromosomes unwind into the chromatin stage.

D. Cytokinesis: The Cytoplasm Divides

1. Cytokinesis divides the cytoplasm of a cell that has undergone division.
   a) A band of proteins circles the cell at the equator.
   b) The band divides the cytoplasm.
   c) The band contracts to pinch apart the daughter nuclei.
2. Each cell enters interphase after cytokinesis.

IV. Mutations Override Cell-Cycle Controls

A. Controls in the Cell Cycle

1. There are a series of checkpoints in the cell cycle.
   a) At each checkpoint, proteins determine whether the cell cycle should continue.
   b) Checkpoints occur during G_1, G_2, and metaphase.
      (1) At the G_1 checkpoint,
         (a) proteins detect the presence of growth factors that stimulate cell division; and
         (b) if enough factors are present, other proteins check to see that the cell is large enough and that enough nutrients are present.
      (2) At the G_2 checkpoint,
         (a) proteins ensure that DNA has replicated properly, and
         (b) proteins ensure that the cell is large enough to divide.
      (3) At the metaphase checkpoint,
         (a) proteins make sure that all chromosomes are attached to microtubules, and
         (b) this ensures that DNA will divide correctly.
   c) If conditions are not favorable for division at any checkpoint, the cell cycle is stopped.
   d) The cell may undergo apoptosis (programmed cell death) if it cannot progress to division.
   e) Proteins that regulate the cell cycle are encoded by genes.
      (1) Functional genes produce functional proteins that properly regulate the cell cycle.
      (2) Mutated genes may fail to produce functional proteins, so unregulated cell division may occur.
         (a) Mutations may be inherited.
         (b) Mutations may occur spontaneously during DNA replication.
         (c) Mutations may be induced by exposure to carcinogens.

B. From Benign to Malignant
   1. Cancer cells can stimulate the growth of surrounding blood vessels through angiogenesis.
      a) Cancer cells secrete a substance that reroutes blood vessels to the tumor.
      b) These vessels supply the tumor with oxygen and nutrients.
   2. A tumor with its own blood supply can grow.
      a) Rapidly dividing cancer cells grow faster than do normal cells.
      b) Cancer cells can fill an organ, causing it to lose functionality.
   3. Unlike normal cells, cancer cells do not have contact inhibition, so they divide and pile up on each other.
   4. Unlike normal cells, cancer cells do not have anchorage dependence.
      a) Cancer cells do not secrete adhesion molecules to glue the cells together.
      b) Cells without anchorage dependence can travel to different tissues through the lymphatic or circulatory systems.
   5. Cancer cells override the programming that controls life span.
      a) Most cells divide 60 to 70 times and then stop dividing.
      b) Cancer cells are immortal and can divide without limit.
         (1) Cancer cells activate a gene that produces telomerase.
         (2) Telomerase prevents the degradation of chromosomes.

C. Multiple Hit Model
   1. Multiple mutations are required for cancer development and progression.
   2. Most cancers are not caused by inheritance of mutant genes alone.
   3. Close to 70% of cancers are caused by mutations that occur during a person's lifetime.

V. Cancer Detection and Treatment

A. Detecting Cancer
   1. Some tumors can be found by self-examination.
   2. X-ray techniques can detect breast cancer or lung cancer.
   3. Blood tests can detect cancers that cause excess production of certain proteins.
   4. A biopsy can be performed to remove tissue to be examined.
      a) Benign tumors show orderly cell growth.
      b) Malignant tumors have an abnormal appearance, due to the rapid growth of cells.
   5. A laparoscope can be used to take a biopsy of tissue within the body.

B. Cancer Treatments: Chemotherapy and Radiation
   1. Chemotherapy
      a) Chemotherapeutic agents kill dividing cells.
      b) Chemotherapy uses many drugs because certain cancer cells may become resistant to one type of drug.
         (1) Mutations allow cancer cells to become resistant.
         (2) Mutations develop quickly due to the rapid rate of cancer cell division.
         (3) An average tumor contains 1 billion cells, 1,000 of which may be resistant to a chemotherapy drug.
      c) Chemotherapy affects normal cells that are dividing rapidly.
         (1) Chemotherapy affects hair follicles, cells that produce red and white blood cells, and cells that line the intestine and stomach.
         (2) Effects of chemotherapy include hair loss, anemia, weakened immune system, nausea, vomiting, and diarrhea.
   2. Radiation Therapy
      a) Radiation therapy uses high-energy particles to injure or destroy cancer cells.
      b) Radiation is typically used on cancers that are located close to the surface of the body.
   3. Remission
      a) A person is in remission from cancer after 5 years of being cancer-free.
      b) A person is considered cured of cancer after 10 years.

VI. Meiosis: Making Reproductive Cells

A. Overview
   1. Meiosis is cell division that takes place within the testes of males and the ovaries of females.
   2. Meiosis produces gametes.
      a) Sperm cells are male gametes.
      b) Egg cells are female gametes.

3. Gametes have half as many chromosomes as somatic cells.
   a) Human somatic cells have two pairs of 23 chromosomes, for a total of 46 chromosomes.
   b) Human gamete cells have 23 chromosomes.
4. Gametes combine during fertilization to produce a zygote with a full suite of chromosomes.
5. A karyotype views chromosomes.
   a) A karyotype is a highly magnified photo of chromosomes from a single cell.
   b) Chromosomes are organized in pairs:
      (1) 22 pairs consist of autosomes, or nonsex chromosomes, and
      (2) 1 pair consists of the sex chromosomes.
   c) Nonsex chromosomes are arranged in homologous pairs.
      (1) Each member of the pair contains the same genes along its length.
      (2) Genes in homologous pairs may have different alleles.
6. Meiosis separates members of each homologous pair from each other.
   a) A cell that contains only one member of each homologous pair is called haploid (*n*).
   b) A cell that contains both members of each homologous pair is called diploid (2*n*).
7. Meiosis is preceded by interphase.
8. Meiosis consists of two stages of division: meiosis I and meiosis II.

B. Interphase
1. The interphase preceding meiosis includes G_1, S, and G_2 phases.
2. This interphase is similar in most respects to the interphase preceding mitosis.

C. Meiosis I
1. Prophase I is the first stage of meiosis I.
   a) The nuclear envelope begins to break down.
   b) Microtubules begin to assemble.
   c) Chromosomes condense.
   d) Homologous chromosomes exchange genetic information by crossing over.
2. Metaphase I is the second stage of meiosis I.
   a) Homologous pairs line up at the cell equator.
   b) Microtubules lengthen and bind to chromosomes near the centromere.
   c) Members of each homologous pair are arranged toward one pole or the other, in random alignment.
3. Anaphase I is the third stage of meiosis I.
   a) Homologous pairs are separated from each other, as microtubules shorten.
   b) Members of each homologous pair move to opposite poles.
4. Telophase I is the final stage of meiosis I.
   a) Nuclear envelopes form around the chromosomes.
   b) Daughter cells are separated by cytokinesis.

D. Meiosis II
1. Prophase II is the first stage of meiosis II.
   a) Microtubules lengthen.
   b) The nuclear envelope breaks down.
2. Metaphase II is the second stage of meiosis II.
   a) Chromosomes line up along the cell's equator.
   b) Chromosomes are not paired.
3. Anaphase II is the third stage of meiosis II.
   a) Sister chromatids separate.
   b) Separated chromatids move toward opposite poles.
4. Telophase II is the final stage of meiosis II.
   a) Nuclear envelopes form around the chromosomes.
   b) Each gamete contains half of each parent's genes.

E. Crossing Over and Random Alignment
1. Crossing over allows gene exchange between homologous chromosomes.
   a) Crossing over likely occurs several times for each homologous pair during meiosis.
   b) Linked genes are affected by crossing over.
   c) Crossing over increases genetic diversity by increasing the combinations of alleles that may be present in gametes.

2. Random alignment increases the number of genetically distinct gametes that can be produced.
   a) As chromosomes randomly align, different combinations of chromosomes can occur.
   b) Eight million different combinations of chromosomes are possible for human gametes.

Practice Questions

Matching

1. anaphase
2. karyotype
3. remission
4. metastasis
5. somatic
6. centriole
7. telophase
8. centromere
9. contact inhibition
10. diploid
11. biopsy
12. angiogenesis
13. microtubule
14. metaphase
15. carcinogen

a. structure that anchors microtubules
b. state of being cancer-free for five years
c. surgical removal of cells to check for cancer
d. substance that can cause damage to cell DNA
e. property that prevents cells from piling on top of each other
f. non-gamete cells
g. a magnified photograph of the chromosomes in a cell
h. fiber that moves chromosomes during cell division
i. stage during which sister chromatids are pulled toward opposite poles
j. a $2n$ cell
k. stage during which chromosomes line up across the equator of a cell
l. region near the middle of a replicated chromosome
m. stage during which nuclear envelopes form around daughter nuclei
n. stimulation of blood vessel growth around a tumor
o. process in which cancer cells from a tumor move to other locations in the body

Fill-in-the-Blank

16. A ____________ tumor can become ____________ if it starts to invade other tissue.
17. The ____________ found in fruit and vegetables may help to neutralize the electrical charge of ____________.
18. A replicated chromosome consists of two identical ____________ that are attached to each other at a ____________.
19. The enzyme ____________ catalyzes the formation of ____________ bonds between nucleotides on a newly forming strand of DNA.
20. A cell's organelles are duplicated during ____________ phase, and its DNA is replicated during ____________ phase of the cell cycle.
21. A cell's nucleus is divided during ____________, and its cytoplasm is divided during ____________.
22. At the G_2 ____________, proteins make sure that the cell is large enough for division and that the cell's DNA has ____________ properly.
23. Cancer cells override the requirement for ____________, so they do not necessarily maintain contact with an underlayer of cells.

24. The enzyme ____________ prevents the degradation of chromosomes, so cancer cells can become ____________.

25. During ____________, homologous pairs of chromosomes are separated from each other and move to opposite ____________ of the cell.

Labeling

Use the statements below to label Figure 17.1. Any statement listed twice should be used twice.

Chromosomes align at cell equator.
DNA condenses into chromosomes and crossing over occurs.
DNA is replicated.
Four haploid daughter cells are produced.

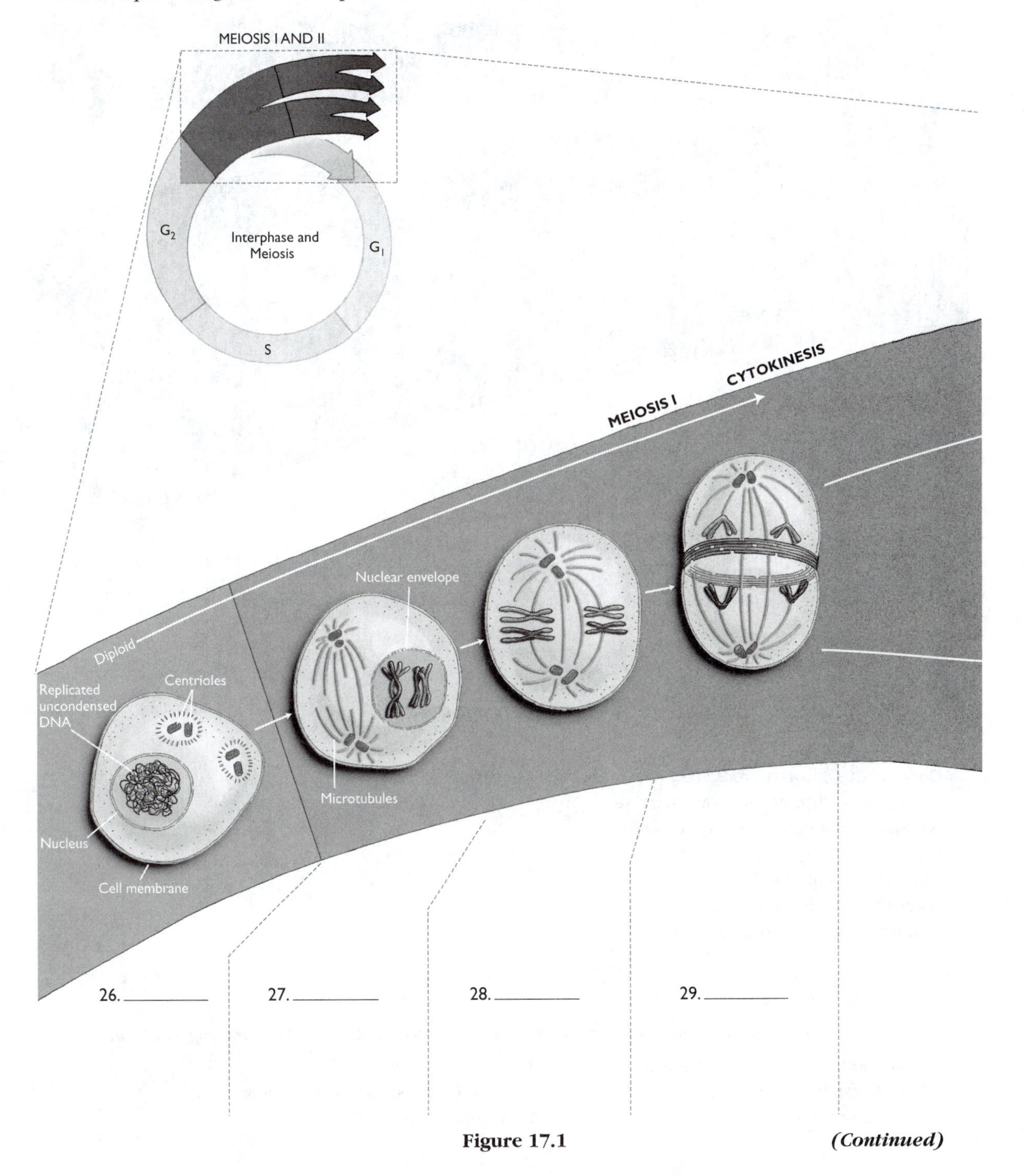

Figure 17.1 *(Continued)*

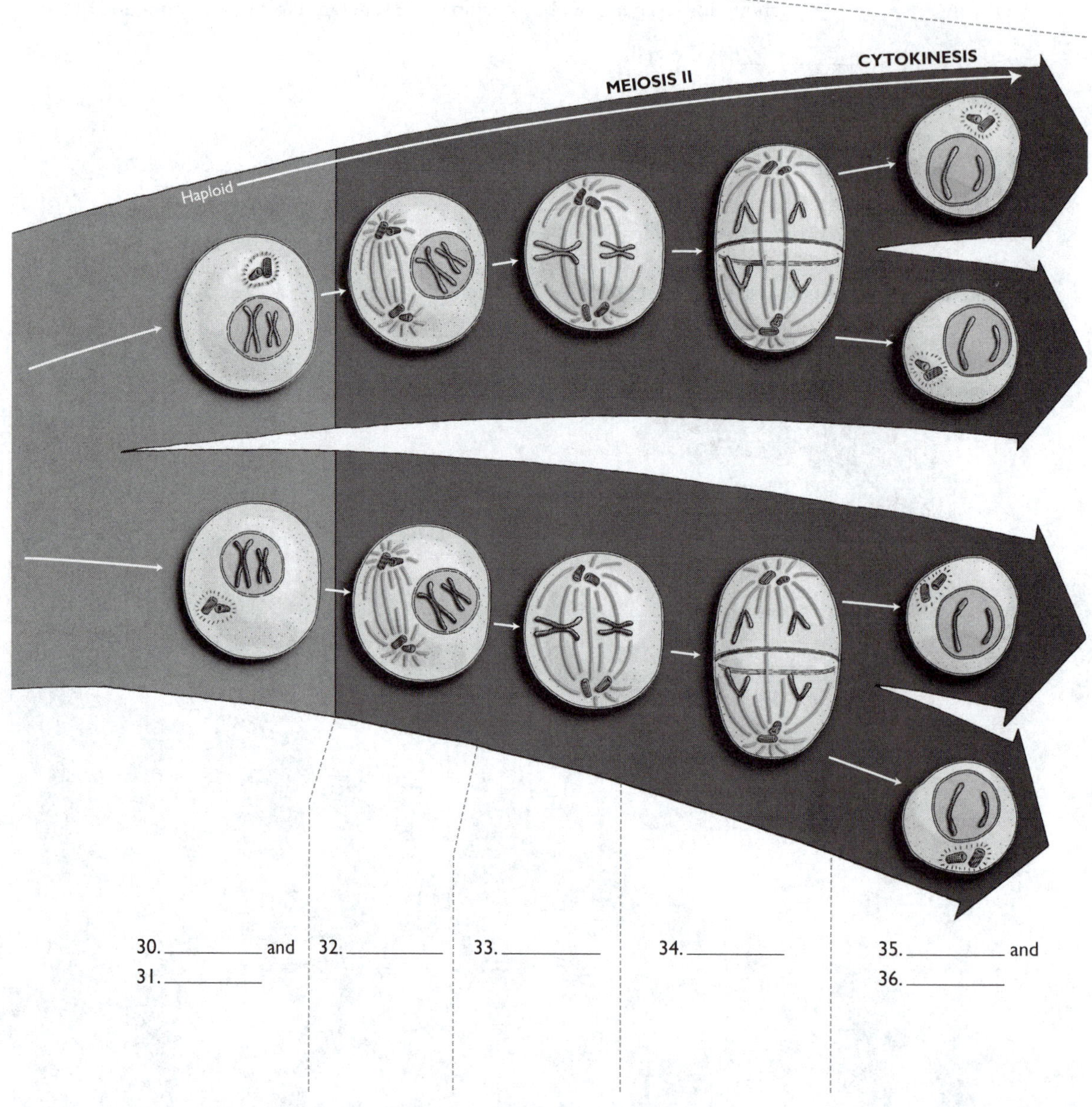

Figure 17.1 *(Continued)*

Homologous chromosomes align randomly at cell equator.
Homologous chromosomes are separated.
Microtubules lengthen in haploid cells.
Nuclear envelope forms.
Nuclear envelope forms.
Sister chromatids are separated.
Two daughter cells are produced.

Roots to Remember

Use your knowledge of the root words presented in this chapter to answer the following questions.

37. According to its roots, what is the literal meaning of *cytokinesis*?
38. How can its word root help you to remember when telophase occurs during mitosis?
39. What process would an oncogene promote if it were switched on?

Crossword Puzzle

Across

40. magnified image of chromosomes
41. cancer cell growth; not normal
42. If a cell is not large enough at a checkpoint, the process of division will ____________.
43. a tumor that invades surrounding tissue
44. structure containing alpha, beta, and delta cells
45. Gametes are ____________ cells.

Down

46. a procedure that removes tissue for examination
47. two chromosomes that carry the same genes
48. during this stage, sister chromatids separate
49. when cancer cells travel from a tumor to other tissue

Word Choice

Circle the word or phrase that correctly completes each sentence.

50. During (metaphase I/metaphase II), homologous chromosomes line up along the cell equator.
51. Linked genes are genes that are located on the same (allele/chromosome).
52. A sperm cell is a (diploid/haploid) cell.
53. At the (G_1/G_2) checkpoint, proteins check for the presence of growth factors.
54. Chromosomes in a cell condense during the (G_2 phase/prophase) of the cell cycle.

Paragraph Completion

Use the following terms to complete the paragraphs. Terms used more than once are listed multiple times.

46
complementary
DNA polymerase
G_2
helix
interphase
mutation
nucleotides
organelles
S
somatic

A human skin cell is a non-gamete, or (55) ____________ cell, that has (56) ____________ chromosomes. Skin cells divide frequently, so they spend less time in (57) ____________ than cells that rarely divide. When a skin cell prepares to divide, it duplicates its (58) ____________ and grows larger during G_1 phase. It replicates its DNA during (59) ____________ phase. DNA replicates with the help of (60) ____________, which moves down the length of the unwound DNA (61) ____________. This enzyme helps to bind free (62) ____________ to each other as they form new DNA strands that are (63) ____________ to their parent strands. A (64) ____________ may occur if DNA replication does not proceed properly. Proteins at the (65) ____________ checkpoint will then detect the error and stop the process of cell division.

Critical Thinking

66. Drugs that reduce or eliminate angiogenesis have been investigated as cancer treatments. How would these drugs work to fight cancer? Why might these drugs have limited usefulness?
67. Nondisjunction is a mistake during cell division that relates to chromosome separation. Nondisjunction may occur during meiosis I, if homologous chromosomes do not separate properly. It may also occur during meiosis II, if sister chromatids do not separate. Nondisjunction results in one or two gametes with an extra chromosome and one or two gametes with one less chromosome than normal. Down syndrome is a result of a nondisjunction that results in an extra chromosome 21 in the cells of the affected individual. What structures in the dividing cell are responsible for nondisjunction? How can a cell protect against nondisjunction?

Practice Test

1. During which of the following phases is the chromosome in a somatic cell unreplicated?
   A. G_2 phase
   B. prophase
   C. metaphase
   D. telophase

2. A male office worker is 6-feet tall and weighs 170 pounds. He goes out with friends once a week and has a couple of beers. He usually smokes when he drinks and has a few cigarettes every day at work. He runs every day and participates in a weekend soccer league. He usually eats a balanced diet with plenty of fruits and vegetables, but he occasionally eats fast food. What is the one lifestyle change that the man could make that would most significantly cut his cancer risk?
   A. stop eating fast food
   B. lose a few pounds
   C. stop smoking
   D. stop drinking

3. Which of the following is *not* a checkpoint in the cell cycle?
   A. G_1 checkpoint
   B. S checkpoint
   C. G_2 checkpoint
   D. metaphase checkpoint

4. Cancer cells are different from normal cells in that they
   A. divide much more rapidly.
   B. grow in orderly rows.
   C. secrete adhesion molecules to glue cells together.
   D. cannot move through the bloodstream.

5. Which statement about the development of cancer is true?
   A. Most cancers are caused by inheritance of mutant genes.
   B. Multiple factors are likely to cause cancer.
   C. Almost all cancers are caused by carcinogens.
   D. Scientists have no idea what causes most cancers.

6. During sperm cell development, sister chromatids separate during
   A. metaphase I
   B. anaphase I
   C. metaphase II
   D. anaphase II

7. A diploid cell in the testes has homologous chromosomes with linked genes for hair color and finger length. One chromosome has an allele for red hair and an allele for long fingers. The other chromosome has an allele for black hair and short fingers? Which of the following sperm cells could *only* be produced if crossing over occurred?
   A. a sperm cell with a chromosome containing an allele for black hair and an allele for short fingers
   B. a sperm cell with a chromosome containing an allele for red hair and an allele for short fingers
   C. a sperm cell with a chromosome containing an allele for blonde hair and an allele for short fingers
   D. a sperm cell with a chromosome containing an allele for red hair and an allele for long fingers

8. When does random alignment of chromosomes occur in a cell going through meiosis?
   A. during anaphase I
   B. during metaphase I
   C. during telophase II
   D. during prophase II

9. A patient with bladder cancer is likely to be treated with
   A. chemotherapy only.
   B. radiation therapy only.
   C. both chemotherapy and radiation therapy.
   D. neither chemotherapy nor radiation therapy.

10. Which statement best describes mitosis?
    A. Mitosis is the process that divides a parent cell into two daughter cells.
    B. Mitosis is the process that replicates cell DNA.
    C. Mitosis is the process that divides a cell nucleus to form two daughter nuclei.
    D. Mitosis is the process that replicates organelles and divides the cell cytoplasm.

11. An X-ray reveals a small growth on a patient's kidney. The growth has not invaded surrounding tissue. Further tests reveal that the growth was a fluid-filled sac. In a few months, the growth goes away. What was the likely identity of the growth?
    A. It was a cyst.
    B. It was a benign tumor.

C. It was a malignant tumor.
D. It was a metastatic cancer.

12. What is angiogenesis?
A. the stimulation of blood vessel growth by cancer cells
B. the process that leads to the death of a cell that has mutations
C. the activation of a gene that produces an enzyme to prevent chromosomal degradation
D. a process that prevents cells from piling on top of each other as they divide

13. How many pairs of nonsex chromosomes does the human karyotype contain?
A. 21
B. 22
C. 23
D. 24

14. Which of the following does *not* happen during prophase?
A. microtubules form and grow
B. chromosomes condense
C. the nuclear envelope breaks down
D. centrioles form

15. Homologous pairs of chromosomes
A. are identical to each other.
B. always carry the same alleles.
C. always carry the same genes.
D. are separated during mitosis.

Answer Key

Matching

1. i; **2.** g; **3.** b; **4.** o; **5.** f; **6.** a; **7.** m; **8.** l; **9.** e; **10.** j; **11.** c; **12.** n; **13.** h; **14.** k; **15.** d

Fill-in-the-Blank

16. benign; malignant; **17.** antioxidants; free radicals; **18.** sister chromatids; centromere; **19.** DNA polymerase; covalent; **20.** G_1; S; **21.** mitosis; cytokinesis; **22.** checkpoint; replicated; **23.** anchorage dependence; **24.** telomerase; immortal; **25.** anaphase I; poles

Labeling

26. DNA is replicated; **27.** DNA condenses into chromosomes and crossing over occurs; **28.** Homologous chromosomes align randomly at cell equator; **29.** Homologous chromosomes are separated; **30.** Nuclear envelope forms; **31.** Two daughter cells are produced; **32.** Microtubules lengthen in haploid cells; **33.** Chromosomes align at cell equator; **34.** Sister chromatids are separated; **35.** Nuclear envelope forms; **36.** Four haploid daughter cells are produced

Roots to Remember

37. cell motion; **38.** The root *telo-* means end or completion, so you know telophase is the last stage of mitosis; **39.** uncontrolled cell growth, cancer development, or tumor formation

Crossword Puzzle

Across

40. karyotype; **41.** abnormal; **42.** stop; **43.** malignant; **44.** sex; **45.** meiosis

Down

46. biopsy; **47.** homologous; **48.** anaphase; **49.** metastasis

Word Choice

50. metaphase I; **51.** chromosome; **52.** haploid; **53.** G_1; **54.** prophase

Paragraph Completion

55. somatic; **56.** 46; **57.** interphase; **58.** organelles; **59.** S; **60.** DNA polymerase; **61.** helix; **62.** nucleotides; **63.** complementary; **64.** mutation; **65.** G_2

Critical Thinking

66. These drugs would reduce the ability of cancer cells to reroute or stimulate the growth of blood vessels around a tumor. This would effectively starve the tumor cells. These drugs may not be effective if the tumor is too large and if it has already damaged surrounding organs and tissues. The drugs may also not prevent metastasis, and they would not be able to destroy metastatic cells that were not part of a tumor.

67. Microtubules that pull apart chromosomes or sister chromatids cause nondisjunctions when they do not attach properly. Proteins that make sure microtubules are properly attached before the cell is allowed to divide can prevent nondisjunction.

Practice Test

1. D; **2.** C; **3.** B; **4.** A; **5.** B; **6.** D; **7.** B; **8.** B; **9.** A; **10.** C; **11.** A; **12.** A; **13.** B; **14.** D; **15.** C

CHAPTER 18

HUMAN REPRODUCTION: FERTILITY AND INFERTILITY

Learning Goals

1. Define *sexual reproduction*.
2. Describe the structure and function of the male and female gonads.
3. Diagram the male reproductive system and describe the function of all associated structures.
4. Diagram the female reproductive system and describe the function of all associated structures.
5. Describe gametogenesis in males and females.
6. Compare and contrast spermatogenesis and oogenesis.
7. Describe events of the ovarian cycle.
8. Describe the hormonal regulation of the menstrual cycle.
9. Describe the four-stage model of human sexual response. Explain how the physiological changes associated with these stages promote reproduction.
10. Compare the biological mechanisms of the various methods of birth control.

Chapter Outline

I. The Human Reproductive Systems
  A. Overview
    1. The process of reproduction allows organisms to produce offspring.
    2. Humans reproduce by sexual reproduction.
      a) Sexual reproduction involves two genetically distinct parents.
      b) Gametes from both individuals combine genetic information at fertilization.
        (1) Male gametes, called sperm, are produced by male gonads called testes.
        (2) Female gametes, called eggs, are produced by female gonads, called ovaries.
      c) Offspring are a genetic mixture of both parents.
      d) Sexual reproduction produces a nearly infinite variety of genetically unique individuals.
    3. Bacteria and other organisms reproduce by asexual reproduction.
      a) One organism subdivides itself during asexual reproduction.
      b) Offspring are genetic clones of the parent.
      c) Asexual reproduction does not produce genetic variety.
  B. The Male Reproductive System
    1. The penis delivers sperm to the female reproductive tract.
      a) The penis is composed of spongy tissue.
      b) When sexually aroused, the tissue of the penis fills with blood.
        (1) Pressure from increased blood volume in the penis seals off veins that drain the penis.
        (2) The penis remains erect and able to be inserted into the vagina to deliver sperm.
    2. The urethra is a tube inside the penis.
      a) The urethra allows sperm or urine to pass out of the body.
      b) A valve prevents these two substances from passing out of the body at the same time.
    3. The glans penis is the head of the penis
      a) The glans penis is highly sensitive.
      b) In uncircumcised males, the glans penis is covered by foreskin.
    4. The scrotum is a pouch below the penis.
      a) The skin of the scrotum is thin, wrinkled, and lacks fatty tissue.
      b) The scrotum contains the testes.
        (1) Cremaster muscle surrounds each testis.

(2) This muscle regulates the position of the testes relative to the body, to maintain a temperature in the testes that produces maximum sperm.
(a) In cold surroundings, the muscle holds the testes closer to the body.
(b) In warm surroundings, the muscle holds the testes away from the body.
(3) A varicose vein in the scrotum, called a varicocele, can disrupt this mechanism.
(a) Blood pools in the scrotum.
(b) Temperature control is disrupted and survival of sperm is prevented.
5. The testes produce sperm and male sex hormones.
a) Each testis is composed of coiled tubes, called seminiferous tubules, where sperm form.
b) Connective tissue surrounds the seminiferous tubules.
(1) This tissue is rich in Leydig cells, which produce androgens.
(2) Androgens are the male sex hormones.
c) Sperm formation takes around 60 days.
(1) Sperm cells pass from the seminiferous tubules into the epididymis.
(2) The cells develop over the 20 days that they spend in the epididymis.
(a) They become able to move.
(b) They become able to fertilize an egg cell.
6. Sperm mixes with secretions from accessory glands during ejaculation.
a) Sperm from the epididymis are propelled through the vas deferens.
b) The seminal vesicles secrete mucus and fructose as an energy source for sperm.
c) The prostrate gland secretes a milky, white fluid into the urethra that contributes to the mobility and viability of sperm.
d) The bulbourethral glands secrete clear mucus that neutralizes acidic urine in the urethra.
e) The combination of sperm and secretions is called semen.
(1) Semen has a pH between 7.1 and 8.0.
(2) This pH is optimal for sperm motility.
(3) This pH helps to neutralize acidic conditions in the female reproductive tract.
f) Semen production may be affected by disease or injury.
(1) Sexually transmitted infections (STIs) can scar the sperm-carrying ducts and impair sperm passage.
(2) Getting mumps after puberty can cause inflammation of the testes and impair sperm production.
(3) Inflammation of the prostate gland can impair sperm passage.
(4) Spinal cord injuries may affect the production of semen.
C. The Female Reproductive System
1. The vulva is part of the female external genitalia.
a) The vulva has two sets of labia.
(1) The outer labia majora are fatty and have a hairy surface.
(2) The inner labia minora do not have fat or hair.
b) The clitoris is located at the front of the vulva.
(1) The labia minora divide around the clitoris.
(2) The clitoris is an important organ for female arousal.
2. Separate openings for the urethra and vagina are found between the folds of the labia.
a) The female urethra is much shorter than the male urethra.
b) Bacteria and viruses have a shorter length to travel, so females are more likely to get urinary tract infections than males are.
3. The ovaries are the female gonads.
a) Ovaries produce gametes (ova) and sex hormones.
b) At birth, ovaries contain about 2 million ova each.
4. The oviducts are tubes that extend from the uterus toward the ovaries.
a) The oviducts are an extension of the top surface of the uterus.
b) The oviducts end in fingerlike projections called fimbria that move over the surface of each ovary.
(1) The oviducts are not directly connected to the ovaries.
(2) The movement of fimbria and suction created by cilia lining the oviduct direct eggs into the oviduct.
5. The uterus is the structure where an embryo implants and grows.
a) The uterus is about the size of a fist in women who are not pregnant.
b) The uterine wall is thick and composed of powerful muscles.

c) The internal surface of the uterus is called the endometrium.
d) The lower third of the uterus is called the cervix.
(1) The cervix is the narrow opening to the uterus.
(2) The cervix dilates during childbirth to allow the baby to be born.
6. The vagina is the passageway into and out of the uterus.
a) The passageway from the vagina, through the uterus, and up the oviducts links the outer body to the abdominal cavity.
b) Because there is a natural passage, females can experience bacterial infections in the abdominal cavity without experiencing injury.
7. Injury or disease can affect fertility.
a) STIs such as gonorrhea or chlamydia can scar the oviducts and prevent sperm from reaching eggs.
b) Scarred or damaged oviducts can produce ectopic pregnancies.
(1) Ectopic pregnancy occurs if a fertilized egg implants in an oviduct.
(2) Ectopic pregnancy does not allow proper fetal development.
(3) Ectopic pregnancy may kill the mother.
c) Uterine fibroids can reduce fertility.
(1) Uterine fibroids are noncancerous growths in the wall of the uterus.
(2) They can reduce the ability of sperm to reach an egg.
(3) They can reduce the ability of an egg to implant in the uterus.

II. Gametogenesis: Development of Sex Cells
A. Overview
1. Gametogenesis is the development of gametes, or sex cells.
2. Gametogenesis involves meiosis.
a) Meiosis reduces the chromosome number by half.
b) In humans, gametes contain 23 chromosomes.
3. Sex cells must go through a maturation process before they are capable of fertilization.
a) Sperm cells lose cytoplasm as they mature to become more agile.
b) Egg cell production maximizes cytoplasm, so an egg will be large enough and contain enough nutrients to divide rapidly if it is fertilized.
4. Men produce gametes from puberty to the end of their lives.
5. Women produce gametes a few days each month from puberty to menopause.
B. Spermatogenesis: Development of Men's Gametes
1. Overview
a) Spermatogenesis begins at puberty.
(1) Parent (stem) cells line the walls of the seminiferous tubules.
(2) Each parent cell duplicates to form two daughter cells.
(a) One daughter cell undergoes spermatogenesis.
(b) The other daughter cell becomes a new parent (stem) cell.
b) Spermatogonia are the stem cells inside the seminiferous tubules that serve as the starting point for cell division.
(1) Spermatogonia continuously divide.
(2) Each spermatogonium produces one primary spermatocyte and one stem cell.
(a) Primary spermatocytes become secondary spermatocytes after meiosis I.
(b) Secondary spermatocytes become spermatids after meiosis II.
(c) Spermatids mature to become spermatozoa or sperm.
c) Sertoli cells in the seminiferous tubules secrete substances that help sperm to develop.
d) Mature sperm are swimming cells with tails.
(1) The head of the sperm contains DNA.
(a) An acrosome is located at the tip of the head.
(b) The acrosome contains digestive enzymes that allow the sperm to penetrate the egg.
(2) A midpiece provides ATP energy.
(3) The tail, or flagellum, propels the sperm forward.
(a) The tail is formed at the end of spermatogenesis.
(b) The tail takes about 10 weeks to form.
2. Hormonal Control of Spermatogenesis
a) Testosterone is secreted by Leydig cells.
(1) Testosterone causes spermatogonia to divide.

(2) Testosterone regulates the growth and development of male reproductive structures.
(3) Testosterone stimulates the development of male secondary sex characteristics.
(a) It stimulates the growth of facial hair.
(b) It deepens the voice.
b) Secretion of testosterone is regulated by other hormones.
(1) When testosterone levels get too low for sperm development, the hypothalamus secretes gonadotropin-releasing hormone (GnRH).
(a) GnRH stimulates the pituitary gland to release follicle-stimulating hormone (FSH) and luteinizing hormone (LH).
(i) FSH stimulates the release of substances from Sertoli cells.
(ii) LH stimulates the release of testosterone from Leydig cells.
(2) When testosterone levels get too high, negative feedback inhibits the release of GnRH from the hypothalamus.
(3) When the concentration of sperm is too high, Sertoli cells release inhibin, which prevents the release of GnRH.
3. Problems with Spermatogenesis
a) Around 90% of all male infertility problems are due to problems with sperm production or formation.
b) Sperm may have defects in motility.
(1) Irregularly shaped sperm cannot swim properly.
(2) Sperm that move too slowly may not reach an egg.
c) Sperm count may be too low.
(1) Sperm count begins to decline around age 35.
(2) Exposure to certain chemicals can decrease sperm count.
(3) Drug use can affect sperm number and quality.
(4) Smoking can lower sperm count.
(5) Use of anabolic steroids can lower sperm count.
C. Oogenesis: Development of Women's Gametes
1. Overview
a) Oogenesis occurs in the ovaries.
b) Oogenesis begins while the female is in her mother's uterus.
(1) At about 8 weeks into development, mitosis "seeds" the ovaries with cells that can later undergo meiosis.
(2) Meiosis begins at 11–12 weeks of development, but stops at prophase I.
(3) Primary oocytes remain in prophase I until puberty.
(4) About 2 million potential egg cells are present at birth.
(5) After birth, cells degenerate until there are about 350,000 left by puberty.
2. The Ovarian Cycle
a) Each ovary contains many follicles.
b) Each follicle contains an oocyte.
c) In the ovarian cycle, a primary follicle develops into a secondary follicle and matures into a Graafian follicle.
(1) The primary follicle secretes estrogen.
(2) The secondary oocyte within the secondary follicle is bathed in estrogen.
(3) The mature Graafian follicle develops.
(a) The Graafian follicle contains a fluid-filled cavity that increases in volume.
(b) When the cavity bursts, it expels the secondary oocyte.
(c) The bursting forth of the secondary oocyte is called ovulation.
d) The remnant Graafian follicle is called the corpus luteum.
(1) The corpus luteum secretes reproductive hormones.
(2) It disintegrates after about 12 days if fertilization does not occur.
e) An ovulated egg moves into the oviduct.
f) If sperm is present within 12 hours, fertilization is likely.
g) Each ovulated egg is produced by off-center meiosis.
(1) Meiosis produces one large cell and three polar bodies.
(a) The large cell gives rise to the female gamete.
(b) Polar bodies are not involved in fertilization.
(2) The large cell carries enough nutrients to nourish the zygote and embryo through early development.

3. Hormonal Control of the Ovarian Cycle
   a) The ovarian cycle is regulated by feedback loops involving the hypothalamus.
      (1) Positive Feedback
         (a) Prior to ovulation, estrogen levels stimulate the hypothalamus to secrete GnRH.
         (b) GnRH stimulates the pituitary gland to secrete FSH and LH.
         (c) FSH and LH stimulate the ovary to produce more estrogen.
      (2) Negative Feedback
         (a) High levels of progesterone (or estrogen during most of the ovarian cycle) cause the inhibition of GnRH secretion by the hypothalamus.
         (b) As GnRH levels decline, FSH and LH levels decline.
   b) The first half of the ovarian cycle is called the follicular phase.
      (1) FSH promotes the development of the follicle.
      (2) The maturing follicle secretes estrogen.
      (3) As estrogen levels increase, FSH production declines.
   c) The second half of the ovarian cycle is called the luteal phase.
      (1) LH promotes the development of the corpus luteum.
      (2) The corpus luteum secretes estrogen and progesterone.
      (3) If pregnancy doesn't occur, an increase in progesterone inhibits the pituitary.
         (a) LH is not secreted.
         (b) The corpus luteum degenerates.
         (c) Menstruation occurs.
4. Problems with Oogenesis
   a) A disruption in the LH surge prevents ovulation.
      (1) A number of factors can prevent the LH surge:
         (a) damage to the hypothalamus,
         (b) damage to the pituitary gland,
         (c) excessive exercise, or
         (d) anorexia.
      (2) If the LH surge does not occur, the follicle keeps growing but does not rupture.
         (a) The follicle may turn into a cyst.
         (b) Follicular cysts generally disappear without intervention.
   b) Polycystic ovarian syndrome can disrupt ovulation.
      (1) This syndrome is caused by abnormal levels of FSH, LH, and androgens.
      (2) This syndrome can cause the following symptoms:
         (a) irregular menstrual cycles,
         (b) excess hair growth, and
         (c) obesity.

III. The Menstrual Cycle
   A. The menstrual cycle refers to periodic changes that occur in the uterus.
   B. Each menstrual cycle is coordinated by hormonal relationships between the brain, ovaries, and uterine lining.
   C. Each menstrual cycle prepares an egg for fertilization and the uterus for pregnancy.
   D. The menstrual cycle lasts for approximately 28 days.
      1. The cycle begins with the first day of bleeding.
      2. Cycle length can vary among women.
   E. The menstrual cycle is maintained by feedback mechanisms.
      1. When a follicle is large enough, it produces estrogen at levels that stimulate the release of GnRH.
         a) FSH and LH levels subsequently spike for around 24 hours.
         b) Mitosis of endometrial cells and regrowth of blood vessels are also stimulated by estrogen.
      2. Ovulation occurs 10–12 hours after the peak of LH.
      3. After ovulation, progesterone secreted by the corpus luteum prepares the endometrium for pregnancy.
         a) Blood flow to the uterine lining is maintained.
         b) LH levels decline.
      4. If fertilization does not occur, the corpus luteum degenerates 12–14 days later.
         a) Progesterone and estrogen levels fall.

b) Arteries that supply the uterine lining spasm.
  (1) Menstrual cramps occur.
  (2) The lining is shed.
  (3) Endometrial tissues die.
  (4) Blood flows out from weakened capillary walls.
c) Decreasing progesterone levels release the hypothalamus from inhibitory control.
  (1) LH and FSH levels rise.
  (2) The cycle starts again.

5. If fertilization has occurred, the early embryo secretes human chorionic gonadotropin (HCG) to extend the life of the corpus luteum.
  a) Progesterone and estrogen levels remain high.
  b) The endometrium is maintained.
  c) The corpus luteum disintegrates after 6–7 weeks of pregnancy, when the placenta produces enough progesterone to maintain the uterine lining.
  d) HCG is the hormone detected by home pregnancy tests.

F. Endometriosis is a condition in which uterine tissues migrate through the oviducts and implant on other organs.
  1. These displaced uterine tissues still respond to hormones of the menstrual cycle.
  2. Blood from these tissues cannot exit the body.
    a) Trapped blood can lead to the growth of cysts, scar tissue, and adhesions.
    b) Growths can cause abdominal pain.
  3. Endometriosis can lead to scarring and inflammation of the oviducts.

IV. The Human Sexual Response
  A. Masters and Johnson introduced a four-stage model for the human sexual response.
    1. Stage one begins with excitement, which is triggered by erotic stimuli.
      a) In both males and females,
        (1) heart and respiratory rate rise, and
        (2) blood pressure rises.
      b) In males,
        (1) the penis becomes erect, and
        (2) muscles in the scrotum tense and draw the testes close to the body.
      c) In females,
        (1) nipples become erect,
        (2) breasts and external genitalia swell,
        (3) lubricating fluid is produced by cells in the vaginal walls, and
        (4) the uterus becomes elevated.
    2. Stage two is a plateau phase of increasing sexual tension.
      a) In both males and females,
        (1) heart and respiratory rate rise, and
        (2) blood pressure rises.
      b) In males,
        (1) the bladder sphincter closes tightly, and
        (2) the penis may secrete seminal fluid.
      c) In females
        (1) breasts and external genitalia swell, and
        (2) the opening of the vagina tightens and its lower area swells.
    3. Stage three is orgasm, or the release of sexual tension.
      a) The muscles of the anus, lower pelvis, vagina, and uterus contract rhythmically.
      b) The diaphragm and vocal cords may contract.
      c) Males ejaculate semen.
    4. Stage four is resolution.
      a) Heart rate, blood pressure, and body structures return to baseline conditions.
      b) This stage may be bypassed by further stimulation.
  B. Male infertility can interfere with the sexual response.
    1. Erectile dysfunction occurs when erectile tissue does not expand enough to compress veins and maintain the erection.
    2. Premature ejaculation is ejaculation that occurs outside of the vagina.
    3. In retrograde ejaculation, semen enters the bladder during orgasm.

V. Controlling Fertility
  A. Principles of Fertility Control
    1. There are many modes of action for birth control.
      a) Some methods block sperm transport.
      b) Some methods inhibit ovulation.
      c) Some methods remove the fertilized egg or embryo.
    2. Different birth control methods have different levels of effectiveness.
  B. Barrier Methods
    1. Withdrawal is erroneously thought of as a barrier method.
      a) In withdrawal, the man removes his penis from the woman's vagina before ejaculation.
      b) This method is ineffective, because sperm can be released before orgasm.
    2. Spermicides inactivate sperm by damaging their cell membranes.
      a) Spermicides come in a cream, jelly, or foam that can be inserted into the vagina.
      b) The active ingredient in spermicides, nonoxynol-9, kills the bacteria that cause chlamydia and gonorrhea.
      c) Spermicide may cause vaginal sores in women that may increase the transmission of STIs.
    3. The contraceptive sponge covers the cervix and blocks the passage of sperm.
      a) The sponge is a polyurethane foam disk infused with spermicide.
      b) The sponge can be very effective when used with other barrier methods.
    4. Latex condoms for males cover the penis and trap sperm.
      a) Condoms effectively reduce the spread of STIs.
      b) The condom has increased in popularity in recent years.
    5. Female latex condoms line the vagina and block the passage of sperm.
      a) The female condom was introduced to help women protect themselves from STIs when their male partner would not wear a condom.
      b) The female condom is more expensive than the male condom and requires practice to use.
    6. Diaphragms and cervical caps cover the cervix to block the passage of sperm.
      a) Both are effective birth control when filled with spermicide.
      b) Neither provides protection from STIs.
  C. Hormonal Birth Control
    1. Combined hormone contraceptives inhibit ovulation and fertilization.
    2. Combined hormone contraceptives contain synthetic estrogen and synthetic progesterone.
      a) Constant levels of estrogen prevent ovarian follicles from developing and inhibit ovulation.
      b) Synthetic progesterone is a backup to estrogen.
        (1) It prevents cervical mucus from thinning to facilitate sperm passage.
        (2) It makes the endometrium unfavorable for embryo implantation.
    3. Hormonal methods include the pill, the patch, and the ring.
      a) The pill is an oral contraceptive taken daily.
        (1) The pill is a highly effective means of birth control.
        (2) Missed pills may result in ovulation.
      b) The patch is applied to the skin weekly.
      c) The ring is inserted in the vagina and removed after three weeks.
    4. Hormonal methods of birth control may have side effects.
      a) Nausea, breast discomfort, weight gain, acne, and headaches may occur.
      b) Side effects may diminish after two to three months.
      c) Serious side effects may include increased risk of blood clots, cervical cancer, liver cancer, and delayed return to fertility.
    5. Hormonal methods may protect women from ovarian and endometrial cancer.
    6. Hormonal methods are generally taken for 21 days, followed by 7 days of a placebo.
      a) The hormone-free week triggers menstruation.
      b) New contraceptive pills are being introduced that reduce or eliminate placebo pills.
        (1) Reduction or elimination of placebo pills reduces or eliminates menstrual periods.
        (2) There is little evidence that eliminating menstrual periods is dangerous.
    7. Progesterone-only pills can be used by breast-feeding women.
      a) Progesterone does not interfere with milk production.

b) A progesterone mimic is used as an injectable form of birth control (Depo-Provera is one example).

8. High doses of birth control pills can be used as emergency contraception.
   a) A progesterone-only pill must be taken within 72 hours of unprotected sex.
   b) Emergency contraception prevents ovulation or fertilization, depending on where the woman is in her cycle.

D. Other Methods of Birth Control

1. Intrauterine Devices (IUDs)
   a) IUDs are implanted in the uterus.
   b) IUDs cause nonspecific inflammation of the endometrium, which makes the uterus less hospitable for egg implantation.
   c) IUDs may also have backup methods of contraception.
      (1) Some release copper, which is toxic to cells.
      (2) Some release progesterone.
   d) IUDs remain effective for 5–10 years.
   e) IUDs must be inserted by a clinician.
   f) Possible complications from IUDs include infection and perforation of the uterine wall.
   g) Pregnancies that occur when an IUD is present can be dangerous to both mother and fetus.
2. Fertility Awareness
   a) The calendar method or rhythm method requires a woman to become familiar with the timing of her cycle by observing signs of fertility.
      (1) If periods are regular, this method can predict ovulation.
      (2) Most women do not have such regular cycles, so this method is risky.
   b) The symptothermal method can help to identify when ovulation is occurring.
      (1) Cervical mucus changes throughout the menstrual cycle.
         (a) Cervical mucus increases and becomes thin and slick prior to ovulation.
            (i) This is a sign that estrogen levels are high.
            (ii) This type of mucus allows sperm to avoid becoming stuck in the folds of the cervix.
            (iii) A woman is fertile when mucus is in this state.
         (b) Cervical mucus becomes whitish and gummy during the second half of the menstrual cycle.
            (i) This is a sign that progesterone levels are high.
            (ii) This type of mucus occurs after ovulation.
            (iii) This type of mucus can prevent sperm from reaching an egg.
      (2) Body temperature changes throughout the menstrual cycle.
         (a) Temperature drops steadily until the day after ovulation.
         (b) On the day after ovulation, temperature rises at least 0.5°C (0.9°F).
3. Sterilization
   a) Surgical sterilization is extremely effective but not easily reversible.
   b) Surgical methods eliminate the possibility of fertilization.
      (1) Men can get a vasectomy.
         (a) In a vasectomy, each vas deferens is severed and tied off.
         (b) This surgery eliminates the passageway in which sperm travels from the testicles to the urethra.
         (c) This surgery does not interfere with the production of semen or the ability to achieve erection or orgasm.
         (d) Vasectomy can be done in an outpatient clinic using local anesthetic.
         (e) 50% of vasectomy surgeries can be successfully reversed.
      (2) Women can get a tubal ligation.
         (a) Most tubal ligations involve cauterization, which melts the tissues of the oviducts.
         (b) A springlike tube can also be inserted into the oviducts to permanently block the movement of eggs.
         (c) Tubal ligation involves abdominal surgery.
         (d) 30% of tubal ligations can be successfully reversed.
4. Abortion
   a) Abortion is the medical termination of a pregnancy.

b) Abortion is legal in the U.S., but it is a divisive issue.
c) Pregnancies in the first seven weeks can be terminated by ingestion of mifepristone, or RU-486.
   (1) Mifepristone blocks the action of progesterone.
   (2) This leads to shedding of the uterine lining and loss of the embryo.
d) Pregnancies between 5 and 12 weeks can be terminated with vacuum aspiration.
   (1) Suction draws the fetus or embryo out of the uterus.
   (2) This is the most frequently used abortion technique.
e) Pregnancies after 12 weeks can only be terminated with techniques that remove both the fetus and placenta.
   (1) These techniques involve dilating the cervix and using forceps, a curette, and vacuum suction to destroy and extract the fetus.
   (2) In partial birth abortion, the body of the fetus is pulled through the cervix and its head is collapsed using pressure from surgical scissors.
      (a) This procedure was banned in the U.S. in 2003.
      (b) The U.S. Supreme Court upheld the ban in 2007.
f) Infection is the most serious complication of surgical abortion.

E. The Future of Birth Control Technology
   1. Some researchers are working on immunocontraception.
      a) Women are vaccinated against biological markers of pregnancy.
      b) Their immune system destroys a new embryo.
   2. Some proposed hormonal birth controls for men interfere with sperm production.

VI. Health, Lifestyle, and Fertility
   A. Taking good care of your health can protect your fertility.
      1. Eat a healthy diet.
      2. Exercise.
      3. Avoid prolonged or intense emotional stress.
      4. Avoid tobacco and other drugs.
      5. Do not abuse alcohol.
      6. Men should avoid exposing the testes to too much heat.
   B. Fertility begins to decline in the mid-30s.

Practice Questions

Matching

1. scrotum
2. endometriosis
3. spermicide
4. follicle
5. oviduct
6. cervix
7. gametogenesis
8. vasectomy
9. urethra
10. oogenesis
11. fimbria
12. glans penis
13. Sertoli cell
14. Leydig cell
15. diaphragm

a. when uterine tissues migrate out of the uterus and implant on other organs
b. fingerlike projections of the oviduct
c. pouch that contains the testes
d. production of sex cells
e. tubes that extend from the body of the uterus to the ovaries
f. a latex dome that blocks sperm from entering the cervix
g. production of eggs
h. substance that damages the cell membrane of sperm
i. cells that secrete substances to aid in sperm development
j. severing and tying off of the vas deferens
k. the head of the penis
l. structure in the ovary that contains an immature egg
m. androgen-producing cell
n. the lower portion of the uterus that dilates during childbirth
o. the tube that allows sperm or urine to leave the male body

Fill-in-the-Blank

16. Male gametes, or ____________, are produced in the ____________ tubules of the ____________.
17. Sperm would lack an energy source if the ____________ in a man's reproductive system were damaged.
18. The top of the uterus splits into ____________, which have ____________ at their upper ends.
19. If an embryo implants in the ____________, it causes a dangerous ____________ pregnancy.
20. When testosterone is low, ____________ cells are stimulated to produce testosterone; when testosterone is high, ____________ cells release inhibin to inhibit hormones that stimulate testosterone production.
21. Mitosis of each ____________ produces one stem cell and one primary ____________.
22. ____________ occurs when a ____________ follicle bursts and releases a secondary ____________.
23. After implantation, the ____________ produces HCG, a hormone that keeps the ____________ from degenerating.
24. During ____________, sexual tension is released as the male ____________.
25. ____________ is a ____________ method of birth control that damages the cell membranes of sperm.

Labeling

Use the statements below to label Figure 18.1.

bulbourethral gland
epididymis
glans penis
prostate gland
seminal vesicle
seminiferous tubules
testis
vas deferens

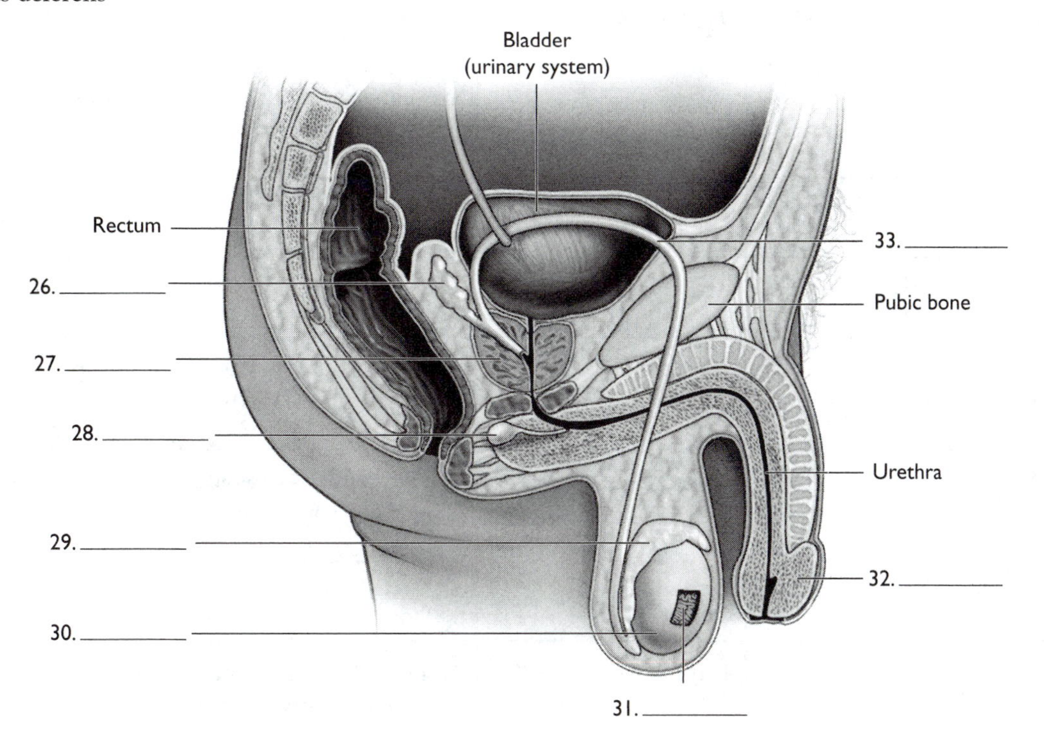

Figure 18.1

Use the statements below to label Figure 18.2.

clitoris
vaginal opening
urethra
vulva

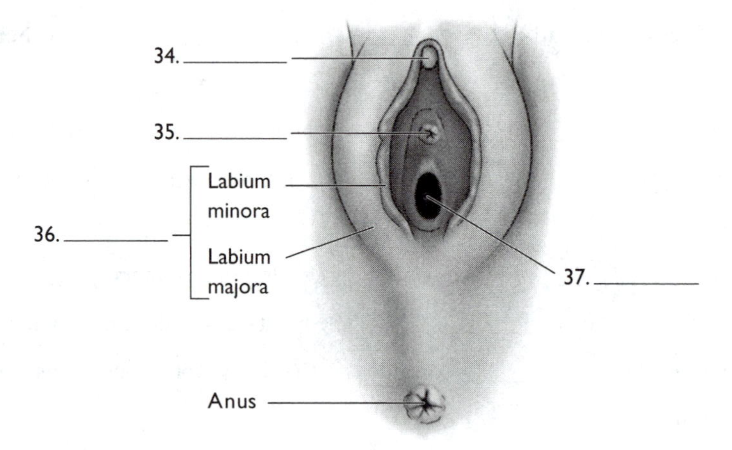

Figure 18.2

Roots to Remember

Use your knowledge of the root words presented in this chapter to answer the following questions.

38. What is the literal meaning of *oogenesis*?
39. The Greek word *tropos* refers to turning. The root *gonado-* refers to seeds or generation. Why are these roots used for the gonadotropin-releasing hormone?
40. Considering its word root, what could you call the cervix?

Table Completion

Birth Control Method	Type of Method	Mode of Action
The patch	41.	Prevents ovulation; makes conditions unfavorable for implantation
IUD	IUD	42.
Mifepristone	Abortion	43.
44.	Sterilization	Severs the vas deferens to prevent sperm from traveling to the urethra
45.	46.	Covers the cervix to block sperm; contains a spermicide to kill sperm
Female condom	47.	48.
49.	Fertility awareness	Monitors body temperature and the condition of cervical mucus to predict ovulation

Word Choice

Circle the word or phrase that correctly completes each sentence.

50. During the (ovarian/menstrual) cycle, changes occur in the endometrium.
51. (Abstinence/Sterilization) in females prevents pregnancy by preventing the transport of the egg through the oviduct.
52. During the (excitement/plateau) phase of the human sexual response, the bladder sphincter in men closes tightly.
53. Spermatocytes become spermatids after (meiosis I/meiosis II).
54. An important organ for female sexual arousal is the (vulva/clitoris).

Paragraph Completion

Use the following terms to complete the paragraphs. Terms used more than once are listed multiple times.

estrogen
falling
follicle
FSH
GnRH
hypothalamus
LH
menstrual
ovulation
progesterone
rising

The (55)____________ cycle begins as the endometrium and blood are shed. This part of the cycle corresponds with (56) ____________ levels of estrogen and progesterone. The (57) ____________ is released from inhibitory control, which allows (58) ____________ and LH levels to start (59) ____________. A new (60) ____________ develops and releases (61) ____________. This stimulates the release of (62) ____________, which causes levels of FSH and LH to spike. (63) ____________ occurs about 10–12 hours after the hormone (64) ____________ spikes. The empty follicle becomes the corpus luteum and secretes (65) ____________ to prepare the body for pregnancy. If fertilization does not occur, the corpus luteum degenerates. Menstruation then occurs.

Critical Thinking

66. Women who begin menstruating early or those who go through menopause at a late age have a higher risk for breast cancer. What might be the cause of this higher risk?
67. A deeply religious person wants to choose a method of contraception that does not result in fertilization or the loss of an embryo. What methods could this person choose?
68. As men grow older, they often experience prostate problems. What are some of the potential consequences of having an enlarged prostate, or a prostate that loses function?

Practice Test

1. How many spermatozoa are produced from one spermatocyte during meiosis?
   A. 1
   B. 2
   C. 3
   D. 4

2. During which stage of the human sexual response do the cells in a woman's vaginal wall first produce lubricating fluid?
   A. excitement
   B. plateau
   C. orgasm
   D. resolution

3. How many chromosomes are in a polar body?
   A. none
   B. 23
   C. 46
   D. 69

4. A runner is training for a long-distance race. She trains for five hours each day and cuts down on calories so that she will be lighter for the race. During her training, she notices that she has not gotten her period in two months. What is happening?
   A. The runner's ovarian follicles are not developing properly because she is not getting enough nutrition.
   B. The excess exercise is causing estrogen and other hormones to break down too quickly in the runner's body.
   C. The runner's body is producing excess testosterone, which is blocking the effects of estrogen in her body.
   D. The excess exercise and weight loss are disrupting homeostasis and preventing the release of GnRH from the hypothalamus.

5. What hormone stimulates the development of a thick layer of endometrial cells and the regrowth of blood vessels in the uterus?
   A. estrogen
   B. FSH
   C. LH
   D. progesterone

6. What happens when the concentration of sperm is too high in the testes and Sertoli cells release inhibin?
   A. The hypothalamus releases GnRH, which triggers the release of FSH, which inhibits testosterone production.
   B. Leydig cells stop producing testosterone, which signals the hypothalamus to stop releasing GnRH.
   C. The hypothalamus stops releasing GnRH, which inhibits the release of LH so that testosterone levels go down.
   D. The pituitary gland begins to release FSH, which inhibits the release of LH and lowers testosterone levels.

7. The fimbria are projections of the
   A. ovaries.
   B. oviducts.
   C. cervix.
   D. vagina.

8. Which glands produce secretions that help sperm to survive in the acidic female reproductive tract?
   A. seminal vesicles
   B. epididymis
   C. bulbourethral glands
   D. prostate gland

9. A fertility doctor recommends that a male patient with fertility problems wear boxer shorts instead of briefs. Why?
   A. Briefs can press on the testes, which destroys sperm cells.
   B. Boxer shorts allow the scrotum to move freely and maintain a temperature that does not damage sperm.

C. Briefs apply pressure to the scrotum, causing varicose veins that increase the temperature of the scrotum, which destroys sperm.
D. Boxers may help to improve the patient's sexual response, which can help to combat his fertility problems.

10. Which method of birth control helps to protect against STIs?
A. IUD
B. vasectomy
C. condom
D. the pill

11. A young woman experiences pelvic pain and visits her gynecologist. The gynecologist finds that the young woman has a follicular cyst. Why might this have occurred?
A. The woman's LH levels did not get high enough during her last menstrual cycle.
B. The woman's estrogen levels were too high during her last menstrual cycle.
C. The woman's progesterone levels went too low during her last menstrual cycle.
D. The woman has polycystic ovarian syndrome.

12. Which of the following produces the corpus luteum?
A. menstruation
B. pregnancy
C. ovulation
D. fertilization

13. Which method of birth control listed below is most effective with typical use?
A. vaginal ring
B. male condom
C. diaphragm
D. IUD

14. On what day of the menstrual cycle does progesterone reach its lowest point?
A. around day 1
B. around day 14
C. around day 21
D. around day 28

15. A 40-year-old woman ovulates during her monthly cycle. She had her first period at age 12. How long has the egg cell been in her ovary?
A. for 20 days
B. for 60 days
C. for 28 years
D. for over 40 years

Answer Key

Matching

1. c; **2.** a; **3.** h; **4.** l; **5.** e; **6.** n; **7.** d; **8.** j; **9.** o; **10.** g; **11.** b; **12.** k; **13.** i; **14.** m; **15.** f

Fill-in-the-Blank

16. sperm; seminiferous; testes; **17.** seminal vesicles; **18.** oviducts; fimbria; **19.** oviduct; ectopic; **20.** Leydig; Sertoli; **21.** spermatogonium; spermatocyte; **22.** Ovulation; Graafian; oocyte; **23.** embryo; corpus luteum; **24.** orgasm; ejaculates; **25.** Spermicide; barrier

Labeling

26. seminal vesicle; **27.** prostate gland; **28.** bulbourethral gland; **29.** epididymis; **30.** testis; **31.** seminiferous tubules; **32.** glans penis; **33.** vas deferens; **34.** clitoris; **35.** urethra; **36.** vulva; **37.** vaginal opening

Roots to Remember

38. the generation or birth of eggs; **39.** This hormone starts the process that directs or turns hormones toward the structures that produce human "seeds" (egg and sperm); **40.** the neck of the uterus

Table Completion

41. Hormonal; **42.** Causes nonspecific inflammation of the uterus; releases ions or hormones to inhibit fertilization, ovulation, or implantation; **43.** Blocks the action of progesterone, so the uterine lining and embryo are shed; **44.** Vasectomy; **45.** Contraceptive sponge; **46.** Barrier; **47.** Barrier; **48.** Lines the vagina to prevent sperm from entering the female reproductive system; **49.** Symptothermal method

Word Choice

50. menstrual; **51.** Sterilization; **52.** plateau; **53.** meiosis II; **54.** clitoris

Paragraph Completion

55. menstrual; **56.** falling; **57.** hypothalamus; **58.** FSH; **59.** rising; **60.** follicle; **61.** estrogen; **62.** GnRH; **63.** Ovulation; **64.** LH; **65.** progesterone

Critical Thinking

66. A longer lifetime exposure to sex hormones, such as estrogen or progesterone, could put them at risk for breast cancer.

67. Oral contraceptives that inhibit ovulation, any of the barrier methods, abstinence, or fertility awareness methods

68. Sperm mobility and viability could go down if the prostate could no longer secrete fluid. An enlarged prostate could prevent sperm from traveling to the urethra.

Practice Test

1. D; **2.** A; **3.** B; **4.** D; **5.** A; **6.** C; **7.** B; **8.** C; **9.** B; **10.** C; **11.** A; **12.** C; **13.** D; **14.** A; **15.** D

CHAPTER 19

HEREDITY: GENES AND INTELLIGENCE

Learning Goals

1. Describe the relationship between genes and chromosomes.
2. Explain how alleles form and describe the consequences of mutation.
3. Define *independent assortment* and explain how it contributes to diversity of gametes.
4. Describe the relationship between the genotype and phenotype.
5. Contrast dominant and recessive mutations.
6. Create and use a Punnett square for a single-gene genetic analysis.
7. Define *quantitative trait* and describe the factors that generate variability in these traits.
8. Define heritability and describe how it is calculated in human populations.
9. Describe the difficulties in using the concept of heritability to explain differences between human groups.

Chapter Outline

I. The Inheritance of Traits
  A. Genes and Chromosomes
    1. Genes are segments of DNA that code for proteins.
    2. All cells in the body have the same genes.
      a) Context matters in the expression of genes or expression of proteins.
        (1) Eye cells and heart cells contain genes that code for rhodopsin.
        (2) Rhodopsin helps to detect light, so it is only produced in eye cells.
        (3) A protein that assists rhodopsin in eyes helps to coordinate muscle contraction in the heart.
      b) The timing of expression and combination of activated genes determine cell activity.
      c) A protein may serve two different functions, depending on its context.
  B. Producing Diversity of Offspring
    1. Gene mutation creates genetic diversity.
      a) There is a chance of mutation every time a cell divides.
      b) Mutations lead to different versions of a gene, called alleles.
      c) Mutations are random.
        (1) Different families are likely to have slightly different alleles for various genes.
        (2) Repeated mutation creates genetic variation in a population.
    2. Segregation and independent assortment create gamete diversity.
      a) Meiosis separates homologous pairs of chromosomes.
      b) Independent assortment arises from the random alignment of homologous chromosomes during metaphase I.
        (1) Homologous chromosomes line up randomly at the cell equator.
        (2) Genes that are on different chromosomes are inherited independently of each other.
      c) When homologous pairs separate, the alleles they carry are separated.
        (1) The separation of pairs of alleles is called segregation.
        (2) A parent with two different alleles of a gene will produce gametes with a 50% probability of containing one of two versions of the gene.
      d) The independent assortment of segregated chromosomes into daughter cells is repeated every time a gamete is produced.
        (1) This results in a different set of alleles for each gamete.
        (2) These alleles are a mix of the alleles inherited by the parent from his or her parents.
    3. Random fertilization results in a large variety of potential offspring.
      a) As a result of independent assortment, each individual can make at least 8 million different types of egg or sperm.

b) In theory, any of these gametes has an equal chance of combining with any of the 8 million different gametes that could be produced from the other parent.

c) The process that allows gametes to combine randomly is called random fertilization.

(1) According to random fertilization, the odds of getting a particular combination of chromosomes are 1 in 8 million times 1 in 8 million, which equals 1 in 64 trillion.

(2) Therefore, each individual is genetically unique, unless produced by the same egg and sperm (such as identical twins).

II. Mendelian Genetics: When the Role of a Gene Is Direct

A. Overview

1. Some genetic traits have a clear pattern of inheritance.
2. These traits are called Mendelian because Austrian monk Gregor Mendel first described their pattern of inheritance correctly.
3. Mendel studied patterns of inheritance using pea plants.

B. Genotype and Phenotype

1. A genotype is the genetic composition of an individual.
2. A phenotype is the pattern of physical traits based on the genotype.
3. Types of alleles determine the genotype.

a) An individual who carries two copies of the same allele for a gene has a homozygous genotype.

b) An individual who carries two different alleles for a gene has a heterozygous genotype.

4. Alleles determine the phenotype.

a) Some alleles are recessive, so their effect can only be seen if the individual has two of these alleles.

(1) Mutations involving the inability to produce functional proteins are often recessive.

(2) The mutation does not influence the phenotype if only one recessive allele is present.

b) Some alleles are dominant, so their effect is seen if the individual has one or two of these alleles.

c) Alleles may also display codominance or incomplete dominance.

C. Genetic Diseases of Human Beings

1. Cystic fibrosis is a recessive condition.

a) Cystic fibrosis is a common genetic disease in European populations.

b) Cystic fibrosis occurs in individuals with two copies of an allele that codes for a nonfunctional protein.

(1) When the protein is functional, it helps transport chloride ions into and out of cells lining the lungs, intestines, and other organs.

(2) Without this functioning protein, the balance between sodium and chloride is disrupted.

(a) Affected cells produce thick, sticky mucus.

(b) This causes deterioration of the lungs and difficulty in nutrient absorption.

c) Carriers of one cystic fibrosis allele and one normal allele do not have symptoms, but they may pass on the disease allele to offspring.

2. Huntington's disease is caused by a dominant allele.

a) Huntington's disease is a fatal condition that is progressive and incurable.

b) The disease occurs when the Huntington's allele produces a protein that clumps up inside the nuclei of cells.

(1) Nerve cells in the brain are most affected.

(2) As these cells die off, the individual loses mental capacity and control of muscle function.

c) Both homozygous dominant and heterozygous individuals are affected by the disease.

d) The disease is passed on because its symptoms generally do not appear until middle age, after the prime reproductive years.

D. Using Punnett Squares to Predict Genotypes of Offspring

1. A Punnett square allows us to determine the probable inheritance of a small number of genes.

a) The Punnett square is a table that shows the alleles in one sperm and one egg.

b) The Punnett square predicts the allelic frequencies from a cross between the egg and sperm.

2. In a Punnett square, a capital letter represents a dominant allele and a lowercase letter represents a recessive allele.

a) In the case of cystic fibrosis, *F* could represent the dominant gene that produces normal proteins.

b) The lowercase *f* could represent the recessive gene that produces abnormal proteins.

c) A genetic cross between two heterozygotes would be symbolized by $Ff \times Ff$.

3. A simple Punnett square contains four boxes—two columns and two rows.
   a) The symbol for one allele carried by the first parent is placed above the first column of the square.
   b) The symbol for the other allele carried by the first parent is placed above the second column of the square.
   c) The symbol for one allele carried by the second parent is placed to the left of the first row of the square.
   d) The symbol for the other allele carried by the second parent is placed to the left of the second row of the square.
4. Alleles are combined within each box of the square.
   a) Each box contains one allele from one parent and one allele from the other parent.
   b) In a cross between *Ff* × *Ff*, there are three possible genotypes.
      (1) *Ff* is the heterozygous genotype.
      (2) *FF* is the homozygous dominant genotype.
      (3) *ff* is the homozygous recessive genotype.
5. A Punnett square can help predict the probability of producing offspring with a particular genotype.
   a) In the cross between *Ff* × *Ff*,
      (1) there is a 25% chance of the *FF* genotype;
      (2) there is a 50% chance of the *Ff* genotype; and
      (3) there is a 25% chance of the *ff* genotype.
   b) In a cross between two people who are carriers for cystic fibrosis, there is a 25% chance that a child will have cystic fibrosis.
   c) Each probability is generated independently for each child.
6. A Punnett square can be used to calculate the likelihood of gene combinations.
   a) As the number of genes being considered increases, the number of boxes in the square increase.
      (1) With two genes, each with two alleles, the number of boxes in the square increases to 16.
         (a) The number of unique gametes a heterozygote can produce is four.
         (b) The number of unique genotypes that can be produced is nine.
      (2) With three genes, each with two alleles, the number of boxes in the square increases to 64.

III. Quantitative Genetics: When Genes and Environment Interact
  A. Overview
    1. Qualitative traits have either one phenotype or the other.
    2. Quantitative traits show continuous variation.
       a) Height, weight, eye color, musical ability, and intelligence are quantitative traits.
       b) Wide variation in quantitative traits contributes to human diversity.
  B. Why Traits Are Quantitative
    1. Multiple genotypes can occur in a population when a trait is influenced by more than one gene.
    2. Traits influenced by many genes are called polygenic traits.
       a) Eye color is influenced by three genes.
       b) These genes are responsible for producing and distributing the pigment melanin in the iris.
          (1) People with dark brown eyes have a lot of melanin.
          (2) People with blue eyes have very little melanin.
          (3) There are many intermediate colors.
    3. Continuous variation in a quantitative trait may be influenced by environmental factors.
       a) For example, skin tone is affected by genes that produce and distribute melanin, as well as amount of sun exposure.
       b) Even individuals with the same genotype can have different phenotypes if they were exposed to different environments.
          (1) These effects can be seen in identical twins that were raised separately.
          (2) If one twin is exposed to damaging environmental factors, such as tobacco smoke and UV rays, that twin's phenotype is likely to suffer more damage than his or her twin's.
  C. The Heritability of Quantitative Traits
    1. Heritability is the genetic component of variation in a given trait.
    2. A strong correlation between a trait and related individuals indicates that a trait has a strong genetic component.
       a) IQ (intelligence quotient) scores of parents and children have an average correlation of 0.42.
          (1) In other words, 42% of the variation between individuals' IQ scores can be explained by the variation among their parents' scores.

(2) This does not mean that 42% of IQ is genetic.
(a) Families share genes and the environment.
(b) Correlations of IQ scores do not identify how much of IQ is based on heritability.
3. Natural experiments are used to separate genetic and environmental influences.
a) Natural experiments provide circumstances that allow scientists to test a hypothesis without intervention.
b) Human twins are a source of natural experiments.
c) Human twin studies indicate that about 50% of the IQ score variations among individuals result from genetic differences.

IV. Genes, the Environment, and the Individual
A. The Use and Misuse of Heritability Calculations
1. Differences between groups may be environmental.
a) Body weight in mice has a strong genetic component.
b) However, diet can strongly affect body weight.
c) In the same way, human IQ can be affected by the enrichment of the environment.
2. A highly heritable trait can still respond to environmental change.
a) Maze running is a highly heritable trait in rats.
b) In a boring environment, maze-bright and maze-dull rats perform the same.
c) In a stimulating environment, all rats did better at maze running, and maze-dull rats improved dramatically.
3. Heritability does not tell us why two individuals differ.
a) Genes may explain 90% of the population variability in a particular environment.
b) However, the variation between one individual and another may be completely a function of the environment.
B. How Do Genes Matter?
1. Genes can have a strong influence on many traits, including risks for disease.
2. Random factors affect the inheritance of alleles that influence these traits.
3. We cannot program the traits of children by selecting the traits of parents.
4. Even if certain traits are more likely, they can be influenced by a physical and social environment that influences gene expression.

Practice Questions

Matching

1. natural experiment
2. recessive
3. genotype
4. continuous variation
5. dominant
6. phenotype
7. independent assortment
8. heterozygous
9. genetic variation
10. polygenic trait
11. homozygous
12. qualitative trait
13. heritability
14. random fertilization
15. quantitative trait

a. the process that randomly combines gametes
b. when random alignment distributes chromosomes to different gametes
c. an individual that carries two copies of the same allele
d. an allele whose phenotypic effects are masked in the presence of a different allele
e. circumstance that allows a hypothesis to be tested without researcher intervention
f. causes a continuous variation of phenotypes
g. differences among genes in a population
h. a large range of phenotypes
i. genetic component of variation in a given trait
j. the physical traits of an individual
k. the genetic composition of an individual
l. causes either one phenotype or another
m. an individual who carries two different alleles for a gene
n. influenced by many genes
o. an allele whose phenotypic effects are seen when it is present

Fill-in-the-Blank

16. ____________ occurs when pairs of alleles on ____________ chromosomes are separated during meiosis.

17. If the allele for red petals on a flower is *R* and the allele for white petals is *r*, a ____________ individual would have white flowers and a ____________ of *rr*.

18. Cystic fibrosis is caused by a ____________ allele whereas Huntington's disease is caused by a ____________ allele.

19. The letter *T* represents the dominant allele for the tongue-rolling trait. A cross between a heterozygous tongue roller and a non-tongue roller would produce the following probabilities: *TT* = ____________; *Tt* = ____________; *tt* = ____________.

20. A ____________ square with 16 boxes would be analyzing ____________ genes.

21. Height is a ____________ trait, so offspring produced by a tall man and a short woman would be ____________.

22. ____________ traits such as eye color are influenced by ____________ genes.

23. The ____________ in the womb can affect the expression of genes in a fetus.

24. Early breast cancer in the women of a family is likely to be a ____________ condition caused by a ____________ that caused a dysfunctional allele.

25. Even when a trait is highly ____________, it can still be influenced by changes in the ____________.

Labeling

Use the words or statements below to label Figure 19.1. Any word or statement listed more than once should be used more than once.

allele 1
allele 1
allele 2
allele 2
codominant
heterozygous
homozygous
homozygous

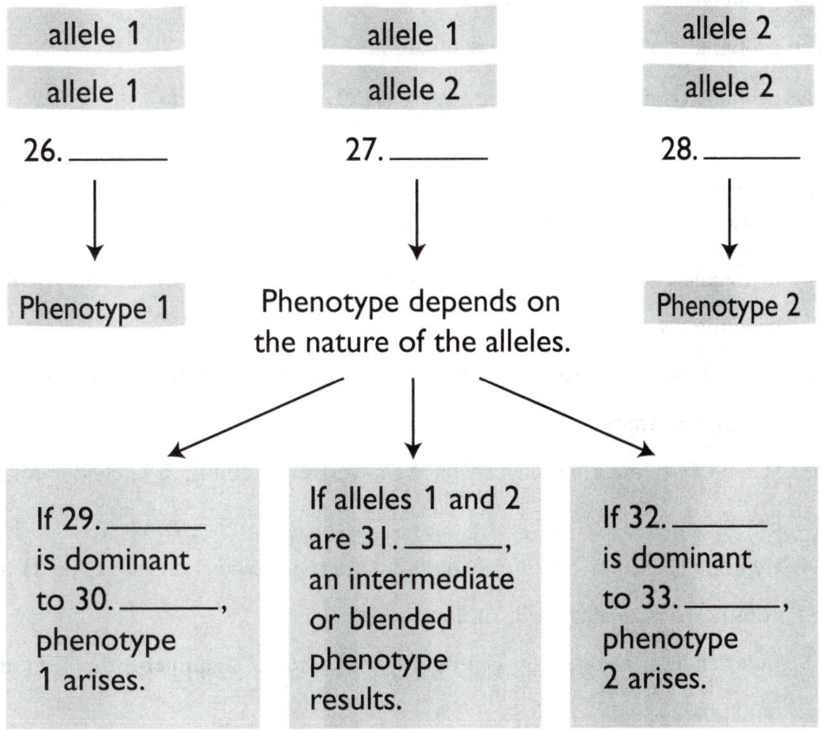

allele 1
allele 1
26. ______
Phenotype 1

allele 1
allele 2
27. ______
Phenotype depends on the nature of the alleles.

allele 2
allele 2
28. ______
Phenotype 2

If 29. ______ is dominant to 30. ______, phenotype 1 arises.

If alleles 1 and 2 are 31. ______, an intermediate or blended phenotype results.

If 32. ______ is dominant to 33. ______, phenotype 2 arises.

Figure 19.1

Roots to Remember

Use your knowledge of the root words presented in this chapter to answer the following questions.

34. Which type of individual has inherited different alleles for a trait?
35. What has happened to cells that have been cryogenically preserved?
36. The Greek root *logos* means relation, reasoning, or computation. How do the word roots of homologous explain what a homologous chromosome is?

Table Completion

Fill in the following Punnett square tables and the information below them.

	a	*a*
A	37.	38.
a	39.	40.

	41.	42.
43.	*Tt*	*tt*
T	44.	*Tt*

The probability of *AA* in the first square is (45)____________.

The probability of *Aa* in the first square is (46)____________.

The probability of *aa* in the first square is (47)____________.

The probability of *TT* in the second square is (48)____________.

The probability of *Tt* in the second square is (49)____________.

The probability of *tt* in the second square is (50)____________.

Word Choice

Circle the word or phrase that correctly completes each sentence.

51. If the allele for Huntington's disease is *H*, an individual with a genotype of *Hh* (would/would not) have the disease.
52. Twin studies are (natural/controlled) experiments that attempt to understand the heritability of quantitative traits.
53. When two slender parents may have an overweight child, it is likely due to (heritable/environmental) factors.
54. Hemophilia occurs only in people who are homozygous recessive for the hemophilia allele, so it is considered a (qualitative/quantitative) trait.
55. Independent assortment is due to the random alignment that occurs during (prophase I/ metaphase I).
56. A cross between two homozygous individuals could (never/possibly) produce a heterozygous individual.

Crossword Puzzle

Across

57. determines probability of certain traits in offspring
58. many
59. one version of a gene
60. separates alleles during meiosis
61. masked by a dominant allele

Down

62. other, another, or different
63. During metaphase I, homologous chromosomes randomly ____________.
64. code for proteins
65. the number of genotypes possible from a cross between *MM* and *Mm*

Critical Thinking

66. A couple gets married and decides to have a baby. They discuss whether they should go to a genetic counselor to determine how likely it is that their child will inherit a disease. They have no Huntington's disease in their family, but they have heard that it is a terrible illness and that it is fatal. Should they be tested to find out if either one of them has the allele for Huntington's disease? Explain your answer.
67. What is the probability that a baby girl will receive an allele for a qualitative trait that her paternal grandmother also possessed? Explain your answer.
68. A man and a woman get married. They are both carriers of cystic fibrosis, and yet none of their six children inherit the disease. Why didn't at least one of their children inherit cystic fibrosis, because the probability of getting the disease from a cross between two carriers is 25%?

Practice Test

Use the following information to answer questions 1–4.

The ability to roll your tongue is determined by dominant and recessive alleles. The *ability* to roll the tongue is the dominant trait. The *inability* to roll the tongue is the recessive trait.

1. A woman who is a non-tongue roller (homozygous recessive) marries a man who is a tongue-roller (homozygous dominant). How many genotypes for the trait of tongue rolling are possible in their offspring?
   A. 1
   B. 2
   C. 3
   D. 4

2. How many phenotypes for the trait of tongue rolling are possible in the offspring of two heterozygous tongue rollers?
   A. 1
   B. 2
   C. 3
   D. 4

3. What is the probability that the daughter of two people who cannot roll their tongues will be able to roll her tongue?
   A. 0%
   B. 25%
   C. 50%
   D. 100%

4. What is the probability that the son of two homozygous tongue rollers will be able to roll his tongue?
   A. 0%
   B. 25%
   C. 50%
   D. 100%

5. Three genes influence eye color. You want to create a Punnett square to find the possible genotypes for eye color from a cross between two individuals who each have two alleles for each gene. How many boxes would you need in the Punnett square?
   A. 16
   B. 32
   C. 64
   D. 128

6. Which process that leads to genetic diversity does *not* occur during the production of gametes?
   A. gene mutation
   B. independent assortment
   C. random fertilization
   D. segregation

7. The dominant allele, *T*, produces a straight thumb phenotype. The recessive allele, *t*, produces a curved thumb phenotype. Which of the following genotypes would produce a straight thumb?
   A. *TT* only
   B. *TT* and *Tt*
   C. *tt* only
   D. *Tt*, and *tt*

8. What factors influence quantitative variation in a trait?
   A. environmental conditions
   B. inheritance of certain alleles
   C. polygenic influences
   D. all of the above

9. Dense breast tissue is a highly heritable trait. This trait increases the risk of breast cancer. Because this trait is highly heritable,
   A. environment cannot influence the expression of this trait.
   B. a woman with this trait will always get breast cancer.
   C. identical twins are both likely to express this trait.
   D. those who do not have the allele for the trait will not have dense breast tissue.

10. A black cat and a tortoiseshell cat have two offspring that are orange, one that is black and white, and one that is tortoiseshell. What does this suggest about fur color in cats?
    A. Fur color in cats is controlled by dominant and recessive alleles.
    B. Fur color in cats is influenced by more than one gene.

C. Fur color in cats is codominant.
D. Fur color in cats is completely unpredictable.

11. Which mutation could be passed on to offspring?
   A. a mutation that occurs in a sperm cell
   B. a mutation that occurs in a skin cell
   C. both A and B
   D. neither A nor B

12. Which of the following traits does *not* show continuous variation?
   A. skin color
   B. height
   C. blood type
   D. intelligence

13. What is the likelihood that a woman who is heterozygous for Huntington's disease and a man who does not carry any allele for the disease will have a child who carries an allele for the disease?
   A. 0%
   B. 25%
   C. 50%
   D. 100%

14. A homozygous dominant plant that has dark purple flowers is crossed with a homozygous recessive plant that has white flowers. What color flowers would the offspring have if the alleles for flower color in this plant are codominant?
   A. white
   B. light purple
   C. dark purple
   D. pink

15. Which of the following describes independent assortment?
   A. Homologous pairs of chromosomes separate randomly into daughter cells.
   B. Alleles on homologous pairs of chromosomes separate during meiosis.
   C. Homologous chromosomes line up randomly during metaphase I.
   D. Gametes combine without regard to the alleles that they carry.

Answer Key

Matching

1. e; **2.** d; **3.** k; **4.** h; **5.** o; **6.** j; **7.** b; **8.** m; **9.** g; **10.** n; **11.** c; **12.** l; **13.** i; **14.** a; **15.** f

Fill-in-the-Blank

16. Segregation; homologous; **17.** homozygous recessive; genotype; **18.** recessive; dominant; **19.** 0%; 50%; 50%; **20.** Punnett; two; **21.** quantitative; medium-height; **22.** Polygenic; many or multiple; **23.** environment; **24.** heritable; mutation; **25.** heritable; environment

Labeling

26. homozygous; **27.** heterozygous; **28.** homozygous; **29.** allele 1; **30.** allele 2; **31.** codominant; **32.** allele 2; **33.** allele 1

Roots to Remember

34. heterozygous individual; **35.** They have been frozen. **36.** The roots mean the same relation, same reasoning, or the same computation. These roots may refer to the fact that homologous chromosomes have the same genes that code for (compute) the same traits.

Table Completion

37. *Aa*; **38.** *Aa*; **39.** *aa*; **40.** *aa*; **41.** *T*; **42.** *t*; **43.** *t*; **44.** *TT*; **45.** 0%; **46.** 50%; **47.** 50%; **48.** 25%; **49.** 50%; **50.** 25%

Word Choice

51. would; **52.** natural; **53.** environmental; **54.** qualitative; **55.** metaphase I; **56.** possibly

Crossword Puzzle

Across

57. Punnett; **58.** poly; **59.** allele; **60.** segregation; **61.** recessive

Down

62. hetero; **63.** align; **64.** genes; **65.** two

Critical Thinking

66. The couple does not need to be tested. Because Huntington's disease is caused by a dominant allele, there would have been family members with the disease for either of them to have a chance of carrying the allele.

67. There is a 50% chance that her father received the allele from his mother and a 50% chance that he passed that allele on to his daughter. Therefore, there is a 25% chance that the baby girl received an allele for a trait that her paternal grandmother also possessed.

68. The probability of being homozygous recessive for cystic fibrosis is assessed separately for each child because this probability is independent of the number of children that the couple has. Therefore, each child has a 25% probability of being homozygous recessive for cystic fibrosis. In this family, none of the children acquired the disease.

Practice Test

1. A; **2.** B; **3.** A; **4.** D; **5.** C; **6.** C; **7.** B; **8.** D; **9.** C; **10.** B; **11.** A; **12.** C; **13.** C; **14.** B; **15.** A

CHAPTER 20

COMPLEX PATTERNS OF INHERITANCE: DNA DETECTIVE

Learning Goals

1. Compare and contrast codominance and incomplete dominance.
2. Use the ABO blood system to explain the concept of multiple allelism.
3. Compare and contrast pleiotropy and polygenic inheritance.
4. Predict the outcome of a dihybrid cross.
5. Explain how autosomes and sex chromosomes differ.
6. Describe how sex is determined in humans.
7. Define *sex-linked genes*, and describe how their expression is different from that of autosomal genes.
8. Analyze genetic pedigrees for various patterns of inheritance.
9. Describe the technique of DNA fingerprinting.
10. Draw a DNA fingerprint that might be generated by two siblings and their parents.

Chapter Outline

I. Extensions of Mendelism
   A. Patterns of inheritance that are not caused by dominant and recessive alleles are called extensions of Mendelian genetics.
   B. Some traits are polygenic, or caused by more than one gene.
   C. Some traits display incomplete dominance.
      1. In incomplete dominance, the phenotype displayed is intermediate to the phenotypes of the parents.
      2. Hair texture in humans displays incomplete dominance.
         a) Straight hair has the genotype *CC*.
         b) Curly hair has the genotype *cc*.
         c) Wavy hair has the genotype *Cc*.
   D. Some traits are caused by multiple allelism.
      1. In multiple allelism, there are more than two alleles of a gene in a population.
      2. The ABO blood system displays multiple allelism.
         a) The three alleles of the blood-group gene are I^A, I^B, and *i*.
         b) Each person inherits two blood-group alleles.
   E. Some traits display codominance.
      1. In codominance, both alleles are expressed in the individual.
      2. The ABO blood system displays codominance.
         a) Both alleles are expressed for each blood type.
            (1) A person who has type AB blood expresses the alleles I^A and I^B.
            (2) A person who has type A blood expresses the alleles I^A and *i* or I^A and I^A.
            (3) A person who has type B blood expresses the alleles I^B and *i* or I^B and I^B.
            (4) A person who has type O blood expresses the alleles *i* and *i*.
         b) Blood typing can sometimes reveal paternity, but not always.
            (1) A child with type O blood could not have a father with type AB blood.
            (2) A child with type B blood who has a mother with type AB blood could have a father with type AB, A, B, or O blood.
   F. Some traits can cause pleiotropy.
      1. Pleiotropy is a phenomenon in which a single gene can have multiple effects on an individual's phenotype.

2. Hemophilia is a genetic disorder that has pleiotropic effects.
   a) A person with hemophilia cannot produce a protein called clotting factor VIII.
   b) Individuals with hemophilia bleed excessively because their blood cannot clot.
   c) Pleiotropic effects of hemophilia include excessive bruising, pain *in* and swelling of joints, vision loss, anemia, fatigue, and possible neurological problems.

II. Dihybrid Crosses
  A. Dihybrid crosses are crosses that involve two traits.
  B. A Punnett square may be used to analyze a dihybrid cross.
  C. Traits of pea plants provide an example for analysis.
    1. Seed color in peas is determined by
       a) a dominant allele, *Y*, that codes for yellow peas in a homozygous dominant or heterozygous plant; and
       b) a recessive allele, *y*, that codes for green peas in a homozygous recessive plant.
    2. Seed shape in peas is determined by
       a) a dominant allele, *R*, that codes for round peas in a homozygous dominant or heterozygous plant; and
       b) a recessive allele, *r*, that codes for wrinkled peas in a homozygous recessive plant.
    3. The genes for seed color and seed shape are on different chromosomes, so they sort independently.
       a) A pea plant that is heterozygous for both traits (*YyRr*) can produce gametes with the following alleles:
          (1) *YR*
          (2) *Yr*
          (3) *yR*
          (4) *yr*
       b) Every possible gamete from one parent is crossed with every possible gamete from the other parent in the Punnett square.
    4. A Punnett square for a dihybrid cross has four rows and four columns (16 boxes).
    5. The phenotypic ratio for a cross between parents that are both heterozygous for seed color and seed shape is 9:3:3:1.
       a) 9/16 of the offspring will have round yellow peas, resulting from a genotype that contains at least one dominant allele for each trait.
       b) 3/16 of the offspring will have round green peas, resulting from a genotype that contains at least one dominant allele for seed shape and two recessive alleles for seed color.
       c) 3/16 of the offspring will have wrinkled yellow peas, resulting from a genotype that contains at least one dominant allele for seed color and two recessive alleles for seed shape.
       d) 1/16 of the offspring will have wrinkled green peas, resulting from a genotype that contains two recessive alleles for each trait.

III. Sex Determination
  A. Chromosomes and Sex Determination
    1. The X and Y chromosomes determine the sex of an individual.
       a) Males produce gametes with 22 autosomes and either an X or Y sex chromosome.
       b) Females produce gametes with 22 autosomes and an X chromosome.
       c) Offspring that receive an X chromosome from the father are females (XX).
       d) Offspring that receive a Y chromosome from the father are males (XY).
    2. Karyotyping can reveal the sex of an individual (as well as genetic abnormalities) from tissue fragments.
  B. Sex Linkage
    1. X-linked Genes
       a) X-linked genes are located on the X chromosome.
       b) Diseases caused by recessive alleles on the X chromosome are more likely to affect males than females.
          (1) Males have only one copy of the gene because they have only one X chromosome.
          (2) Females have two copies of the gene because they have two X chromosomes.
       c) Only females can be carriers of X-linked recessive traits.
          (1) A carrier shows no sign of the disease.
          (2) A carrier can pass the disease on to her male offspring.

- d) There are a number of well-known X-linked traits.
  - (1) Hemophilia is an X-linked trait that affects over 20,000 men in the U.S.
  - (2) Red-green color blindness is an X-linked trait that affects around 4% of men.
    - (a) Red blindness is the inability to see red as a distinct color.
    - (b) Green blindness is the inability to see green as a distinct color.
    - (c) A lack of opsin proteins causes insensitivity to red or green light wavelengths in affected individuals.
  - (3) Duchenne muscular dystrophy is a fatal, X-linked disease that affects 1 in 3,500 males.
    - (a) This disease causes muscle wasting.
    - (b) Disease onset occurs between the ages of 1 and 12.
    - (c) The recessive gene that causes this disease fails to produce dystrophin—a protein that stabilizes cell membranes during muscle contraction.

2. Y-linked Genes
   - a) Y-linked genes are located on the Y chromosome.
   - b) Y-linked genes are passed from fathers to sons.
   - c) Very few genes have been localized to the Y chromosome.
     - (1) The *SRY* gene is located exclusively on the Y chromosome.
     - (2) When the *SRY* gene is expressed, it triggers a series of events that lead to the development of testes and other cells required for male sexual characteristics.

IV. Pedigrees

- A. A pedigree is a family tree that follows the inheritance of a trait over many generations.
- B. Pedigrees allow scientists to study inheritance by analyzing matings that have already occurred.
- C. Pedigrees show if a trait is autosomal dominant or recessive, or if the trait is sex linked.
- D. A pedigree of European royal families traces the inheritance of hemophilia.
  1. The fertilized egg that produced Queen Victoria carried the mutation for hemophilia.
  2. Some females descended from Queen Victoria carried and passed on the mutated gene.
  3. Some males descended from Queen Victoria were affected with hemophilia.

V. DNA Fingerprinting

- A. Overview
  1. DNA fingerprinting allows unambiguous identification of individuals.
  2. DNA fingerprinting can be used to solve crimes or establish family relationships.
- B. Copying DNA Through Polymerase Chain Reaction
  1. The polymerase chain reaction (PCR) can be used to amplify DNA.
     - a) Double-stranded DNA is placed in a test tube with nucleotides.
     - b) The *Taq* polymerase enzyme is added to the test tube.
       - (1) The *Taq* polymerase enzyme was isolated from the single-celled organism *Thermus aquaticus*.
       - (2) *Thermus aquaticus* lives in hydrothermal vents and can withstand very high temperatures.
       - (3) The *Taq* polymerase enzyme acts as a DNA polymerase.
     - c) The DNA molecule is first denatured with high heat to produce single-stranded templates.
     - d) As the solution cools, *Taq* polymerase adds complementary nucleotides to each strand.
     - e) The cycle of heating and cooling is repeated many times, with each round doubling the amount of DNA.
  2. Specific nucleotide sequences are cut out of the DNA using restriction enzymes.
     - a) Restriction enzymes are like highly specific molecular scissors.
     - b) The DNA of different people produces fragments of different sizes when cut with the same enzymes.
- C. Size-Based Separation Through Gel Electrophoresis
  1. DNA fragments are separated from each other using gel electrophoresis.
     - a) Fragments are added to an agarose gel.
     - b) When electric current is applied, the fragments move.
       - (1) Small fragments move more quickly because they are less impeded by the gel.
       - (2) Larger fragments move more slowly because they are more impeded by the gel.
  2. DNA fragments are then stained so they can be visualized.
- D. DNA Fingerprints Can Be Compared to Solve Mysteries
  1. Matching banding patterns can show a genetic relationship, or lack of relationship, between two individuals being tested.

2. A perfect match, or lack of a match, between banding patterns on two samples may help to solve crimes.

Practice Questions

Matching

1. codominance
2. carrier
3. gel electrophoresis
4. incomplete dominance
5. polymerase chain reaction
6. pleiotropy
7. phenotypic ratio
8. multiple allelism
9. denatured
10. sex-linked gene
11. autosome
12. hemophilia
13. sex determination
14. DNA fingerprinting
15. X-linked gene

a. gene carried on the X chromosome
b. gene carried on a sex chromosome
c. causes both alleles to be expressed in the phenotype
d. the process by which the sex of an individual is determined
e. technique to separate DNA fragments using an electric current
f. the ratio of different phenotypes produced in a cross
g. technique that allows unambiguous identification of individuals
h. technique to amplify DNA
i. X-linked clotting disorder
j. nonsex chromosome
k. an individual who has one normal allele and one allele for a recessive disease
l. produces a phenotype that is intermediate to that of either parent
m. multiple effects from a single gene
n. split apart DNA strands
o. when more than two alleles for a gene exist in a population

Fill-in-the-Blank

16. A cross between a curly-haired mother and a straight-haired father produces a ____________ -haired son. This cross illustrates ____________.
17. Blood type displays ____________ as well as multiple allelism. An individual's blood type is determined by ____________ alleles, but there are three possible alleles in the human population.
18. A daughter who has type AB blood has a sister who has type O blood. If her mother has type A blood, her father must have type ____________ blood and be ____________ for the trait.
19. A ____________ cross that examines the flower color and seed shape of pea plants has the possibility of producing ____________ phenotypes if heterozygotes are crossed.
20. Chromosomes called ____________ code for most cell functions. ____________ chromosomes determine the sex of an individual.
21. A couple has a child who has red-green color blindness. The child is a ____________ who got the allele for color blindness from his/her ____________, who was a ____________ of the trait.
22. A ____________ trait is not likely to show up in every generation represented in a ____________.
23. The technique called ____________ amplifies DNA so that it can be used to produce a DNA ____________.
24. Ten cycles of PCR would produce ____________ identical ____________ of DNA.
25. Large DNA fragments travel more ____________ than smaller DNA fragments do during the process of ____________.

Labeling

Use the words or statements below to label Figure 20.1. Any word or statement listed more than once should be used more than once.

carrier female
carrier female
hemophiliac male
unaffected female
unaffected male
unaffected male

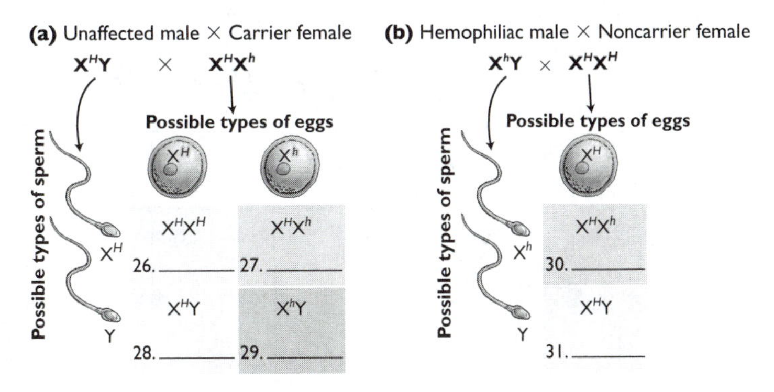

Figure 20.1

Use the words or statements below to label Figure 20.2. Any word or statement listed more than once should be used more than once.

denaturing
double-stranded DNA
identical DNA molecules
nucleotides
replication
Taq polymerase

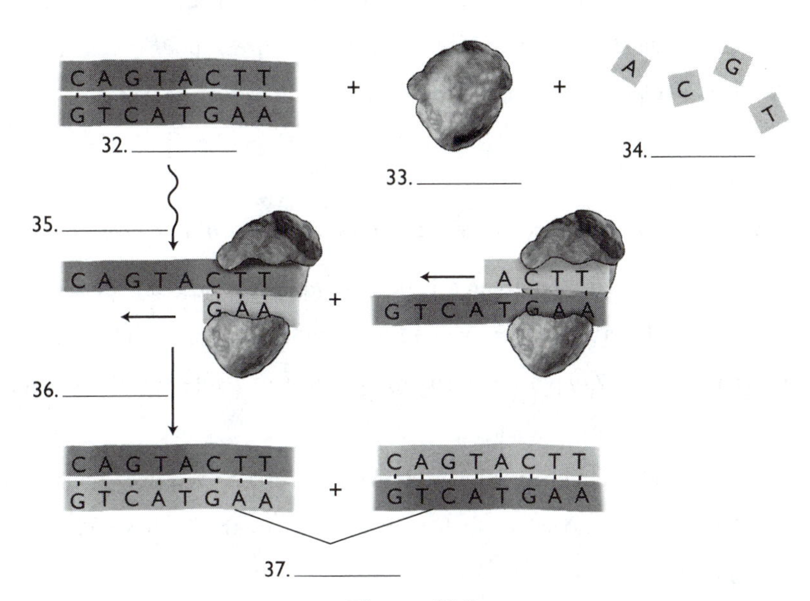

Figure 20.2

Roots to Remember

Use your knowledge of the root words presented in this chapter to answer the following questions.

38. Why would it make sense for the word *forensic* to have the root *forum*?
39. The root *-tropy* means a turn or change in response to a stimulus. What is the meaning of *pleiotropy*?
40. Which word root could be used instead of *pleio-* to describe multiples of something?

Table Completion

Fill in the following Punnett square table and the information below it.

The following Punnett square represents a cross between two pea plants. The traits being tested are height of plant and pod color. The dominant allele for height, *T*, produces a tall plant. The recessive allele, *t*, produces a short plant. The dominant allele for pod color, *G*, produces a green pod. The recessive allele, *g*, produces a yellow pod.

	T 41.__	42.__*g*	*t g*	43.__*G*
t 44.__	45.__	*Tt gg*	46.__	*tt Gg*
47.__	48.__	*Tt gg*	49.__	*tt Gg*
t G	*Tt GG*	50.__	*tt Gg*	51.__
52.__	*Tt GG*	*Tt Gg*	53.__	54.__

The frequency of each phenotype is predicted by the Punnett square. Frequencies can be represented by fractions, with the total number of boxes (16) as the denominator.

55. The predicted frequency of tall plants with green pods is ____________.
56. The predicted frequency of tall plants with yellow pods is ____________.
57. The predicted frequency of short plants with green pods is ____________.
58. The predicted frequency of short plants with yellow pods is ____________.

Word Choice

Circle the word or phrase that correctly completes each sentence.

59. A woman with blood type A is heterozygous for the trait. She has a child with a man with blood type B who is also heterozygous for the trait. The child (could/could not) have blood type O.
60. The enzyme (*Taq* polymerase/DNA polymerase) is able to withstand very high temperatures without becoming nonfunctional.
61. The daughter of a hemophiliac is (likely to/certain to) receive an allele for hemophilia.
62. The son of a carrier of Duchenne muscular dystrophy has a (25%/50%) chance of inheriting the allele that causes the disease.
63. Of the two sex chromosomes, the (X chromosome/Y chromosome) has more genetic material.
64. A dihybrid cross between two heterozygous individuals can produce (four/nine) different genotypes.
65. (Multiple allelism/Pleiotropy) is when a single gene causes multiple effects.

Critical Thinking

66. Why would you expect the DNA fingerprint of a young girl to be a mix of the DNA fingerprint of her mother and the DNA fingerprint of her father?
67. Could you use a karyotype to identify inherited conditions or diseases? Specifically, could it be used to identify a recessive allele for an X-linked disease? Explain your answer.
68. Could a female inherit hemophilia? Explain your answer.

Practice Test

1. A man with blood type AB marries a woman with blood type A. Their children could *not* have blood type
   A. A.
   B. B.
   C. AB.
   D. O.

2. Two wavy-haired parents (genotype *Cc*) have a child. What is the probability that their child will have wavy hair?
   A. 0%
   B. 25%
   C. 50%
   D. 100%

3. Two alleles are both expressed fully in an individual through
   A. codominance.
   B. incomplete dominance.
   C. multiple allelism.
   D. dominant and recessive alleles.

4. In pea plants, the allele for tall plants (*T*) is dominant over the allele for short plants (*t*). The allele for purple flowers (*P*) is dominant over the allele for white flowers (*p*). A pea plant that is homozygous recessive for both traits is crossed with a pea plant that is homozygous dominant for both traits. What will be the phenotypic ratio of the offspring?
   A. 0:8:8:0
   B. 6:2:6:2
   C. 9:3:3:1
   D. 16:0:0:0

5. Two pea plants that are heterozygous for seed shape and seed color are crossed. How many of their offspring will have wrinkled, green seeds (the homozygous recessive genotype)?
   A. 1/16
   B. 3/16
   C. 4/16
   D. 9/16

6. Which of the following genotypes belongs to a carrier of an X-linked recessive disorder?
   A. X^HY
   B. X^HX^h
   C. X^hY
   D. X^HX^H

7. The pedigree below shows four generations of a family. Shaded boxes indicate an individual affected by a particular trait.

The trait shown in the pedigree is most likely a

A. dominant trait.
B. recessive trait.
C. sex-linked trait.
D. codominant trait.

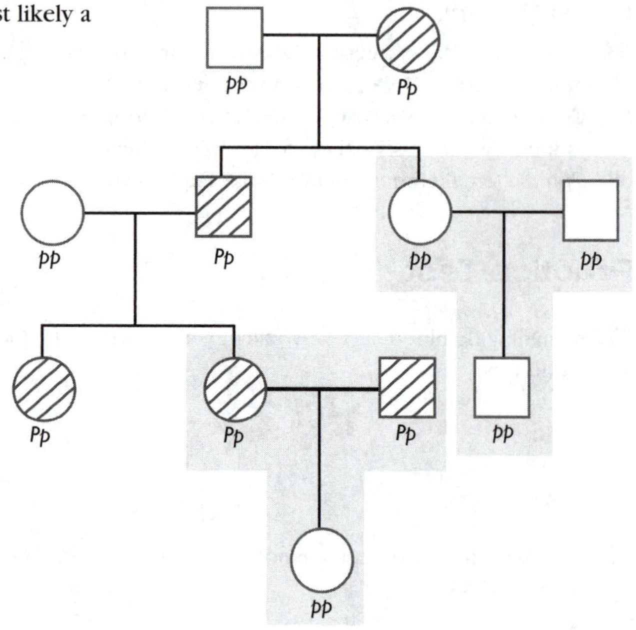

8. Which of the following is *not* added to a test tube before PCR is done?
   A. DNA
   B. *Taq* polymerase
   C. restriction enzymes
   D. nucleotides

9. The following is a DNA fingerprint of six adults and three children. Which adults are most likely to be the parents of child 2?
   A. adults 1 and 3
   B. adults 2 and 4
   C. adults 3 and 5
   D. adults 4 and 6

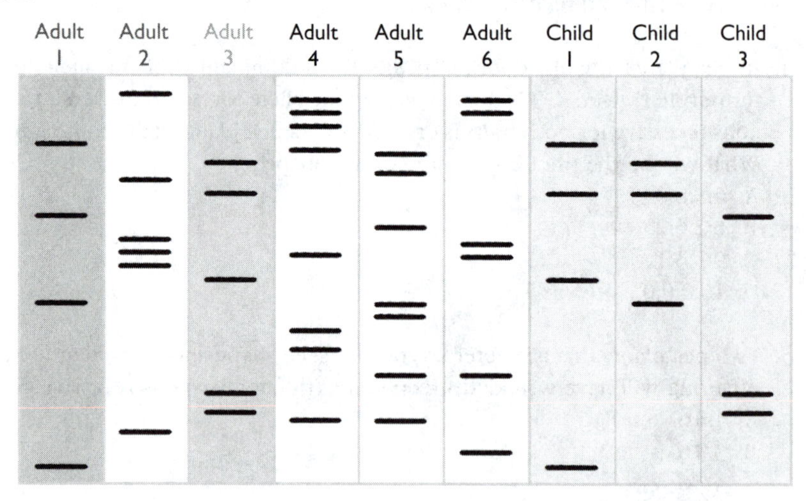

10. How many PCR cycles would it take to get over 1 million copies of DNA from one original molecule?
    A. 17 cycles
    B. 18 cycles
    C. 19 cycles
    D. 20 cycles

11. What is the genotype of a human male?
    A. XX
    B. XY
    C. YY
    D. none of the above

12. The recessive allele in the ABO blood system is
    A. I^A.
    B. I^B.

C. I^O.
D. *i*.

13. Attached earlobes are a recessive trait. What percentage of offspring is likely to have attached earlobes if both parents are heterozygous for the trait?
A. 0%
B. 25%
C. 50%
D. 100%

14. Which phenotype would *not* be seen in a cross between a woman with straight hair and dark eyes (*ccDD*) and a man with wavy hair and blue eyes (*Ccdd*)?
A. wavy hair
B. straight hair
C. blue eyes
D. dark eyes

15. Which of the following produces fragments of DNA?
A. polymerase chain reaction
B. gel electrophoresis
C. treatment with restriction enzymes
D. treatment with agarose gel

Answer Key

Matching

1. c; **2.** k; **3.** e; **4.** l; **5.** h; **6.** m; **7.** f; **8.** o; **9.** n; **10.** b; **11.** j; **12.** i; **13.** d; **14.** g; **15.** a

Fill-in-the-Blank

16. wavy; incomplete dominance; **17.** codominance; two; **18.** B; heterozygous; **19.** dihybrid; four; **20.** autosomes; Sex; **21.** boy; mother; carrier; **22.** sex-linked; pedigree; **23.** polymerase chain reaction; fingerprint; **24.** 1,024; molecules; **25.** slowly; gel electrophoresis

Labeling

26. unaffected female; **27.** carrier female; **28.** unaffected male; **29.** hemophiliac male; **30.** carrier female; **31.** unaffected male; **32.** double-stranded DNA; **33.** *Taq* polymerase; **34.** nucleotides; **35.** denaturing; **36.** replication; **37.** identical DNA molecules

Roots to Remember

38. The forum was a place where legal disputes and lawsuits were heard. Forensics gathers information that might be used to settle legal disputes or lawsuits. **39.** many changes in response to a stimulus (where the stimulus is the production of proteins, or inability to produce proteins, by an allele); **40.** *poly-*

Table Completion

41. *G*; **42.** *T*; **43.** *t*; **44.** *g*; **45.** *Tt Gg*; **46.** *tt gg*; **47.** *tg*; **48.** *Tt Gg*; **49.** *tt gg*; **50.** *Tt Gg*; **51.** *tt GG*; **52.** *tG*; **53.** *tt Gg*; **54.** *tt GG*; **55.** 6/16 or 3/8; **56.** 2/16 or 1/8; **57.** 6/16 or 3/8; **58.** 2/16 or 1/8

Word Choice

59. could; **60.** *Taq* polymerase; **61.** certain to; **62.** 50%; **63.** X chromosome; **64.** nine; **65.** Pleiotropy

Critical Thinking

66. The girl received half of her genetic material from her mother and half from her father. Therefore, her DNA fingerprint would match half of her mother's fingerprint and half of her father's.
67. A karyotype can identify nondisjunctions that may be the cause of certain inherited syndromes or diseases. It cannot, however, identify specific alleles on any chromosome.
68. Yes, if she inherited the allele for hemophilia from each parent.

Practice Test

1. D; **2.** C; **3.** A; **4.** D; **5.** A; **6.** B; **7.** A; **8.** C; **9.** A; **10.** D; **11.** B; **12.** D; **13.** B; **14.** C; **15.** C

CHAPTER 21

DEVELOPMENT AND AGING: THE PROMISE AND PERILS OF STEM CELLS

Learning Goals

1. Describe the differences between adult, fetal, and embryonic stem cells.
2. Describe the process of fertilization in detail.
3. Summarize the development of a preembryo from zygote through blastocyst.
4. Illustrate the structure of the gastrula, and compare and contrast the eventual fate of its three layers.
5. Describe the early stages of organogenesis of the central nervous system.
6. Explain the role of the *SRY* gene in the development of sex-specific organs.
7. Describe the function of the placenta.
8. Compare and contrast fetal circulation and adult circulation.
9. Describe the three stages of labor.
10. Describe the changes in males and females during puberty.
11. Describe two hypotheses for the causes of aging, and provide evidence for each.
12. Describe some of the effects of aging on organ systems.

Chapter Outline

I. The Production of Embryonic Stem Cells
  A. Overview
    1. When cells go through differentiation, they obtain a specific function and become a fixed type of cell.
    2. Stem cells are undifferentiated and totipotent.
      a) Undifferentiated cells do not perform a specific function.
      b) Totipotent cells can become any other type of cell.
    3. There are three types of stem cells.
      a) Embryonic stem cells build tissues in the embryo.
        (1) The embryo stage is the stage of development from the second week after fertilization until the ninth week.
        (2) Embryonic stem cells are pluripotent, which means they can develop into any of the more than 200 cell types in the human body.
        (3) Due to their pluripotency, embryonic stem cells are of interest to researchers in regenerative medicine.
      b) Fetal stem cells build tissues in the fetus.
        (1) The fetal stage is the stage of development from the end of the ninth week until birth.
        (2) Fetal stem cells are multipotent.
          (a) They are partially differentiated.
          (b) They are able to produce only a few types of cells.
      c) Adult stem cells build some tissues in the adult.
        (1) Adult stem cells are less multipotent than fetal stem cells.
        (2) Adult stem cells are uncommon and difficult to find in the body.
  B. Fertilization: Forming the Ultimate Stem Cell
    1. The fertilized egg is the single cell that gives rise to all cells in the body.
    2. Fertilization occurs after sexual intercourse.

a) A man ejaculates roughly 200 million sperm into the woman's reproductive tract during sexual intercourse.
b) Prostaglandin released in semen triggers muscular contraction of the uterus.
c) Sperm are drawn into the upper portion of the oviduct.
d) Chemicals in the female reproductive tract lead to capacitation of sperm, which weakens the plasma membrane surrounding the acrosome at the tip of each sperm cell.
e) Once capacitation is complete, a sperm that encounters an egg can fertilize it.
   (1) Fertilization can take place six hours after intercourse.
   (2) Eggs can only be fertilized within 24 hours of their release from an ovary.
   (3) Sperm can live for 24 hours or longer, so sperm that enter the woman's reproductive tract up to five days before ovulation can fertilize an egg once the egg is ovulated.

3. Couples who have trouble conceiving may use in vitro fertilization (IVF).
   a) A woman receives drug therapy that stimulates excess follicle development and the ovulation of multiple eggs.
   b) These eggs are removed from the ovaries by a needle inserted through the wall of the vagina.
   c) Eggs are examined to make sure they have reached metaphase II.
   d) Sperm are collected from the man.
   e) Sperm are treated using the process of capacitation.
   f) Sperm and eggs are incubated together in a glass dish for 48 hours.
   g) Fertilization is almost assured using IVF.
4. The process of fertilization creates a zygote.
   a) The sperm cell meets the egg cell.
      (1) As the flagellum beats, it forces the sperm cell through the mass of follicle cells that nourish the egg.
      (2) A specific protein receptor on the head of the sperm binds to a receptor on the zona pellucida of the egg cell.
         (a) The zona pellucida is a translucent covering that shields the egg from mechanical damage.
         (b) The receptors on the zona pellucida only bind to sperm of the same species.
      (3) Binding of the sperm cell triggers the release of enzymes from the acrosome.
      (4) Enzymes digest a tunnel through the zona pellucida, through which the sperm travels to the egg's plasma membrane.
   b) The plasma membranes of sperm and egg fuse.
      (1) The egg's plasma membrane depolarizes to produce a reduced electric gradient across it.
      (2) Depolarization causes a release of enzymes that detach the zona pellucida from the egg and destroy its sperm receptors.
         (a) Any more sperm are inhibited from attaching to and fertilizing the egg.
         (b) The zona pellucida remains for several days as a shell of protection around the egg.
   c) The egg completes meiosis II.
      (1) A single mature egg, or ovum, is produced.
      (2) A tiny polar body is produced that quickly degrades.
   d) The sperm nucleus fuses with the egg nucleus.
      (1) Each gamete has n chromosomes.
      (2) This fusion produces a $2n$ zygote.
   e) Follicle cells are shed.
   f) The zygote begins cell division.
5. Dizygotic, or fraternal, twins can be produced if two eggs are released at ovulation and both are fertilized by different sperm.
6. Multiple zygotes can be produced by IVF.
   a) One to four preembryos will be implanted in the mother's uterus.
   b) Remaining preembryos are frozen and stored, in case the pregnancy is not successful.
   c) If the pregnancy is successful, frozen preembryos may be discarded, donated to other couples, or used for stem cell research.

C. Preembryonic Development
   1. Development that occurs during the preembryonic stage produces a multicellular embryo that is capable of producing adult tissues.
   2. Preembryos undergo cell division and morphogenesis, but they do not grow in size.

3. Rapid mitotic cell division, called cleavage, begins when the zygote is in the oviduct.
   a) The first cell division occurs 24 hours after fertilization.
   b) The second and third divisions happen the following day.
   c) Gap junctions form between adjacent cells, so that the cell mass contracts into a solid ball.
      (1) If a preembryo completely breaks apart before the cell mass contracts, monozygotic, or identical, twins can result.
      (2) If a preembryo fragments but does not split completely, conjoined twins may result.
   d) Cells divide again to form a 16-cell mass called a *morula*.
   e) A fluid-filled sphere called a blastocyst forms over the next two days.
      (1) The inner cell mass is a cluster of cells inside the blastocyst.
         (a) The inner cell mass contains 30–34 cells.
         (b) Some of these cells will give rise to the body of the embryo.
      (2) The trophoblast is the outer layer of blastocyst cells.
         (a) The trophoblast is made up of about 200 cells.
         (b) The trophoblast includes the early chorion.
            (i) The chorion is one of the extraembryonic membranes that are not part of the developing individual.
            (ii) The chorion combines with tissues from the mother to form the placenta.
      (3) Most IVF embryos are introduced into the uterus at blastocyst stage.
      (4) Cells from the inner cell mass of a blastocyst are harvested to create stem cell lines.
         (a) Cell lines are colonies of cells that can replicate themselves when provided with nutrients and hormones.
         (b) Cells can be removed from the colony for experimentation.
         (c) Harvesting cells from a blastocyst to begin a cell line destroys the preembryo.

II. Early Embryonic Development
   A. Development continues when a blastocyst implants in the uterus.
      1. Implantation occurs about one week after fertilization.
         a) The embryo breaks out of the zona pellucida.
         b) The trophoblast secretes enzymes that digest part of the mother's endometrium.
         c) The embryo sinks deeply into the endometrial tissue and becomes completely engulfed.
         d) The trophoblast secretes human chorionic gonadotropin (HCG).
            (1) HCG inhibits the breakdown of the corpus luteum.
               (a) The corpus luteum produces progesterone and estrogen.
               (b) Progesterone and estrogen maintain the endometrium.
               (c) The increase in these hormones during the first five weeks of pregnancy may cause nausea.
            (2) The presence of HCG indicates pregnancy.
      2. Not all blastocysts implant successfully.
         a) One-third to one-half of all embryos fail to implant, probably due to genetic abnormalities.
         b) The rate of implantation for IVF preembryos is around 15%.
            (1) The timing of development and suitability of the uterus must be ideal for implantation to occur.
            (2) Rate of implantation can be improved by transferring two or more preembryos.
      3. When a blastocyst implants in the wall of the oviduct, it is called an ectopic pregnancy.
         a) Ectopic pregnancies cannot be successful.
         b) Ectopic pregnancies must be surgically removed to prevent injury to the mother.
   B. Structures form in the uterus after implantation.
      1. Two additional extraembryonic membranes form from the trophoblast.
         a) The yolk sac will eventually produce blood cells.
         b) The amnion surrounds the amniotic cavity, which fills with amniotic fluid.
            (1) Amniotic fluid acts as an insulator against temperature changes.
            (2) Amniotic fluid provides a cushion that absorbs shock from movements of the mother.
   C. The trophoblast and inner cell mass separate.
      1. The inner cell mass, now termed the embryonic disk, changes to produce the gastrula.
         a) The gastrula contains the three primary germ layers.
            (1) The ectoderm (outer primary germ layer) gives rise to
               (a) skin,
               (b) the nervous system, and
               (c) sensory organs.

(2) The mesoderm (middle primary germ layer) gives rise to
(a) muscles,
(b) kidneys,
(c) heart,
(d) blood vessels,
(e) sex organs, and
(f) skeleton.
(3) The endoderm (innermost primary germ layer) gives rise to
(a) lining of the gastrointestinal tract,
(b) gastrointestinal organs,
(c) glands,
(d) respiratory organs, and
(e) excretory organs.
b) The three germ layers are the first differentiation among body cells.
2. Different chemical signals likely produce differentiated layers.
a) Chemicals are unequally distributed in a zygote.
b) Chemical signals turn some genes on in a cell and leave others off.

III. Organ Formation
A. Overview
1. Cell differentiation is the first step in organogenesis.
2. Organogenesis is the formation of organs and organ systems.
3. Organogenesis is nearly complete by the end of the embryonic period.
4. Environmental exposure to toxic compounds can disrupt organogenesis.
B. Cell Migration and Death
1. Organogenesis results from three factors:
a) Cell differentiation
b) Cell migration
(1) Chemical signals at the final destination and along the route to that destination facilitate cell migration.
(2) Environmental toxins can disrupt chemical signals that facilitate cell migration.
(a) Nicotine interferes with the migration of bone marrow stem cells to the fetal liver.
(b) Babies of heavy smokers may have cardiovascular and immunological problems.
c) Apoptosis, or programmed cell death
(1) The role of programmed cell death in embryonic and fetal development is not well understood.
(2) Programmed cell death helps to shape fingers and toes from paddlelike embryonic feet and hands.
(3) Greater understanding of apoptosis could help scientists develop therapies that selectively kill tumor cells.
C. Early Organogenesis: Development of the Nervous System
1. The central nervous system (CNS) appears first as a primitive streak during the gastrula stage.
2. The primitive streak develops into a neural plate by the third week of pregnancy.
3. The plate partially folds to form a hollow neural tube that is covered with ectoderm.
4. The neural tube develops into the brain and spinal cord.
a) Cells that connected the tube to ectoderm migrate to form peripheral nerves or undergo apoptosis.
b) The neural tube becomes separate from other tissues of the embryo.
c) Mesoderm on either side of the tube migrates to form somites, which eventually make up some of the muscles, skin, and bone.
5. Spina bifida is a birth defect that occurs when the neural tube fails to close and separate completely from the ectoderm.
a) Part of the spinal cord is exposed in a cyst outside the spine.
b) Infants with spina bifida have symptoms ranging from numbness to paralysis.
c) Lack of folic acid during pregnancy can cause the development of spina bifida.
D. Later Organogenesis: The Reproductive Organs
1. Gonadal development begins around seven weeks after fertilization.
a) Cells migrate to the gonadal ridge in the abdomen of the embryo.
(1) Cells divide to form two indifferent gonads.

(2) Germ cells inside the gonads could become sperm or eggs.
(3) Steroid-producing cells inside the gonads could become either structural and estrogen-secreting cells or structural and testosterone-secreting cells.

b) Two sets of ducts are found in each embryo.
(1) The Wolffian duct may develop into male structures.
(2) The Mullerian duct may develop into female structures.

c) Male reproductive structures develop if the *SRY* gene on the Y chromosome is present.
(1) The *SRY* gene produces a protein that stimulates transcription of other genes.
(2) The SRY protein turns on the genes required for testicular development.
(3) Embryonic testes produce male sex hormones.
(a) Testosterone stimulates the development of the vas deferens, epididymis, and urethra from the Wolffian duct.
(b) Anti-Mullerian hormone causes the Mullerian duct to regress, thereby preventing development of female reproductive organs.

d) Female reproductive structures develop if the *SRY* gene is absent.
(1) Indifferent gonads develop into ovaries.
(2) The Mullerian duct develops into the oviducts, uterus, cervix, and vagina.
(3) Without testosterone, Wolffian duct cells die.

2. External genitalia develop around 12 weeks after fertilization.
a) Each embryo contains an undifferentiated genital tubercle and labioscrotal swellings.
b) In the presence of dihydroxytestosterone (DHT), the tubercle differentiates into a penis and the swellings differentiate into the scrotum.
(1) The testes descend into the scrotum late in development.
(2) The scrotum provides cooler temperatures that facilitate sperm production.
c) In the absence of DHT, the tubercle differentiates into the clitoris and the swellings differentiate into the vulva.

3. About 2 in every 1,000 children born in the U.S. have ambiguous genitalia.
a) Ambiguous genitalia can result from exposure to abnormal levels of sex hormone during development.
(1) Hormone levels can be affected by genetics.
(a) A mutation in the enzyme that converts testosterone to DHT can cause the affected person to have internal male reproductive structures and external female genitalia.
(b) Testosterone gives this individual other external male characteristics.
(2) Hormone levels can be affected by the environment.
(a) Maternal exposure to DES, a synthetic estrogen prescribed to prevent miscarriages, caused malformation of daughter uteruses and ovaries.
(b) Exposures to some environmental hormones present in plastics or pesticides have correlated with abnormalities in male reproductive organs.

4. Embryonic stem cell research may find treatments for or help to reduce birth defects.

IV. Fetal Development and Birth

A. The Purpose of the Placenta in Pregnancy

1. The placenta contains both fetal and maternal tissue.
2. Placental development begins at implantation.
a) The trophoblast digests part of the uterine lining.
b) Blood vessels in the endometrium leak blood, forming a pool around the trophoblast.
c) Oxygen, nutrients, and wastes are passed between embryo and mother through diffusion within the pool.
d) The chorion grows fingerlike projections into uterine tissue.
(1) These projections are called chorionic villi.
(2) Chorionic villi increase the surface area for substance exchange.
3. The placenta produces hormones that help to sustain the pregnancy and prepare the mother for nursing.
a) Production of estrogen and progesterone allows the placenta to take over the functions of the corpus luteum.
(1) The placenta maintains endometrium.
(2) The placenta prevents follicle formation.

b) Production of progesterone relaxes smooth muscles in the uterus and gastrointestinal tract.
  (1) The uterus is able to expand in size.
  (2) The mother experiences acid reflux and constipation.
  (3) Blood pressure drops.
    (a) The kidneys release hormones to increase blood volume.
    (b) Estrogen levels also promote water retention.
    (c) Blood volume increases by 40% in pregnant women.
c) Production of estrogen and progesterone causes darkening of the areolae and other areas of the skin.
d) Production of an additional hormone makes the mother's cells resistant to insulin.
  (1) Higher blood-sugar levels are needed to supply the fetus with energy for growth.
  (2) This hormone may induce pregnancy-related diabetes.
e) Production of somatomammotropin stimulates development of the mother's mammary glands and breasts.
  (1) This hormone works together with estrogen, progesterone, and the pituitary hormone, prolactin.
  (2) These hormones prepare the mammary glands to give milk.

B. Fetal Circulation
1. Dissolved substances in the blood move down a concentration gradient and across the placental membrane.
  a) Glucose that is elevated in the mother's bloodstream flows continually to the fetus.
  b) Oxygen flows to the fetus due to special fetal hemoglobin.
    (1) Fetal hemoglobin is produced from the end of the embryonic period until birth.
    (2) Fetal hemoglobin has a higher affinity for oxygen than does adult hemoglobin.
    (3) After birth, adult hemoglobin replaces fetal hemoglobin over the first year.
  c) Carbon dioxide flows from fetus to mother.
    (1) The thoracic cavity increases by 40% as the fetus grows.
    (2) This leads to an increase in maternal lung capacity and a decrease in maternal CO_2 levels.
    (3) High fetal heart rate also maintains higher CO_2 levels in the fetus.
2. Damaging substances, such as alcohol, can pass across the placenta to the fetus.
3. The umbilical cord connects the bloodstream of the fetus to the placenta.
  a) Umbilical arteries carry blood from arteries in the fetus to the placenta.
  b) The umbilical vein carries blood from the placenta to the fetus.
  c) The ductus venosus allows returning blood to bypass the fetal liver.
4. Special structures in the heart allow blood to bypass the undeveloped fetal lungs.
  a) A gap between atria, called the foramen ovale (oval window), allows blood entering the right atrium to flow directly to the left atrium.
  b) The ductus arteriosus allows blood that reaches the right ventricle to empty into the aorta.
  c) These structures close after birth.

C. Stem Cells and the Fetal Period
1. Almost all birth defects occur during the embryonic period.
2. Almost all testing for birth defects occurs during the fetal period.
  a) Amniotic fluid is removed for genetic testing using amniocentesis.
  b) Organs can be directly observed using ultrasound.
3. Some defects can be repaired by fetal surgery.
4. Fetal stem cells may be used to repair some birth defects.
  a) Fetal stem cells are more multipotent than adult stem cells.
  b) Fetal stem cells may be collected from amniotic fluid.

D. The Process of Childbirth
1. Parturition, or childbirth, is triggered by a number of factors.
  a) At the end of its development in the uterus, the fetus releases hormones from its hypothalamus, pituitary, and adrenal glands.
  b) Fetal hormones stimulate the placenta to release more estrogen.
  c) Increased estrogen triggers an increase in oxytocin and prostaglandin.
  d) Oxytocin stimulates the smooth muscle in the uterus to contract.
  e) Contractions stimulate the cervix to dilate.
  f) The mucous plug in the cervix is expelled.
  g) The amniotic sac bursts and releases fluid when the child's head presses against the cervix.

2. Positive feedback accelerates labor.
   a) Oxytocin causes uterine contractions.
   b) Contractions force the child's head against the cervix.
   c) Stretching of the cervix stimulates the release of more oxytocin.
3. Labor is divided into three stages.
   a) During stage one, the cervical opening dilates to 10 centimeters.
      (1) This stage is known as effacement, or thinning of the cervix.
      (2) This stage takes several hours or days.
   b) During stage two, the baby moves into the birth canal and is born.
      (1) Contractions are frequent.
      (2) The mother has an urge to push with abdominal muscles.
      (3) An episiotomy may be made at the mouth of the birth canal.
   c) During stage three, the placenta is delivered.
4. Some women may require a Caesarean section (C-section).
   a) In a C-section, an incision is made through the upper abdomen and the infant is removed.
   b) C-section may be necessary when a small-hipped woman is carrying a large baby.
   c) C-section is necessary if the baby is breech (positioned with its feet toward the cervix).
5. After birth, the umbilical cord is cut.
   a) Carbon dioxide builds up in the baby's bloodstream.
   b) A breathing reflex is triggered.
   c) The first inhalation inflates the lungs and initiates gas exchange.
   d) Premature babies with underdeveloped lungs may require a respirator.

V. Development After Birth

A. Growth and Maturation

1. Critical development occurs in the neonatal period.
   a) The child's digestive system begins to function.
   b) Fat deposits develop under the skin to assist with thermoregulation.
   c) Brain development allows adjustment to stimuli and the ability to interact.
   d) Bony skull plates begin to fuse together.
      (1) Fontanels are soft spots where the bone has not fused.
      (2) Fontanels remain until the infant is about two years old.
2. Breast milk nourishes the infant.
   a) During the first few days, the mammary glands produce colostrum—a low-fat yellow fluid that is rich in proteins, antibodies, minerals, and vitamin A.
   b) Prolactin stimulates lactation, the production of mature milk.
   c) Suckling triggers the release of oxytocin, which stimulates milk release.
3. Development continues in stages until early adulthood.
   a) Infancy is the stage from 1 to 15 months.
   b) Childhood is the stage from 15 months to around 12–15 years of age.
   c) Pubescence occurs at different ages for boys and girls.
      (1) Puberty occurs between ages 10 and 16 for girls.
      (2) Puberty occurs between ages 13 and 16 for boys.
   d) Adolescence is the stage from 16 to around 19–21 years of age.
   e) Early adulthood is the stage from 19 to 21 years of age to around age 25.
4. Patterns of growth are irregular after birth.
   a) Some body parts grow faster than others.
   b) This differential pattern is called allometric growth.

B. Puberty

1. Sexual organs mature and secondary sex characteristics develop during puberty.
2. Hormones trigger puberty.
   a) The hypothalamus begins to secrete gonadotropin-releasing hormone (GnRH).
   b) GnRH triggers the pituitary to release luteinizing hormone (LH) and follicle-stimulating hormone (FSH).
   c) LH and FSH stimulate the gonads to produce sex hormones.
      (1) Estrogen is more abundant in females.
      (2) Testosterone is more abundant in males.
      (3) Both males and females release both estrogen and testosterone.

3. Girls have their first menstruation (menarche) at puberty.
4. Bodies develop secondary sex characteristics.
   a) Girls
      (1) Breasts begin to develop.
      (2) Pubic hair growth occurs after some breast development.
      (3) Body hair becomes thicker and darker.
      (4) Hips widen as the lower pelvis increases in size and more fat is stored.
      (5) Girls grow to adult height.
   b) Boys
      (1) The testes and penis enlarge.
      (2) Libido increases.
      (3) Erections become common.
      (4) Pubic hair develops.
      (5) Growth of the larynx causes the voice to deepen by an octave.
      (6) Hair grows on the face and body.
      (7) Boys experience a long growth spurt in height.

VI. Aging
  A. Why Do We Age?
    1. Senescence is the process of physical change in the body that increases the risk of disease, disability, and death.
       a) Senescence begins around age 40.
       b) On average, death occurs between 72 and 79 years in the U.S.
       c) Maximum life span is 125 years.
    2. Maximum life span may be encoded in our genes.
       a) Adult cells can divide only a certain number of times before dying.
          (1) Each time a cell divides, a portion of telomere at the end of the chromosome is lost.
          (2) The cell dies when only a small portion of telomere remains.
          (3) Telomerase, an enzyme produced by stem cells and cancer cells, restores telomere after division and makes a cell immortal.
       b) Genetic mutations can lead to unusually shortened or lengthened lives.
    3. Environmental conditions may affect life span.
       a) Free radicals can cause damage to DNA.
       b) Suppression of free radical production can cause organisms to live longer.
    4. Failure of one body system affects others.
       a) Body systems are integrated.
       b) Damage to one system can cause damage to others.
  B. Effects of Aging
    1. Patterns of aging depend on genetic and environmental factors for each individual.
    2. Human health habits affect aging.
    3. Chemical cross-links that interfere with normal protein function affect aging.
    4. Effects of aging on organ systems include the following:
       a) Integumentary
          (1) Skin loses elasticity.
          (2) Reduced fat under the skin causes sagging and wrinkling.
          (3) A reduction in sweat glands, hair follicles, and fat impairs thermoregulation.
          (4) A reduction in sebaceous glands causes skin to become dry.
          (5) Pigment cells decline, causing hair to gray.
          (6) Pigment cells increase in size, causing skin blotchiness.
          (7) The cornea may become cloudy, reducing visual acuity.
       b) Cardiovascular
          (1) Cardiac cell size declines, reducing heart size and cardiac output.
          (2) Arteries become less elastic and contain more plaque, increasing heart attack and stroke risk.
          (3) Blood flow to the liver declines, interfering with drug metabolism.
       c) Digestive
          (1) Organs lose muscle tone, which may cause heartburn and constipation.
          (2) Metabolic rate declines.
          (3) Teeth may decay or be lost.

d) Excretory
   (1) Kidneys become less efficient in maintaining water balance, which may result in dehydration.
   (2) Reduced muscle tone may increase the risk of bladder infection.
   (3) Reduced pelvic muscle tone may lead to incontinence in women.
   (4) Enlarged prostate may cause urinary problems in men.
e) Musculoskeletal
   (1) The number of muscle cells declines with lack of use.
   (2) Bones lose calcium and mass, resulting in a decrease in height.
   (3) Joint cartilage degrades, causing arthritis.
   (4) Cell division and repair decreases, resulting in slower healing.
f) Reproductive
   (1) Sex hormone levels decline.
   (2) Women experience menopause.
   (3) Men may take longer to achieve an erection.
g) Nervous
   (1) Brain mass declines.
   (2) Reduction of blood flow to the brain causes nerve cell death and reduction in brain function.
   (3) Senses decline.
   (4) The lens of the eye becomes less flexible, impairing near vision.

C. Restoring the Brain
1. Alzheimer's disease (AD) affects around 3% of people over age 65 and around 50% of those over age 85.
   a) AD occurs when tangled clumps of fibers form in brain neurons.
   b) AD causes progressive memory loss, dementia, and death.
   c) Stem cell treatments are unlikely to reestablish previous neural connections, so they are unlikely to restore memory.
2. Parkinson's disease (PD) occurs when brain cells that produce dopamine stop functioning.
   a) PD is typically associated with aging.
   b) Stem cell treatments may be able to replace damaged brain cells and restore dopamine levels.
3. Restrictions on federal funding for stem cell research have impeded medical progress in the U.S.
   a) Some states are funding their own research.
   b) Donated IVF embryos have gotten around some moral objections.
4. Research to reprogram differentiated cells to become multipotent or pluripotent has shown some promise.

Practice Questions

Matching

1. Mullerian duct	a. stage of development from weeks 0–2 after fertilization
2. embryo	b. germ layer that gives rise to muscles and the heart
3. parturition	c. stage of development from weeks 9–38 after fertilization
4. gastrula	d. stage of development from weeks 2–9 after fertilization
5. fetus	e. changes in form
6. Wolffian duct	f. germ layer that gives rise to skin and sensory organs
7. trophoblast	g. childbirth
8. ectoderm	h. may develop into female reproductive structures
9. foramen ovale	i. fluid-filled sphere that implants in the uterus
10. mesoderm	j. first menstruation
11. lactation	k. allows fetal blood to flow between atria
12. menarche	l. may develop into male reproductive structures
13. preembryo	m. structure containing three primary germ layers
14. blastocyst	n. production of mature breast milk
15. morphogenesis	o. cells that give rise to the chorion, yolk sac, and amnion

Fill-in-the-Blank

16. Embryonic stem cells are ____________, which means they have great flexibility to develop into a wide variety of cell types. Adult stem cells are ____________, which means they are less flexible.
17. A sperm cell that has not undergone ____________ cannot burrow through the ____________ to reach the ____________ membrane of the egg.
18. When the egg completes ____________ it produces a(n) ____________ and a smaller polar body.
19. ____________ twins can be different sexes, but ____________ twins must be the same sex.
20. A mutation that prevented the release of HCG from the ____________ would prevent successful ____________ of the blastocyst.
21. All adult body cells are formed from the three ____________ of the gastrula that were formed from the ____________ of the trophoblast.
22. Structures of the ____________ system are the first to form in a developing ____________.
23. The Wolffian duct only develops into reproductive structures in the ____________ of testosterone. The Mullerian duct only develops into reproductive structures in the ____________ of testosterone.
24. Blood ____________ increases in the pregnant mother due to the action of progesterone and estrogen.
25. The release of ____________ causes uterine contractions during labor. Contractions cause the movement of the child to stretch the ____________. This stretching triggers the release of ____________.

Labeling

Use the words or statements below to label Figure 21.1.

blastocyst
inner cell mass
morula
trophoblast
zona pellucida
zygote

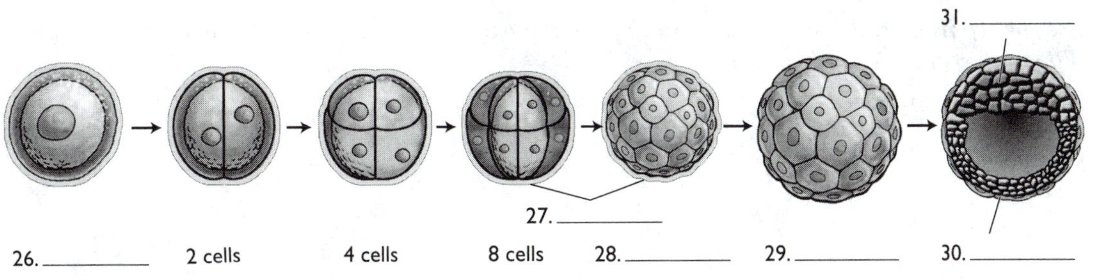

Figure 21.1

Use the words or statements below to label Figure 21.2.

fetal capillary
maternal blood vessels
placenta
pool of blood
trophoblast
uterine wall

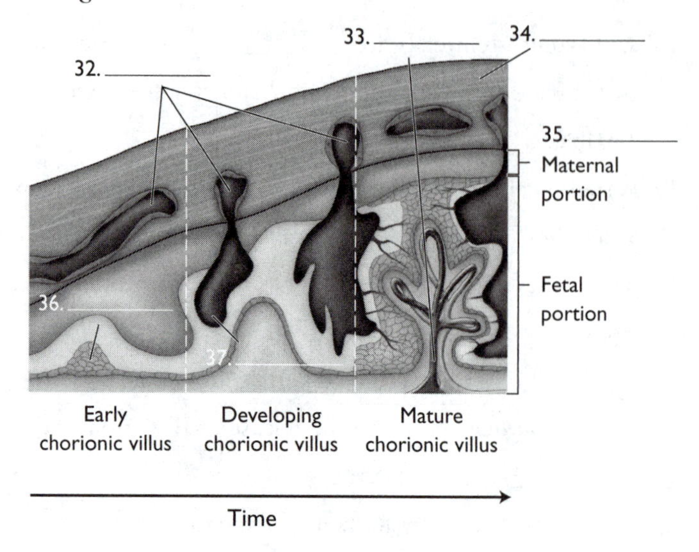

Figure 21.2

Roots to Remember

Use your knowledge of the root words presented in this chapter to answer the following questions.

38. Human body types are sometimes classified into three forms. An ectomorph body is generally long and lean. An endomorph body is generally soft and round. What would be the name of a body type between these two?
39. Someone who is lactose intolerant has a hard time digesting what types of food?
40. Why could the process of sexual development be called a *pluriglandular* process?

Sequencing

Put the following events related to fertilization in order by numbering each event, from 1 to 8.

41. ____________ Sperm and egg nuclei fuse.
42. ____________ Sperm releases acrosomal enzymes.
43. ____________ Enzymes destroy sperm receptors in zona pellucida.
44. ____________ Sperm and egg plasma membranes fuse.
45. ____________ Sperm head binds to zona pellucida.
46. ____________ Sperm penetrates follicle cells.
47. ____________ Sperm nucleus enters egg cell.
48. ____________ Egg completes meiosis II.

Paragraph Completion

Use the words or statements below to complete the paragraph. Any word or statement listed more than once should be used more than once.

anti-Mullerian hormone
clitoris
DHT
estrogen
estrogen
FSH
GnRH
gonads
indifferent
ovaries
pituitary
puberty

scrotum
sexual
testes
testosterone
testosterone

Hormones play a key role in the (49) ____________ development of an individual. During the embryo stage, an individual develops (50) ____________ gonads that become (51) ____________ in the presence of testosterone. In the absence of testosterone, these same structures become (52) ____________. The hormone (53) ____________ determines whether Wolffian or Mullerian ducts will develop into sexual structures. If this hormone is present, the testes will also produce (54) ____________, to inhibit the formation of female reproductive organs. Hormones also influence external genitalia. The hormone (55) ____________ triggers the development of the penis and the (56) ____________. In the absence of this hormone, the (57) ____________ and vulva develop. Hormones take on a key developmental role again during (58) ____________. When the hypothalamus begins to secrete (59) ____________, the hormones LH and (60) ____________ are released from the (61) ____________. This triggers the release of estrogen and testosterone from the (62) ____________. In girls, (63) ____________ leads to the development of breasts and pubic hair. (64) ____________ also helps to trigger menarche. In boys, (65) ____________ leads to the enlargement of the testes and penis, as well as the development of other secondary sexual characteristics.

Critical Thinking

66. An embryo contains a Y chromosome that has a mutation in the *SRY* gene that inhibits the production of SRY protein. How would this impact the developing embryo?
67. How are the substances carried in an umbilical vein different from substances carried in an adult vein?
68. Are all conjoined twins identical? Explain your answer.

Practice Test

1. Which of the following is formed from cells that began in the inner cell mass of a trophoblast?
   A. muscle
   B. yolk sac
   C. allantois
   D. chorion

2. An infant is born with a major birth defect affecting the heart. When did the birth defect likely develop?
   A. during the first two weeks after fertilization
   B. during weeks 3–6 after fertilization
   C. during weeks 7–9 after fertilization
   D. during weeks 10–16 after fertilization

3. Gonads in an embryo develop in the presence of SRY protein. The genital tubercle and labioscrotal swellings develop in the presence of DHT. When this individual is born, which adult reproductive structures will the individual possess?
   A. male internal reproductive structures and female external genitalia
   B. female internal reproductive structures and female external genitalia
   C. male internal reproductive structures and male external genitalia
   D. female internal reproductive structures and male external genitalia

4. The last stage of a preembryo is called a
   A. zygote.
   B. blastocyst.
   C. trophoblast.
   D. morula.

5. What happens right after the plasma membrane of a sperm fuses with the plasma membrane of an egg?
   A. The egg's plasma membrane becomes depolarized.
   B. The sperm and egg nuclei fuse together.
   C. The zona pellucida detaches from the egg and floats away.
   D. The sperm releases acrosomal enzymes.

6. Which of the following is an effect of senescence that can be minimized by diet and exercise?
   A. Cell division declines, so damaged muscles are slower to recover.
   B. Reduction in sebaceous glands results in dry skin.
   C. The number of muscle cells decreases and average muscle mass declines.
   D. Brain mass slowly decreases.

7. Which substances flow down a concentration gradient across the placenta, from mother to fetus?
   A. carbon dioxide and oxygen
   B. oxygen and glucose
   C. carbon dioxide and metabolic waste
   D. metabolic waste and glucose

8. The placenta produces all of the following hormones *except*
   A. estrogen.
   B. somatomammotropin.
   C. progesterone.
   D. lactin.

9. The embryonic structure that develops into the vulva in females can develop into what structure in males?
   A. testes
   B. penis
   C. scrotum
   D. epididymis

10. Which embryonic structure shunts blood from the right ventricle to the aorta?
    A. foramen ovale
    B. ductus venosus
    C. umbilical artery
    D. ductus arteriosus

11. Which body system develops over the longest period of time in the embryo and fetus?
    A. nervous system
    B. respiratory system
    C. reproductive system
    D. integumentary system

12. What happens when a cell undergoes apoptosis?
    A. It migrates to a different part of the body.
    B. It dies at a prescribed time.
    C. It differentiates into a specialized cell.
    D. It divides without regulation.

13. Where does fertilization occur?
    A. in an oviduct
    B. in the uterus

C. in an ovary
D. in the placenta

14. How many rounds of cell division has a preembryo undergone when it becomes a morula?
A. 3
B. 4
C. 5
D. 6

15. Bones usually stop growing in an individual during
A. childhood.
B. pubescence.
C. adolescence.
D. early adulthood.

Answer Key

Matching

1. h; **2.** d; **3.** g; **4.** m; **5.** c; **6.** l; **7.** o; **8.** f; **9.** k; **10.** b; **11.** n; **12.** j; **13.** a; **14.** i; **15.** e

Fill-in-the-Blank

16. pluripotent; multipotent; **17.** capacitation; zona pellucida; plasma; **18.** meiosis II; ovum; **19.** Dizygotic; monozygotic; **20.** trophoblast; implantation; **21.** germ layers; inner cell mass; **22.** nervous; embryo; **23.** presence; absence; **24.** volume; **25.** oxytocin; cervix; oxytocin

Labeling

26. zygote; **27.** zona pellucida; **28.** morula; **29.** blastocyst; **30.** trophoblast; **31.** inner cell mass; **32.** maternal blood vessels; **33.** fetal capillary; **34.** uterine wall; **35.** placenta; **36.** trophoblast; **37.** pool of blood

Roots to Remember

38. mesomorph; **39.** dairy products; **40.** It requires production of hormones from several glands.

Sequencing

41. 8; **42.** 3; **43.** 6; **44.** 4; **45.** 2; **46.** 1; **47.** 5; **48.** 7

Paragraph Completion

49. sexual; **50.** indifferent; **51.** testes; **52.** ovaries; **53.** testosterone; **54.** anti-Mullerian hormone; **55.** DHT; **56.** scrotum; **57.** clitoris; **58.** puberty; **59.** GnRH; **60.** FSH; **61.** pituitary; **62.** gonads; **63.** estrogen; **64.** Estrogen; **65.** testosterone

Critical Thinking

66. Without the SRY protein, the genes for testicular development would not be turned on. The indifferent gonads would not develop into testes, so they would not produce testosterone. In the absence of testosterone, the embryo would develop female reproductive structures.

67. An umbilical vein carries substances from the mother to the fetus. Therefore, it would contain high concentrations of oxygen and nutrients. It would contain low concentrations of waste materials, such as carbon dioxide. Veins in adults (with the exception of the pulmonary vein) contain low concentrations of oxygen and nutrients and high concentrations of wastes.

68. Yes. Conjoined twins result from a preembryo that splits but does not break apart completely, thus they are formed from the same zygote.

Practice Test

1. A; **2.** B; **3.** C; **4.** B; **5.** A; **6.** C; **7.** B; **8.** D; **9.** C; **10.** D; **11.** A; **12.** B; **13.** A; **14.** B; **15.** D

CHAPTER 22

EVOLUTION: WHERE DID WE COME FROM?

Learning Goals

1. Define *biological evolution*.
2. Describe the theory of common descent.
3. Describe how Linnaeus's classification system supports the theory of common descent.
4. Define *vestigial structures*.
5. Describe the evidence for evolution provided by biogeography.
6. Explain how fossils form, and illustrate how they support the hypothesis of common descent.
7. Summarize the process of speciation and the biological species concept.
8. Define *fitness* as used in the context of evolution and natural selection.
9. Explain why scientists think there is a single common ancestor for all living organisms.
10. List the three steps required for life to form from nonliving precursors.
11. Describe how evolution can occur via genetic drift.

Chapter Outline

I. Evidence of Evolution
   A. What Is Evolution?
      1. Biological evolution is a change in the characteristics of a population of organisms that occurs over the course of generations.
      2. Evolutionary change results in a change in the frequency of particular alleles in a population.
      3. The theory of evolution is a principle for understanding
         a) how species originate, and
         b) why species display certain characteristics.
      4. The theory of evolution states that
         a) all species were descended from a common ancestor, and
         b) all species represent the product of millions of years of evolutionary change.
      5. The part of the theory of evolution that states the origin of species is called the theory of common descent.
   B. Charles Darwin's Revolution
      1. Darwin served as assistant naturalist on the HMS *Beagle*.
      2. The expedition's most influential stop was on the Galápagos Islands.
         a) The variety of organisms was remarkable.
         b) Each island seemed to have its own unique species.
      3. Darwin realized that his observations fit a theory of common descent.
      4. Darwin amassed evidence to support his theory over 20 years.
      5. Darwin's ideas were published in 1859 in the book *On the Origin of Species by Means of Natural Selection, or the Preservation of Favoured Races in the Struggle for Life.*
   C. Alternative Hypotheses: Scientific and Religious
      1. Most scientists today accept the theory that modern species evolved from extinct ancestors as fact.
      2. In the 1850s, most Europeans believed in special creation—that God created organisms as described in Genesis.
         a) They believed that species have not changed significantly since this creation.
         b) They believed that the creation occurred within the last 10,000 years.

3. Alternatives to the theory of evolution are often not scientific hypotheses.
    a) Theories such as special creation and intelligent design include supernatural aspects.
    b) The existence of a supernatural creator cannot be tested using the scientific method.
4. Alternatives to the theory of evolution contain some testable hypotheses.
    a) The static model states that species were recently derived and they are unchanging.
    b) Other hypotheses fall between the static model and common descent.
5. Observations from the natural world are necessary to evaluate hypotheses related to human origins.

D. Evidence from Biological Classification
1. Classification systems allow scientists to organize and understand biological diversity.
2. Linnaeus developed a classification system that is still used today.
    a) Shared physical traits are used to place organisms into a hierarchy.
        (1) Organisms that shared the most traits are placed in the same narrow classification group.
        (2) Organisms that shared fewer, broader traits are placed in more comprehensive classification groups.
    b) The hierarchy used today consists of the following groups, from broadest to narrowest:
        (1) Domain
        (2) Kingdom
        (3) Phylum or Division
        (4) Class
        (5) Order
        (6) Genus
        (7) Species
    c) Sublevels or superlevels have been added between each major category.
    d) The levels in Linnaean classification can be interpreted as different degrees of relationship.
        (1) All species in the same family share a relatively recent common ancestor.
        (2) All families in the same class share a more distant common ancestor.
3. Linnaeus placed humans, monkeys, and apes in the same order (Primates).
    a) All have forward-facing eyes and coordinated hands.
    b) Humans and apes (in the family Hominidae) have relatively large brains, erect posture, lack of a tail, and flexibility of the thumb.
    c) Humans and the African great apes (in the subfamily Homininae) have elongated skulls, short canine teeth, and reduced hairiness.

E. Evidence from Homology: Related Species Are Similar
1. Homology is the similarity in characteristics that has resulted from common ancestry.
    a) For example, forelimbs have different functions in mammals.
    b) However, each limb has a common set of bones.
    c) The most likely explanation is that each species inherited the basic forelimb structure from a common ancestor.
2. Homology can be observed in anatomy, behavior, DNA, and other features of organisms.
3. Vestigial traits provide support for the hypothesis of common ancestry.
    a) A vestigial trait is a nonfunctional trait or feature.
    b) Similarities between vestigial and functional traits can provide clues to ancestry.
        (1) Great apes and humans have a tailbone like their primate relatives, but no tail.
        (2) Humans have arrector pili like their mammalian relatives.
            (a) In other mammals, arrector pili raise each hair to improve insulation of the coat or increase the perceived size of the animal.
            (b) In humans, arrector pili cause goose bumps.
    c) Common descent helps to explain the presence of useless traits in organisms.
4. Structures present during the development of organisms can support common ancestry.
    a) All chordates have pharyngeal slits as early embryos.
    b) Most chordates have tails as early embryos.
    c) These similarities suggest that chordates were derived from an ancestor that developed along a similar pathway.
5. Similarities in genes can identify common ancestry.
    a) Some genes are shared by all organisms.
    b) Other genes are only shared by organisms that have similar physical structures.

c) DNA sequencing can reveal relationships between species.
    (1) Organisms with DNA sequences that are highly similar are likely to share a recent common ancestor.
    (2) Organisms with DNA sequences that are less similar are likely to share an ancient common ancestor.
    (3) Using DNA sequencing, humans appear more closely related to chimpanzees than to gorillas.
        (a) DNA sequences of humans and chimpanzees are 99.01% similar.
        (b) DNA sequences of humans and gorillas are 98.9% similar.

F. Evidence from Biogeography
1. Biogeography, or the distribution of species on Earth, can support the hypothesis of common ancestry.
    a) Species in a geographic location are generally descended from ancestors that lived in that geographic location.
    b) Darwin found evidence of this in the Galápagos Islands.
        (1) Species of mockingbirds found on the islands were similar to a species in nearby Ecuador.
        (2) Different varieties of tortoises were found on each island, but all were the same species.
2. Darwin predicted that, due to biogeography, fossils of human ancestors would be found in Africa.
    a) Humans are highly mobile.
    b) Great apes are less mobile.
    c) If these organisms shared a common ancestor, you would be likely to find evidence of that ancestor where the less-mobile species are found.

G. Evidence from the Fossil Record
1. Fossils are the remains of an organism preserved in soil or rock.
    a) Most fossils of large animals are rocks.
    b) These rocks formed as organic material in bone decomposed and minerals filled the spaces left behind.
2. The process of fossilization requires special conditions.
    a) Most organisms decay before they can be fossilized.
    b) The fossil record is incomplete.
3. Fossils of hominins (humans and human ancestors) show the structural developments that led to bipedalism.
    a) The foramen magnum moved from the back of the skull to the base of the skull.
    b) The pelvis and knee were modified for an upright stance.
    c) The foot changed from grasping to weight bearing.
    d) The lower limbs elongated relative to front limbs.
4. Fossil hominins were first found in Africa beginning in the early 20th century.
    a) *Australopithecus africanus* was found in 1924.
    b) Lucy, the 3.2 million-year-old *Australopithecus afarensis* skeleton, was found in Ethiopia in 1974.
    c) The oldest hominin fossil, *Ardipithecus ramidus*, is 5.2–5.8 million years old and was found in Ethiopia.
5. The age of fossil remains can be determined by radiometric dating.
    a) Radioactive elements in rock decay at known rates.
    b) When radioactive elements decay, they produce daughter products.
    c) The amount of time it takes one-half of a radioactive element to decay into daughter products is called the element's half-life.
    d) Scientists can determine the ratio of the radioactive element to daughter product in a rock sample.
    e) By comparing that ratio to the element's half-life, they can determine the age of the fossil.
6. Fossil evidence has allowed scientists to construct a human pedigree.
    a) Radiometric dating determined the ages of fossils.
    b) Anatomical similarities were used to arrange fossils of similar ages within the pedigree.
7. The fossil record of hominins shows a progression from apelike to humanlike features.
8. The fossil record and radiometric dating allow us to reject the static model.

a) Radiometric dating shows that Earth and its organisms have been around for much longer than 10,000 years.
b) The fossil record shows that organisms have changed over time.

II. The Origin of Species
A. Speciation: How One Becomes Two
1. According to the biological species concept, a species is a group of individuals that can interbreed and produce fertile offspring.
2. Reproductive isolation is the inability of pairs of individuals from different species to produce fertile offspring.
3. Speciation is the evolution of one or more species from an ancestral form.
4. Speciation can occur under the following conditions.
a) Populations become isolated.
b) Evolutionary changes occur in both populations, causing them to diverge.
c) Divergence is so great that reproductive isolation evolves to prevent further breeding.
5. Populations may become isolated for several reasons.
a) A population may migrate to a distant location.
b) A geological barrier may occur.
c) Timing of mating and reproduction may differ in each population.
6. Divergence occurs through a number of mechanisms.
B. The Theory of Natural Selection
1. Natural selection is a mechanism for evolution.
2. The theory of natural selection is based on four observations.
a) Individuals within a population vary.
(1) Each different type of individual in a population is called a variant.
(2) Variants differ due to differences in alleles.
(3) Gene mutation can generate new variants.
b) Some of the variation among individuals can be passed to their offspring.
c) Populations of organisms produce more offspring than will survive.
d) Survival and reproduction are not random.
(1) Some variants in a population have a higher likelihood of survival and reproduction.
(2) The relative survival of one variant over another is called its fitness.
(3) Traits that increase an individual's fitness are called adaptations.
C. Critical Thinking About Natural Selection
1. Fitness is a relative trait, not an absolute one.
a) The term *survival of the fittest* is misleading.
b) The individuals that survive and reproduce are the ones that are best adapted to their current environment.
2. Natural selection does not result in "ideal" organisms.
a) Natural selection leads to organisms that are better survivors in their current situation.
b) Traits that allow better survival and reproduction in one environment may be detrimental in another.
3. Natural selection does not result in the "progress" of a population toward a predetermined goal.
a) Natural selection is situational.
b) Traits that are selected for increase fitness in the particular environment where the population is living at the time.
4. Natural selection does not cause the emergence of new variants.
a) Natural selection acts only on variants that are already in the population.
b) Variants are produced by the random process of mutation.
c) New variants are typically modifications of the organism's underlying structure, not completely new traits.
5. Natural selection cannot change the characteristics of individuals, only populations.
a) Evolution can only occur when traits that influence survival are present in a population and have a genetic basis.
b) Natural selection cannot cause an individual to change the alleles it carries.
D. Evolution: A Robust Theory
1. The theory of evolution is robust, which means that it explains a wide variety of observations and is well supported by a large body of evidence from many scientific disciplines.

a) The most compelling evidence for the single origin of life is the universality of both DNA and the relationship between DNA and proteins.
(1) Genes from cows can be transferred to bacteria.
(2) Modified bacteria create a functional cow protein.
(3) If bacteria and cows arose separately, they would be unlikely to translate genetic information similarly.
b) Shared cellular processes point to a common ancestor.
2. Intelligent design and other alternatives to evolution are not robust.
a) They lack experimental support.
b) They rely on misreading of data and false statistical arguments.
c) They are based on the idea of irreducible complexity.
(1) Irreducible complexity is the idea that certain structures or processes are too complex to have arisen without a conscious designer.
(2) Evolution provides a framework for understanding how complex designs could have arisen naturally.

III. Human Evolution
A. Overview
1. The binomial name *Homo sapiens* describes humans.
2. Scientists have added a subspecies name, *Homo sapiens sapiens*, to distinguish modern humans from earlier humans.
3. Humans are one of the most successful species on the planet.
4. All modern human populations descended from an African population beginning about 200,000 years ago.
5. Recent shared ancestry supports the hypothesis that human races are not very different from each other.
a) There are no genes that are unique to a single race of humans.
b) There are no alleles that can clearly identify someone as belonging to one race or another.
B. Why Human Groups Differ: Selection and Genetic Drift
1. Natural Selection Causes Some Differences Between Groups
a) The skin color of a native human population appears to be a function of the environmental conditions to which the population is exposed.
b) Darker skin is an adaptation to block UV light in high-UV environments.
(1) UV light interferes with the body's ability to retain folic acid.
(2) Folic acid is required for proper fetal development.
(3) Folic acid is required for adequate sperm production.
c) Lighter skin is an adaptation to absorb UV light in low-UV environments.
(1) Absorption of UV light is required for the synthesis of vitamin D.
(2) Vitamin D is essential for the development of strong bones.
2. Sexual Selection Also Creates Differences
a) Sexual selection occurs when a trait influences the likelihood of mating.
b) The difference in overall size between men and women may be a result of sexual selection.
3. Differences Can Arise by Chance
a) Evolutionary changes can occur by chance through genetic drift.
(1) The Founder Effect
(a) When small groups colonize new areas, they bring a subset of the alleles from their larger population.
(b) The genetic makeup of the smaller population will diverge from the makeup of the larger population over generations.
(c) Genetic diseases that are unusually common in certain populations result from the founder effect.
(2) Population Bottlenecks
(a) A population may experience a rapid decline in size due to environmental conditions.
(b) Alleles are lost when the population declines.
(c) The new population may evolve to have common traits that are uncommon in neighboring populations.
(3) Chance Events
(a) An allele may be low in frequency within a small population.

(b) If the individuals who carry that allele do not successfully reproduce, the allele will not spread to the next generation.
(c) Many alleles may be lost in small, isolated populations.
b) Early human populations were likely small and isolated, so they were susceptible to genetic drift.
4. Differences Are Maintained by Assortative Mating
a) Positive assortative mating occurs when individuals choose to mate with partners who have similar traits.
b) When different populations differ in physical features, the number of matings between them may be small.
c) Positive assortative mating maintains physical differences between groups.
C. Evolution in the Classroom
1. Evolution is a foundation of biological science.
2. Evolution helps us to understand fundamental biological principles.
3. Students who are not taught evolution may lack an appreciation of the basic unity and diversity of life.
4. Students who are not taught evolution may fail to understand the effects of evolution on the natural world and humankind.

Practice Questions

Matching

1. diverge
2. speciation
3. homology
4. hominins
5. fitness
6. genetic drift
7. biogeography
8. species
9. founder effect
10. positive assortative mating
11. adaptation
12. population bottleneck
13. vestigial traits
14. sexual selection
15. theory of common descent

a. choosing a mate that has similar characteristics to yours
b. similarity in characteristics due to common ancestry
c. well-supported hypothesis that all life evolved from one ancestor
d. group of individuals that can interbreed and produce fertile offspring
e. change in a way that is different from another population
f. process in which evolutionary changes occur by chance
g. humans and human ancestors
h. process in which a trait influences the likelihood of mating
i. a reduction in population size followed by an increase
j. evolution of a new species from an ancestral form
k. trait that improves survival and reproduction in a given environment
l. distribution of species on Earth
m. change in allele frequencies due to the formation of a subpopulation
n. relative survival and reproduction of one variant compared to others
o. nonfunctional features

Fill-in-the-Blank

16. In Linnaeus's hierarchy, _____________ is a broader level of classification than family, but it is narrower than _____________.
17. In the five-kingdom system, the kingdom _____________ contains single-celled prokaryotic organisms and the kingdom _____________ contains single-celled eukaryotic organisms.
18. Bones in the _____________ of a bat are homologous with bones in the fingers of a chimpanzee and bones in the _____________ of a sea lion.

19. Cats and dogs are more likely to share ____________ genes than do wolves and dogs.
20. Patterns of ____________ show that species in a certain location are generally ____________ from ancestors in that location.
21. Anatomical changes, such as the movement of the ____________ from the back of the skull to the base, trace the development of ____________ in hominins.
22. If the ____________ of a radioactive element is 1 million years, a rock sample that contains 6.25% of the radioactive element would be ____________ million years old.
23. The immediate predecessor to *Homo sapiens* is ____________ and the most ancient known hominin ancestor is ____________.
24. Human skin color is likely a result of ____________ selection, and the difference in body size between human males and females is likely a result of ____________ selection.
25. A community that rebuilds after a lethal volcanic eruption is likely to experience a ____________ that will make formerly rare ____________ more common.

Labeling

Use the words or statements below to label Figure 22.1.

Ardipithecus ramidus
Australopithecus afarensis
Australopithecus africanus
Homo erectus
Homo habilis
Paranthropus aethiopicus

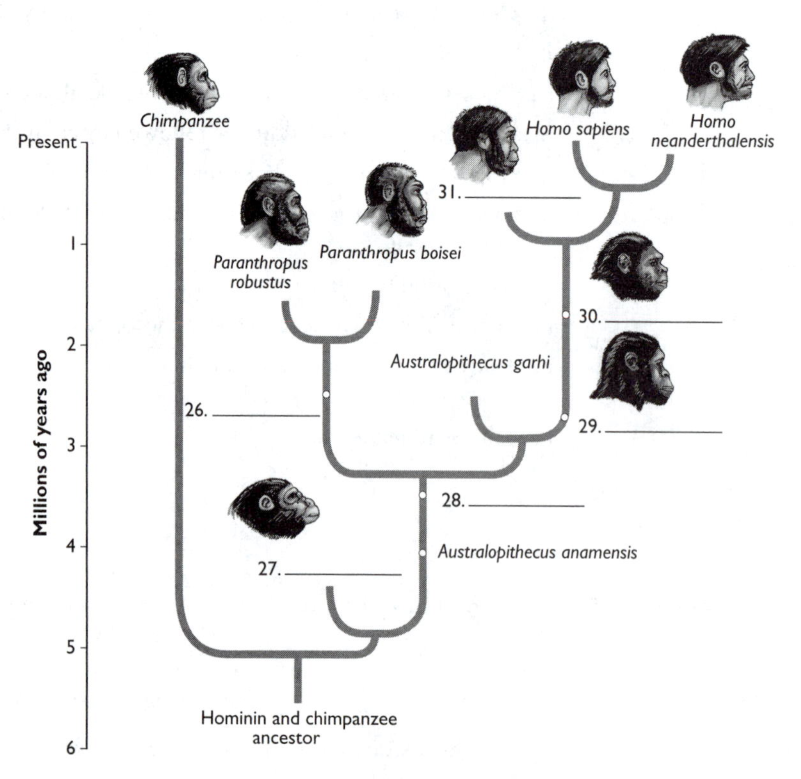

Figure 22.1

Use the words or statements below to label Figure 22.2. Words or statements that appear more than once should be used more than once.

Adequate folate
Adequate vitamin D
Adequate vitamin D
Low folate
Low vitamin D

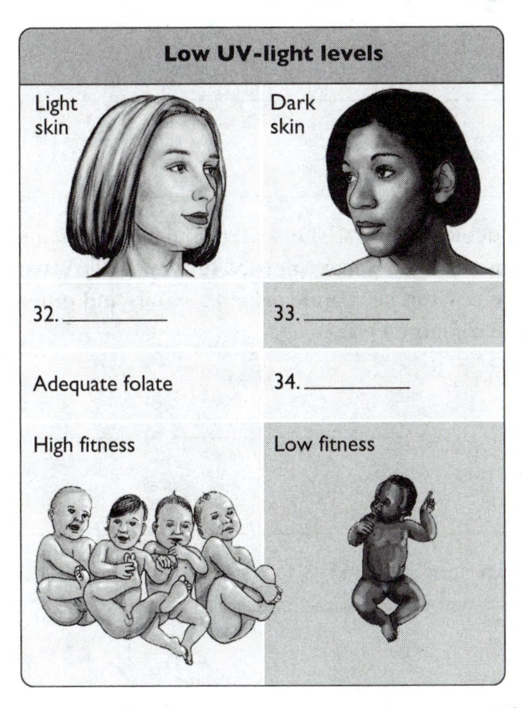

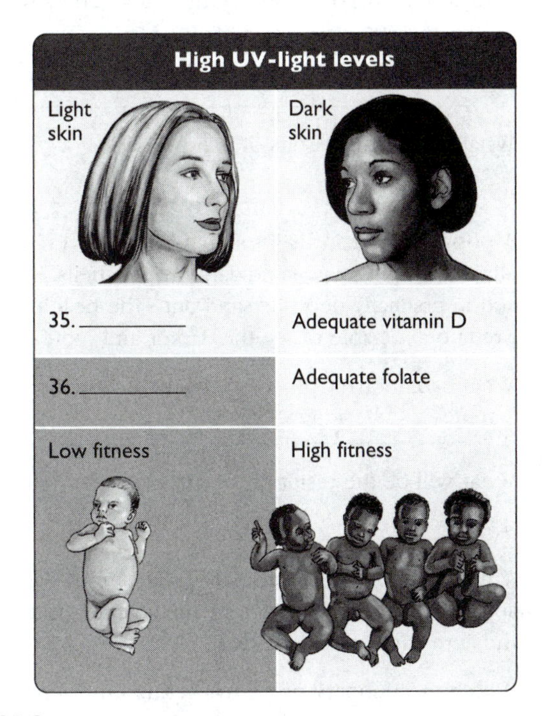

Figure 22.2

Roots to Remember

Use your knowledge of the root words presented in this chapter to answer the following questions.

37. The Latin root *-cidere* means "to kill." What does *homicide* mean?
38. The suffix *-logy* means "the study of." It comes from the Greek word *logos*, which means "word." Using this information, define the term *homology*.
39. The word *bear* is a homograph. You can bear a heavy load, but you can also see a bear in the forest. How would you define this word?

Cause and Effect

Answer the questions after reading each scenario below.

A population contains yellow flowers and red flowers. Some of the flowers in the population are orange—an intermediate between yellow and red. The bees that visit the population are attracted to yellow flowers. Hummingbirds are attracted to red flowers. Few pollinators are attracted to orange flowers.

What evolutionary process is likely to act on the population?

40. ____

What will be the result of this process after many generations?

41. ____

A small population of gray squirrels moves into a neighborhood. Some of these squirrels carry an allele for black fur. Over time, squirrels with black fur become more common in this neighborhood than in a neighborhood across town.

Which evolutionary process is acting on the population?

42. ____________________

Describe how the evolutionary change in this population occurred.

43. ____________________

Refugees from Sudan find jobs in a city in the United States. Relatives and friends from Sudan are gradually able to join them in this city. The Sudanese people establish neighborhoods, with businesses that sell Sudanese food, goods, and crafts. They generally marry within their community.

What process is taking place in the community described above?

44. ____________________

What will be the results of that process?

45. ____________________

A population of snails lives on a beach covered with multicolored rocks. The snail population contains alleles for light brown and dark brown shells. Heterozygous individuals have mottled brown shells. Wave action gradually deposits sand onto the beach, making the environment appear lighter. Gulls and other predators are able to see the darker and mottled brown snails more clearly.

What evolutionary process is likely to act on the population?

46. ____________________

What will be the result of this process after many generations?

47. ____________________

A new housing development replaces most of a forest. An acre of forest is left standing. Some of the spruce trees in that acre carry an allele for golden needles. Over generations, most of the spruce trees in the acre have golden needles.

Which evolutionary process is acting on the population?

48. ____________________

Describe how the evolutionary change in this population occurred.

49. ____________________

Paragraph Completion

Use the words or statements below to complete the paragraph. Any word or statement listed more than once should be used more than once.

natural selection
fitness
folic acid
vitamin D
darker
lighter
sexual selection
positive assortative mating
maintained
genetic drift
founder effect
alleles

Many factors play a role in causing or reinforcing differences in human groups. Differences in skin color appear to be caused by (50) ____________. UV light level can affect the (51) ____________ of individuals by influencing reproduction. Too much UV light can interfere with the body's ability to retain (52) ____________. This vitamin is essential to proper fetal development and adequate sperm production. Too little UV light can interfere with the synthesis of (53) ____________. This vitamin is essential for proper bone development. In areas where UV light is intense, human populations have (54) ____________ skin, in order to block the harmful rays. In areas where UV light is weak, human populations have (55) ____________ skin, in order to absorb enough of the light.

The process of (56) ____________ may reinforce certain physical traits. In this process, men or women choose mates who have particular physical features. This process is similar to (57) ____________, in which individuals choose mates who resemble themselves. In both of these processes, certain traits are (58) ____________ or become more common in the population, and certain traits become less common.

Differences in human populations can also arise by chance through (59) ____________. When a small population migrates to a new area, the (60) ____________ may cause certain alleles to become more common in that population than in the larger population to which these migrants used to belong. Smaller populations are also subject to chance events in which certain rare (61) ____________ disappear from the population.

Word Choice

Circle the word or phrase that correctly completes each sentence.

62. Natural selection (causes/does not cause) the emergence of new variants.
63. A thin, delicate bill is likely an adaptation to a large supply of (small seeds/trumpet-shaped flowers).
64. Variations within a population must have a (genetic/physical) basis in order for their frequency to be affected by natural selection.
65. Both *Homo sapiens* and (*Homo neanderthalensis*/*Homo habilis*) lived during the same time period.
66. Humans and apes both belong to the class (Mammalia/Primates).

Critical Thinking

67. Why is it possible for a scientist to believe in both God and evolution?
68. The black-footed ferret is a highly endangered species of weasel that lives in the grasslands of the western United States. In 1986, only 18 black-footed ferrets were known to exist in the wild. These ferrets were taken into captivity to try to increase the population and save the species. They are gradually being reintroduced into their former habitats. Describe what is likely to have happened genetically to this population as it declined and then increased due to captive breeding.
69. What evidence helps to disprove the hypothesis that humans should be separated into races?

Practice Test

1. Which level of hierarchical classification would contain tigers and fish but not insects?
   A. kingdom
   B. phylum
   C. class
   D. order

2. Which of the following is *not* a vestigial trait in humans?
   A. tailbone
   B. thumb

C. erector pili
D. appendix

3. Males of many species of birds have evolved elaborate mating dances. The evolution of these dances is likely to be a result of
A. natural selection.
B. positive assortative mating.
C. sexual selection.
D. genetic drift.

4. Which of the following organisms are likely to have the most similar DNA sequences?
A. goat and horse
B. lily and moss
C. whale and sea anemone
D. algae and bacteria

5. A scientist uses radiometric dating to test the rock layer below a fossilized bone. The radioactive element measured in the rock has a half-life of 245,000 years. About how old is the fossil if approximately 0.4% of the radioactive material remains?
A. 1.23 million years old
B. 1.47 million years old
C. 1.72 million years old
D. 1.96 million years old

6. In the mid-20th century, a small group of wolves migrated to Isle Royale in Lake Superior by walking across the ice during one cold winter. Evolutionary change in this population has most likely been influenced by
A. assortative mating.
B. a population bottleneck.
C. the founder effect.
D. natural selection.

7. Environmental changes on an island cause a reduction in plant populations that produce small seeds and an increase in plant populations that produce large seeds. A population of birds on the island has variants with different-sized beaks. Over time, which of the following variants would become most common in the population?
A. individuals with large, broad beaks
B. individuals with long, thin beaks
C. individuals with small, strong beaks
D. individuals with small, pointed beaks

8. Which of the following is true about natural selection?
A. Natural selection results in a population that has better traits than the previous population.
B. Natural selection causes new traits to appear.
C. Natural selection produces individuals that are best adapted to their current environment.
D. Natural selection produces individuals that are the best fighters.

9. Which early hominin gave rise to *Homo habilis*?
A. *Australopithecus africanus*
B. *Paranthropus robustus*
C. *Homo erectus*
D. *Homo neanderthalensis*

10. A large lake decreases in size until it becomes two separate lakes. The population of trout in one of the lakes diverges from the population in the other lake until these two populations appear to be different species. How could you test to see if they really are different species?
A. Try to get individuals from the two populations to mate to see if they produce fertile offspring.
B. Examine the anatomy of individuals from both populations to see if there are significant differences.

C. Examine the DNA sequences of individuals from each population to see if there are significant differences.
D. Observe individuals from each population to see if their reproductive behaviors differ significantly.

11. Which of the following is *not* part of the theory of natural selection?
A. Variation among individuals in a population can be passed to offspring.
B. Individuals with adaptations are more likely to survive and reproduce.
C. Certain traits give greater fitness to an individual in any environment.
D. More offspring are produced than will survive in any population.

12. In modern society, many people have poor eyesight and need to wear glasses or contact lenses. This shows that natural selection
A. still greatly affects modern human populations.
B. has less of an effect on modern human populations, due to technological advances.
C. no longer has any effect on modern human populations.
D. has less of an effect on modern human populations than does sexual selection.

13. What is the best evidence for a single origin for all life on Earth?
A. the presence of homologous structures
B. the universality of DNA
C. the presence of vestigial traits
D. the presence of carbon in all living organisms

14. Which of the following organisms would be most closely related to green algae?
A. yeast
B. sulfur bacteria
C. *Paramecium*
D. moss

Answer Key

Matching

1. e; **2.** j; **3.** b; **4.** g; **5.** n; **6.** f; **7.** l; **8.** d; **9.** m; **10.** a; **11.** k; **12.** i; **13.** o; **14.** h; **15.** c

Fill-in-the-Blank

16. order; class; **17.** Monera; Protista; **18.** wings; flippers; **19.** fewer; **20.** biogeography; descended; **21.** foramen magnum; bipedalism; **22.** half-life; 4; **23.** *Homo erectus*; *Ardipithecus ramidus*; **24.** natural; sexual; **25.** population bottleneck; alleles

Labeling

26. *Paranthropus aethiopicus*; **27.** *Ardipithecus ramidus*; **28.** *Australopithecus afarensis*; **29.** *Australopithecus africanus*; **30.** *Homo habilis*; **31.** *Homo erectus*; **32.** Adequate vitamin D; **33.** Low vitamin D; **34.** Adequate folate; **35.** Adequate vitamin D; **36.** Low folate

Roots to Remember

37. to kill a human; **38.** the study of traits with a shared or similar origin; **39.** A homograph is a word that can have different meanings but is spelled the same.

Cause and Effect

40. natural selection; **41.** The population will diversify into a yellow variety and a red variety; **42.** genetic drift; **43.** Due to the founder effect, an allele that may have been rare in the larger population of gray squirrels becomes more common over time in a smaller population; **44.** positive assortative mating; **45.** Alleles that are common in the Sudanese community will be maintained. Physical differences between the Sudanese population and other populations in the area will remain distinct; **46.** natural selection; **47.** The population will evolve in the direction of light brown shells; **48.** population bottleneck;

49. A formerly rare allele in the larger population increased in frequency, due to the dramatic reduction in the population. This allele then became common in the population over generations.

Paragraph Completion

50. natural selection; **51.** fitness; **52.** folic acid; **53.** vitamin D; **54.** darker; **55.** lighter; **56.** sexual selection; **57.** positive assortative mating; **58.** maintained; **59.** genetic drift; **60.** founder effect; **61**. alleles

Word Choice

62. does not cause; **63.** trumpet-shaped flowers; **64.** genetic; **65.** *Homo neanderthalensis*; **66.** Mammalia

Critical Thinking

67. Sample answer: Evolution is a well-supported scientific theory that explains the process by which organisms change and diversify on Earth. It is the best explanation for the diversity of life, based on human logic and the scientific method. Scientific methods cannot prove or disprove the existence of God or a higher power. Belief in God or a higher power is separate from science and often based on personal or community experiences. Therefore, a scientist can accept the theory of evolution and still believe in God or a higher power.

68. Because their numbers got so low, chance events likely eliminated alleles that were low in frequency in the population. The black-footed ferret also went through a population bottleneck, so some alleles that were rare in the former, larger population may have become common in the population that emerged from the bottleneck.

69. No conclusive pattern of allele frequencies among groups that are racially similar versus between groups that are racially different; lack of correlation between differences in DNA sequences and skin color; lack of specific genes or alleles that correspond to a specific race of people; evidence that environmental factors, such as UV light intensity, can affect skin color.

Practice Test

1. B; **2.** B; **3.** C; **4.** A; **5.** D; **6.** C; **7.** A; **8.** C; **9.** A; **10.** A; **11.** C; **12.** B; **13.** B; **14.** C

CHAPTER 23

ECOLOGY: IS EARTH EXPERIENCING A MASS EXTINCTION?

Learning Goals

1. Describe what the study of ecology entails.
2. Describe exponential growth.
3. Define *carrying capacity* and its effect on a population.
4. Explain why species at higher trophic levels are more vulnerable to extinction.
5. Describe the four causes of extinction to which human beings contribute.
6. Compare and contrast the four ecological interactions within a community—predation, parasitism, mutualism, and competition—and give an example of each.
7. Describe how energy flows in an ecosystem and why there is less energy available at higher trophic levels than at lower trophic levels.
8. Give two reasons large populations are needed to remove species from threatened status.

Chapter Outline

I. Limits to Population Growth
  A. Overview
    1. Ecology is the study of interactions between organisms and their environment.
    2. Ecology can be studied at many levels.
  B. Principles of Population Ecology
    1. A population is all of the individuals of a species within a given area.
    2. Populations are structured according to their distribution, or spacing of individuals.
    3. Populations are structured by their abundance.
      a) Abundance depends on growth rate.
      b) Growth rate is a function of the birth rate minus the death rate.
      c) Human populations are growing exponentially.
        (1) Exponential growth is growth that occurs in proportion to the total.
        (2) The quantity of new offspring each year is an ever-growing number.
        (3) Exponential growth results in a J-shaped curve.
        (4) Exponential human growth puts pressure on other species.
  C. Carrying Capacity and Logistic Growth
    1. Population growth is limited by resources that individuals need in order to survive and reproduce.
    2. Carrying capacity is the maximum population that can be supported indefinitely in a given environment.
      a) Populations stabilize around their carrying capacity.
      b) This pattern of growth is called logistic growth.
    3. Density-dependent factors are population-limiting factors that depend on population size.
      a) Density-dependent factors include food supply, risk of infectious disease, and an increase in toxins caused by increased waste.
      b) Density-dependent factors influence both birth and death rates.
    4. Density-independent factors are population-limiting factors that do not depend on population size.
      a) Density-independent factors include drought, dangerous storms, and other weather-related factors.
      b) Density-independent factors can interact with density-dependent factors.

D. Population Crashes
   1. A population crash is a steep decline in population numbers.
   2. Population crashes can occur when population numbers increase above the carrying capacity.
      a) A population may operate in a boom-and-bust cycle.
      b) A population may stabilize at carrying capacity.
   3. Future population growth can be estimated by looking at a population pyramid.
      a) A population pyramid shows the numbers and proportions of each sex and age group.
         (1) When the age structure shows a larger proportion of young people, the population will continue to increase.
         (2) When the age structure shows an even distribution, the population will remain stable.
         (3) When the age structure shows a larger proportion of adults than young people, the population will decline.
      b) The United States shows a stable or declining population.
      c) South Africa shows an increasing population.

II. The Sixth Extinction
A. Overview
   1. Extinction is the complete loss of a species.
   2. The Endangered Species Act (ESA) was passed in 1973 to protect threatened and endangered species.
   3. The ESA was passed to protect and enhance biodiversity, or the variety of all living organisms.
B. Measuring Extinction Rates
   1. The history of extinctions in the past can be estimated by the fossil record.
      a) A wide variety of animal groups appeared around 580 million years ago.
      b) Five mass extinctions—species loss on a global scale—have occurred since then.
   2. Many scientists argue that we are seeing a sixth mass extinction, brought on by global human activity.
      a) The historical rate of extinction is about one species per million per year, or 0.0001% per year.
      b) The rate of known extinctions, calculated over the last 400 years, is about 0.01%.
         (1) The current rate is about 100 times higher than the historical rate.
         (2) The rate has dramatically risen over the past 150 years.
         (3) The World Conservation Union (IUCN) estimates that 12% of birds, 20% of mammals, and 29% of amphibians are in danger of extinction.
C. Causes of Extinction
   1. Habitat Destruction
      a) A habitat is the place where a species lives and obtains food, shelter, water, and space.
      b) Habitat damage or destruction threatens 83% of endangered mammals, 89% of endangered birds, and 91% of endangered plants.
      c) Habitat destruction is the outright loss of habitat.
         (1) Agricultural, industrial, and residential development has destroyed habitat throughout the 20th century.
         (2) Habitat decreases the carrying capacity of Earth for large numbers of species.
      d) A species-area curve shows the relationship between the size of a natural area and the species it can support.
         (1) The number of species increases rapidly in an area as size increases.
         (2) The rate of increase slows as the area becomes very large.
         (3) Using a species-area curve, the predicted habitat loss of 90% of the Amazonian tropical rain forest over the next 35 years would result in an extinction of about 50% of the species living there.
         (4) The IUCN predicts that, at the present rate of habitat destruction, nearly 25% of all living species will be lost in the next 50 years.
   2. Habitat Fragmentation
      a) In habitat fragmentation, large areas of natural habitat are broken up.
      b) Fragmentation threatens large predators, who need large hunting areas.
         (1) Energy flows in one direction along a food chain, which contains many trophic levels.
            (a) Producers (plants and other photosynthetic organisms) convert energy from the sun.
            (b) Primary consumers obtain energy by eating producers.
            (c) Secondary consumers, such as predators, obtain energy by eating primary consumers.

(2) Most of the calories taken in at each trophic level go to support the activities of the individual at that level.
(a) A major part of the solar energy that was originally captured is given off as heat at each level.
(b) Only about 10% is passed along to the next level.
(c) A trophic pyramid shows the bottom-heavy relationship between the biomass (total weight) of populations at each level of the chain.
(i) If the population of producers decreases, fewer consumers can be supported by the ecosystem.
(ii) Habitat destruction and fragmentation can reduce producer populations, thereby reducing consumer populations.

3. Other Human Causes of Extinction
a) Introduced Species
(1) Introduced species are organisms that are brought by humans to a region where they had previously not been found.
(2) Introduced species have not evolved with native species.
(a) They may compete with native species for resources.
(b) They may kill native species.
b) Overexploitation
(1) Overexploitation occurs when the rate of human use of a species outpaces the species' reproduction.
(2) Overexploitation may occur when particular parts of an organism are prized by humans.
c) Pollution
(1) Pollution occurs when poisons, excess nutrients, or other wastes are released into the environment.
(a) Fertilizer can cause pollution.
(i) Fertilizer may have high levels of nitrogen and phosphorus.
(ii) These nutrients increase algal growth when they run off into streams and lakes.
(iii) Increased algal growth leads to an increase in the bacteria that feed on the algae.
(iv) As the bacteria feed, they use up the available oxygen in the water.
(v) Oxygen-depleting eutrophication causes fish kills.
(2) Pollution can affect air, water, and land.

D. The Consequences of Extinction
1. Loss of Resources
a) Wild areas can provide food resources.
(1) The loss of wild species has a direct economic effect.
(2) Wild products obtained in the U.S. are worth an estimated $87 billion a year.
b) Wild species can provide unique biological chemicals.
(1) Some plants have biological chemicals that may be used to develop new drugs.
(2) Drugs derived from the rosy periwinkle of Madagascar help to improve survival from leukemia and Hodgkin's disease.
c) Alleles from wild species that are related to domestic species may be used to develop better crops.
d) Wild organisms that reduce pest damage on wild crops may also be used to prevent pest infestations on related domestic crops.
2. Disruption of Ecological Communities
a) Overview
(1) A community consists of all the organisms that live together in a particular habitat area.
(2) Organisms in a community are linked together in a food web.
(3) Any disruption in one portion of the food web will affect the other portions.
b) Species Interactions
(1) In predation, one species (the predator) feeds on another species (the prey).
(2) In a mutualism, two species benefit from a long-term interaction.
(3) Two species are in competition when they rely on the same limited resources.
(4) In some communities, a single species (called a keystone species) can play a role that is critical and larger than expected.
(a) Effects of a keystone species may be indirect.
(b) Removal of the keystone species can cause a collapse in a food web.

3. Damaged Ecological Systems
   a) Energy Flow
      (1) An ecosystem consists of all of the organisms in a given area, along with their nonliving environment.
      (2) Energy flows through an ecosystem.
         (a) The primary energy source for nearly all ecosystems is the sun.
         (b) Producers convert solar energy to chemical energy through the process of photosynthesis.
         (c) Energy is passed through the trophic levels of the food chain.
      (3) The biodiversity in an ecosystem can have strong effects on energy flow.
         (a) Overall plant biomass tends to be greater in more diverse areas.
         (b) A decline in diversity can decrease biomass, making less energy available to organisms higher on the food chain.
   b) Nutrient Cycling
      (1) Essential mineral nutrients pass through a food web through nutrient cycling.
         (a) Complex molecules move through the food web with relatively minor changes until they reach the soil.
         (b) In soil, complex molecules are broken down into simpler ones by the action of decomposers.
      (2) Nitrogen is one nutrient that cycles through food webs.
      (3) Changes in the soil community can greatly affect nutrient cycling.
         (a) Earthworms have been introduced in forests throughout the northeastern U.S.
         (b) These worms have caused dramatic reductions in the diversity and abundance of plants on the forest floor.

III. Saving Species
   A. Protecting Habitat
      1. Overview
         a) The most effective way to prevent loss of species is to preserve as many habitats as possible.
         b) According to the species-area curve, species diversity declines slowly as habitat area declines.
         c) If the rate of habitat destruction is slowed or stopped, extinction rates will slow as well.
      2. Protecting the Greatest Number of Species
         a) Conservation International has identified 34 worldwide biodiversity "hotspots."
         b) The biodiversity hotspots make up less than 3% of Earth's land surface, yet they contain up to 50% of all plant, amphibian, reptile, bird, and mammal species.
         c) These hotspots are severely threatened by human activity.
         d) By saving these spots, we could greatly reduce the global extinction rate.
      3. Protecting Habitat for Critically Endangered Species
         a) The ESA requires the designation of critical habitats for endangered species.
         b) These critical habitats are areas necessary for survival of the species.
         c) Critical habitat designation results in the restriction of human activities.
         d) Some critical habitats may not be protected if that protection interferes with significant economic benefits.
      4. Decreasing the Rate of Habitat Destruction
         a) Daily personal actions can help to reduce habitat destruction.
            (1) Reducing consumption of meat and dairy products can reduce the conversion of land to agricultural production.
            (2) Reducing the use of wood and paper products, and buying wood products from sustainably managed forests, can reduce the loss of forest habitat.
         b) Aid to developing countries can help these countries to invest in technologies that decrease the use of natural resources.
         c) Programs can help to slow the rate of human population growth, thereby protecting habitat.
         d) Political and educational activities sponsored by conservation groups can help to save habitats.
   B. Ensuring Adequate Population Size
      1. Threatened populations need to be large enough to survive environmental threats.
      2. The heath hen is an example of an animal whose population became too small.
         a) The heath hen once numbered in the hundreds of thousands on the East Coast of the U.S.
         b) Habitat loss and heavy hunting reduced their numbers until the species was endangered.

c) A nature reserve was established, and heath hen numbers increased.
d) Unpredictable events caused the species to become extinct in the 1930s.
3. Larger populations can weather events that kill 90% of their population.
4. Small populations lose genetic variability due to genetic drift.
a) When alleles are lost, individuals become homozygous for more genes.
(1) Increased homozygosity leads to lower reproduction.
(2) Increased homozygosity leads to higher death rates.
b) Populations with low genetic variability cannot evolve in response to environmental changes.
(1) Irish potatoes in the 19th century were widespread but had low genetic variability.
(2) When potato blight came to Ireland in 1845, almost all the potatoes were infected and rotted in the fields.
(3) The entire potato crop failed in 1846 and again in 1848, causing the deaths of almost 1 million Irish people and the emigration of 1.5 million from Ireland to North America.

C. Meeting the Needs of Humans and Nature
1. Actions to save endangered species can cause economic and emotional hardship for many people.
2. A political solution that causes some economic hardship while ensuring short-term survival of the species will often prevail.
3. Risk to the long-term survival of a species often balances the cost to human populations.
4. Restoration ecology can restore habitats to a state where natural processes function effectively to protect an endangered species.
5. The best strategy for preserving biodiversity is to prevent species from becoming endangered.

Practice Questions

Matching

1. producer
2. extinction
3. biodiversity
4. predator
5. logistic growth
6. trophic level
7. consumer
8. keystone species
9. ecology
10. exponential growth
11. mutualism
12. decomposer
13. overexploitation
14. carrying capacity
15. community

a. organism that converts energy from the sun
b. level in a food chain
c. maximum population that can be supported indefinitely in an environment
d. interaction between two species that benefits both species
e. all organisms living together in a given habitat
f. variety of living organisms
g. species with an ecosystem role that is critical
h. complete loss of a species
i. growth that occurs in proportion to the current total
j. organism that eats other organisms to obtain energy
k. organism that breaks down complex molecules into simpler ones
l. field that focuses on interactions between organisms and their environment
m. when human use of a species outpaces reproduction of that species
n. organism that survives by hunting and eating another organism
o. growth that declines to zero as it approaches a carrying capacity

Fill-in-the-Blank

16. In a given area, there may be many ____________, each composed of one species, that make up the ____________, a group that includes all of the species in the area.
17. A population is structured by its spacing, or ____________, and its number, or ____________, of individuals.

18. A population of bacteria may grow __________ in a Petri dish until their food supply becomes __________.

19. Warm weather in early spring may allow robins to nest early and produce more offspring throughout the summer. Warm weather is a density __________ factor that influences __________ growth.

20. A __________ predicts that as size of a habitat __________, the biodiversity of the area increases.

21. In a lake food chain, algae are __________ that are eaten by small fish, which are __________ consumers.

22. Fertilizer that runs off into lakes or streams can increase the growth of __________ and bacteria, which can lead to oxygen-depleting __________, resulting in fish kills.

23. Some species of tropical ants protect certain plants from predators. In turn, these "ant plants" produce small structures in which the ants can hide. The interaction between ants and ant plants is a __________ that benefits __________ species.

24. A food __________ has only one species at each trophic level while a food __________ may have many.

25. One species of warbler may feed at the top of a tree while another species feeds toward the center of the tree. In this way, the two species avoid __________ for __________ resources.

Labeling

Use the words or statements below to label Figure 23.1. Words or statements that appear more than once should be used more than once.

carrying capacity
exponential growth
logistic growth
no growth
population size
time

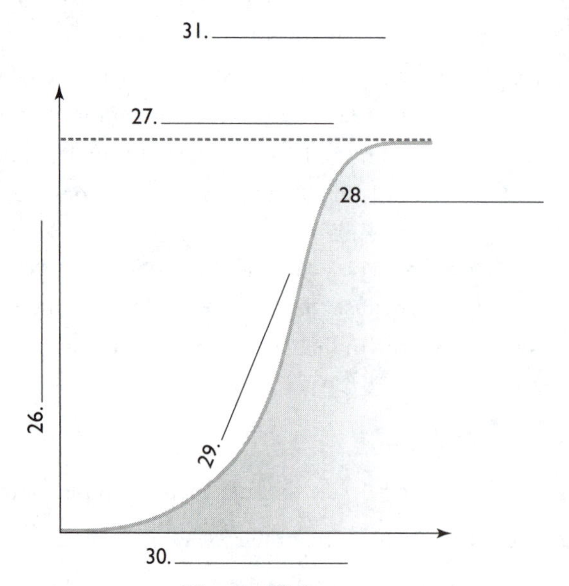

Figure 23.1

Use the words or statements below to label Figure 23.2.

ammonia (NH_3)
decomposers
nitrogen (N_2)
nitrogen-fixing bacteria
primary consumer
producers
secondary consumer

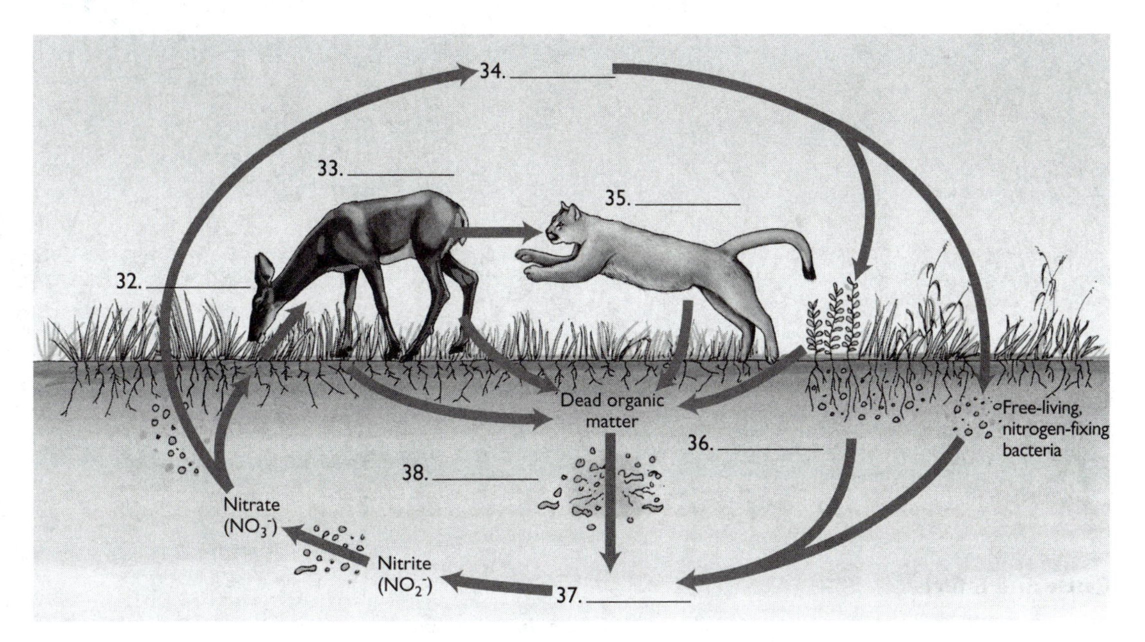

Figure 23.2

Roots to Remember

Use your knowledge of the root words presented in this chapter to answer the following questions.

39. The Greek word *heteros* means "other, different." What would the word *heterotroph* mean?
40. Plants and other photosynthetic organisms are autotrophs. What does the word *autotroph* mean?
41. Translated literally, the word *ecology* means the study of the house or habitation. How does this literal translation relate to the discipline of ecology?

Word Choice

Circle the word or phrase that correctly completes each sentence.

42. A robin hunts for worms in a lawn. The robin could be classified as a (predator/parasite).
43. In the nitrogen cycle, nitrogen in the form of ammonia is (released/taken up) by decomposers.
44. Biodiversity hotspots are overwhelmingly located in (coastal/mid-continental) areas.
45. A tornado is a (density-dependent/density-independent) event that can reduce populations.
46. A(n) (community/ecosystem) includes both biological and nonbiological factors.
47. An antelope in one food chain is likely to be at a (higher/lower) trophic level than a warbler in a different food chain.

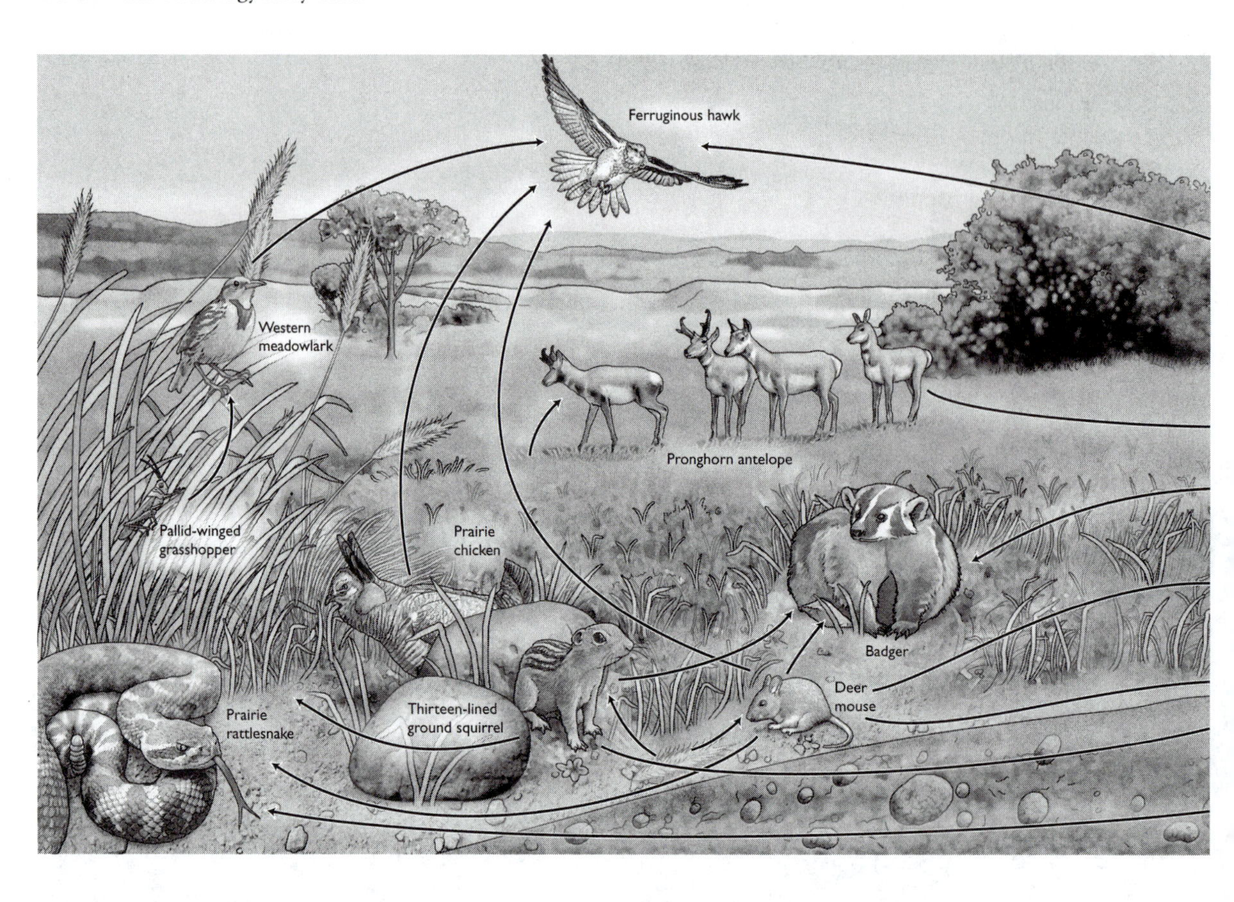

Cause and Effect

Examine each scenario presented below and answer the associated questions.

The Eurasian ruffe is a fish species that traveled across the Atlantic Ocean to the Great Lakes in North America in the ballast water of a ship. The Eurasian ruffe has become established in two of the Great Lakes.

What would you call this fish species?

48. ______________________________

How is this fish species likely to affect native species in its Great Lakes habitats?

49. ______________________________

The yellow-bellied sapsucker is a woodpecker that creates a series of holes in tree bark. Sap flows from these holes, to feed the bird. This sap attracts ants and other insects. The insects that feed on the sap attract other birds. Hummingbirds feed on the sap from these holes, as do squirrels and other mammals.

What would you call this bird species?

50. ______________________________

What would likely occur if yellow-bellied sapsucker populations declined in an area?

51. ______________________________

Roadless areas have been created in some national forests in the United States in order to protect certain threatened species.

What is a roadless area likely to prevent?

52. ______________________________

What would likely happen to large predators in these areas if many roads were constructed and these areas were turned into more active tourist parks?

53. ______________________________

The following illustration shows interactions between North American prairie species.

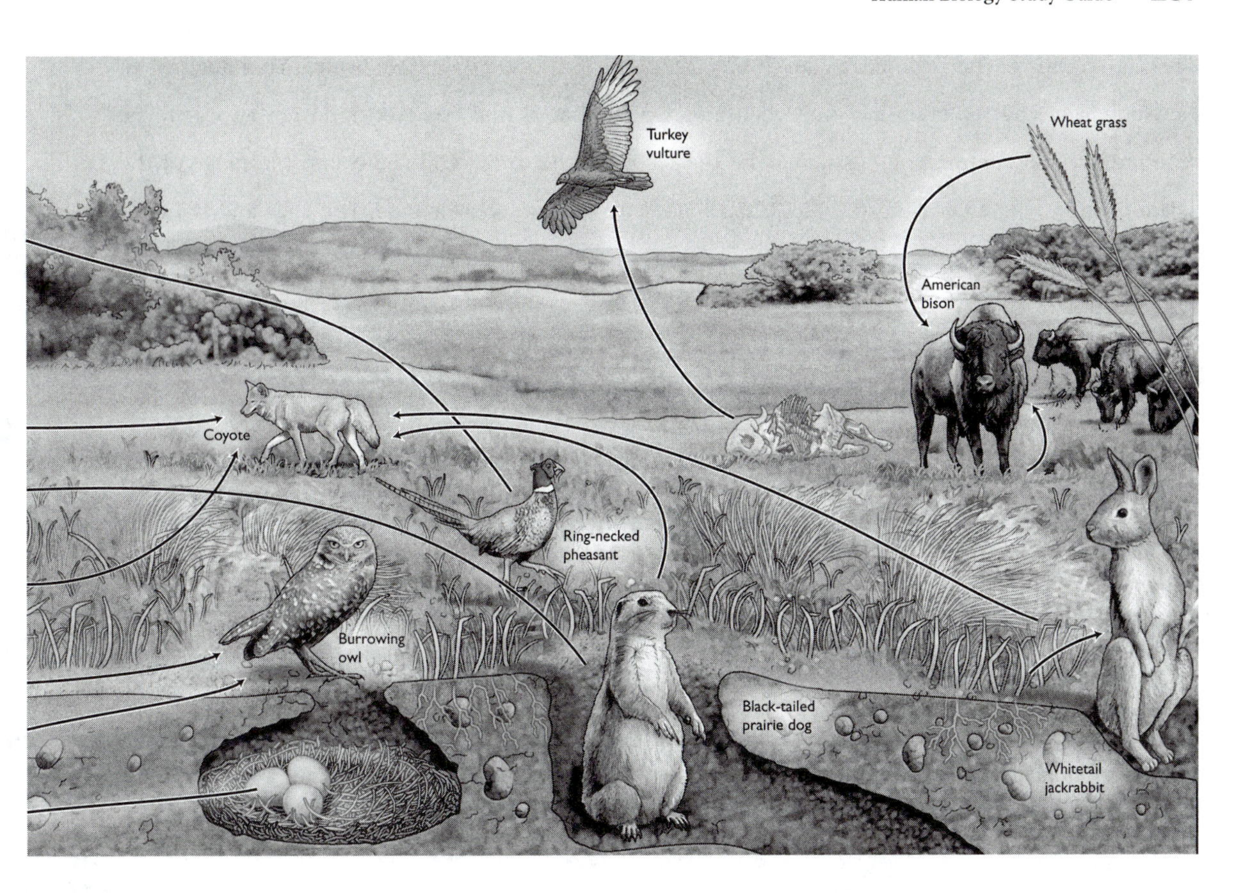

What kind of relationship exists between the 13-lined ground squirrel and the deer mouse?

54. ______________________________

What kind of relationship exists between the black-tailed prairie dog and the coyote?

55. ______________________________

What would likely happen to the ring-necked pheasant population if both the western meadowlark and deer mouse populations decreased? Explain your answer.

56. ______________________________

Why are coyote populations likely to remain relatively stable?

57. ______________________________

What would likely happen to the burrowing owl population if the prairie rattlesnake population increased? Explain your answer.

58. ______________________________

Paragraph Completion

Use the words or statements below to complete the paragraph. Any word or statement listed more than once should be used more than once.

compete
crash
economic
extinction
overexploitation
predators
prey

Atlantic cod populations once numbered in the millions in the waters off of Newfoundland in Canada. However, (59) ______________ caused the Atlantic cod population to (60) ______________ in the late 20th century. This has led to an (61) ____________ decline in the fishing communities of Newfoundland and a moratorium on cod fishing. So far, the Atlantic cod stocks do not appear to be rebounding. One contributing factor may be the movement of arctic cod south into the range of the Atlantic cod. Young arctic cod (62) ____________ with Atlantic cod for food. The eggs and larvae of Atlantic cod are also (63) ____________ for the arctic cod. Another factor may be the abundance of harp seals. Harp seals are (64) ____________ of cod. As years pass and the Atlantic cod fails to return in significant numbers, many scientists worry that this species may be facing (65) ____________.

Critical Thinking

66. The zebra mussel causes major changes to ecosystems into which it has been introduced. Why wouldn't this species cause such dramatic changes in its native habitat?
67. Are humans separate from Earth's ecosystems or part of them? Explain your answer.
68. What are three things that you can do to reduce habitat destruction?

Practice Test

1. In the nitrogen cycle, decomposers convert dead organic material to
   A. nitrate (NO_3^-).
   B. ammonia (NH_3).
   C. nitrite (NO_2^-).
   D. nitrogen (N_2).

2. A food chain consists of grass that is eaten by grasshoppers. The grasshoppers are eaten by grasshopper mice. The grasshopper mice are eaten by red fox. There are 4,000 units of energy in the grass population in this food chain. How many units of energy are there in the grasshopper mouse population?
   A. 0.4 unit
   B. 4.0 units
   C. 40 units
   D. 400 units

3. A population of chickadees reaches its carrying capacity in a forest. Which of the following is true about this population, if it experiences no migration of members in or out?
   A. There are more births than there are deaths.
   B. There are more deaths than there are births.
   C. The number of births and deaths are equal.
   D. There is no way to know how births relate to deaths.

4. An introduced weed would be most likely to colonize and take over an area that
   A. has high biodiversity.
   B. has low biodiversity but very little open ground.
   C. is covered with many other introduced species.
   D. has low biodiversity and open ground.

5. Which of the following is most likely to be most negatively affected by chemical pollution?
   A. a frog, covered with moist, porous skin
   B. a reptile, covered with dry, scaly skin
   C. a bird, covered with feathers
   D. a mammal, covered with fur

6. Each of the following events could cause population declines. Which of these describes a density-dependent factor?
   A. A tsunami washes away crabs in a beach ecosystem.
   B. Disease spreads through a moose population.

C. Climate change destroys frog eggs in a pond ecosystem.
D. Drought causes the death of a soybean crop.

7. Some tropical bats drink nectar from large, white flowers. As the bats fly from flower to flower, they spread pollen. What would you call the relationship between the bats and the flowering plant species?
A. competition
B. parasitism
C. predation
D. mutualism

8. A seed from a maple tree sprouts and grows just inches away from a seed from a cherry shrub. This interaction has
A. a positive effect on both species.
B. a positive effect on the maple but a negative effect on the cherry.
C. a positive effect on the cherry but a negative effect on the maple.
D. a negative effect on both species.

9. Which of the following would be classified as a primary consumer?
A. a rabbit that eats clover
B. a bat that eats mosquitoes
C. a large fish that eats smaller fish
D. a raspberry bush that takes up nitrogen

10. The logistic growth curve is shaped like the letter
A. J.
B. C.
C. S.
D. N.

11. Which population is likely to experience the largest growth rate in the next 20 years?
A. A population in which the age structure resembles a square.
B. A population in which the age structure resembles a pyramid.
C. A population in which the age structure resembles a hexagon.
D. A population in which the age structure resembles an upside-down pyramid.

12. If the current annual extinction rate is 0.01%, how many organisms would likely become extinct this year out of a total of 1.8 million species?
A. 18
B. 180
C. 1,800
D. 18,000

13. A black bear eats grasses, fruits, nuts, insects, and some vertebrates. A black bear is all of the following *except*
A. a primary consumer.
B. a producer.
C. a secondary consumer.
D. a predator.

14. Which of the following substances is *not* a pollutant?
A. phosphorus
B. carbon dioxide
C. pesticide
D. dihydrogen oxide

15. Wild species can be resources for humans in which of the following ways?
A. They can be used to create medicines.
B. They can serve as a food source for human populations.

C. They can be used to improve pest resistance in crops.
D. All of the above.

Answer Key

Matching

1. a; **2.** h; **3.** f; **4.** n; **5.** o; **6.** b; **7.** j; **8.** g; **9.** l; **10.** i; **11.** d; **12.** k; **13.** m; **14.** c; **15.** e

Fill-in-the-Blank

16. populations; community; **17.** distribution; abundance; **18.** exponentially; limited; **19.** independent; population; **20.** species-area curve; increases; **21.** producers; primary; **22.** algae; eutrophication; **23.** mutualism; both; **24.** chain; web; **25.** competition; food

Labeling

26. population size; **27.** carrying capacity; **28.** no growth; **29.** exponential growth; **30.** time; **31.** logistic growth; **32.** producers; **33.** primary consumer; **34.** nitrogen (N_2); **35.** secondary consumer; **36.** nitrogen-fixing bacteria; **37.** ammonia (NH_3); **38.** decomposers

Roots to Remember

39. one who eats another; **40.** makes its own food; **41.** Ecology is the study of the environment, which is the house or habitation of all of Earth's organisms.

Word Choice

42. predator; **43.** released; **44.** coastal; **45.** density-independent; **46.** ecosystem; **47.** lower

Cause and Effect

48. introduced species; **49.** It is likely to compete with native species of fish and reduce the populations of the native species; **50.** keystone species; **51.** Populations of the other species that use sapsucker holes, or feed on species that use sapsucker holes, would also decline; **52.** habitat fragmentation; **53.** Intact hunting areas would be reduced, which would likely reduce large predator populations. Encounters with humans would become more common, which would also reduce large predator populations; **54.** competition; **55.** predator–prey; **56.** It would likely decrease. The deer mouse and western meadowlark are prey for the ferruginous hawk. If their populations decreased, the hawk will need to eat more pheasants and prairie chickens; **57.** Coyotes feed on many different prey species, so if the population of one prey species decreased, they would likely be able to eat more of another prey species; **58.** The burrowing owl population would likely increase, because fewer of its eggs would be preyed upon.

Paragraph Completion

59. overexploitation; **60.** crash; **61.** economic; **62.** compete; **63.** prey; **64.** predators; **65.** extinction

Critical Thinking

66. The zebra mussel has evolved with the other species in its native habitat, so all of these species have developed relatively stable interrelationships. These interrelationships keep any one of the populations from taking over and dramatically changing the ecosystem.

67. Humans are part of Earth's ecosystems. Many human groups have evolved over hundreds or thousands of years in a particular location on Earth. Even if a human group is relatively new to a location, it still interacts with and affects the other organisms in the area.

68. Answers will vary but could include: reduction of resource use; involvement in political or educational campaigns; reduction in the consumption of meat and dairy products; reduction in the use of toxic cleaners or nonorganic agricultural products; recycling of materials.

Practice Test

1. B; **2.** C; **3.** C; **4.** D; **5.** A; **6.** B; **7.** D; **8.** D; **9.** A; **10.** C; **11.** B; **12.** B; **13.** B; **14.** D; **15.** D

CHAPTER 24

BIOMES AND NATURAL RESOURCES: WHERE DO YOU LIVE?

Learning Goals

1. Compare and contrast weather and climate. Describe the major components of climate.
2. List the three factors that influence local temperatures, and briefly describe their effects.
3. Explain how global factors and nearness to water and mountains affect precipitation patterns.
4. Compare and contrast tropical forests, temperature forests, boreal forests, and chaparral.
5. Summarize the characteristics of grasslands, desert, and tundra.
6. Explain the differences between freshwater and marine habitats.
7. Define *ecological footprint*, and describe how individual choices can change footprints.
8. Describe the evidence that shows how increased carbon dioxide in the atmosphere will lead to global climate change.

Chapter Outline

I. Terrestrial Biomes
  A. Overview
    1. Climate and weather are different.
      a) Climate describes the average weather as measured over many years.
      b) Weather describes current conditions, such as temperature, cloud cover, and precipitation.
    2. Temperature and precipitation patterns determine climate.
    3. Features on Earth influence temperature and precipitation.
      a) Latitude
        (1) Temperatures are cooler near the poles and warmer near the equator.
        (2) There is greater temperature variation near the poles.
        (3) Precipitation is high at the equator and reduced at 30° north and south latitudes.
      b) Proximity of Surface Water
        (1) Water moderates air temperature.
        (2) Ocean currents may heat or cool coastal areas.
        (3) More precipitation occurs on the side of a water body opposite the prevailing wind.
      c) Mountains
        (1) Temperatures are cooler at high altitudes.
        (2) Locally, temperatures are cooler in valleys where denser air drains.
        (3) Precipitation is higher on the side of a mountain facing prevailing winds and lower on the sheltered side.
      d) Characteristics of land surface
        (1) Ice and snow reflect sunlight, so temperatures stay cooler.
        (2) Dark surfaces absorb sunlight, so temperatures increase.
    4. Climate influences the physical appearance of vegetation in geographic areas.
      a) Plants native to the region are adapted to particular patterns of precipitation and temperature.
      b) Animals native to the region are adapted to the particular plant community as well as to the climate.
    5. There are four basic categories of land biomes, or vegetation types.
    6. Each basic category may contain several biome types.
  B. Forests and Shrublands
    1. Overview
      a) Forests are communities dominated by trees and other woody plants.
      b) Forests occupy one-third of Earth's land surface.

c) Forests contain about 70% of the biomass found on land.
d) Forests are categorized based on distance from the equator.

2. Tropical Forests
   a) Tropical forests are located at or near the equator.
   b) Tropical forests contain great biological diversity.
   c) Some areas of tropical forest may have annual wet and dry periods.
3. Temperate Forests
   a) Temperate forests are located from 23° to 50° north and south of the equator.
   b) Temperate forests experience seasonal temperature changes.
   c) Temperate forest soils are rich in nutrients.
      (1) Most forest areas in the eastern United States were converted to cropland during the 18th and early 19th centuries.
      (2) These farms were abandoned in the late 19th and early 20th centuries.
      (3) Many of these abandoned areas reverted to forest.
      (4) Urban and suburban development again threatens these forest areas.
4. Boreal Forests
   a) Boreal forests are located toward the poles.
   b) Boreal forests comprise the largest land biome on Earth.
   c) Coniferous plants that produce seed cones dominate boreal forests.
   d) Boreal forests may be threatened by unsustainable logging.
5. Chaparral
   a) Chaparral is located in coastal areas around the Mediterranean Sea, in southern California, South Africa, and southwestern Australia.
   b) Chaparral is dominated by spiny evergreen shrubs.
   c) Chaparral plants are adapted to fire.
   d) Rapid population growth into chaparral threatens the ecosystem due to fire control policies.

C. Grasslands
1. Grasslands are dominated by nonwoody grasses.
2. Grasslands are present in regions where precipitation is too limited to support trees.
3. Tropical grasslands are known as savannas.
4. Temperate grasslands include tallgrass prairies and shortgrass steppes.
   a) Greater precipitation supports taller grasses in these areas.
   b) These landscapes contain few trees.
   c) Cool temperatures in these biomes cause decomposition to slow.
      (1) Soils contain partially decayed roots.
      (2) These rich soils are prized for agriculture.
      (3) In North America, less than 1% of native prairie remains.

D. Desert
1. Deserts occur where rainfall is less than 50 centimeters per year.
2. Most great deserts are located close to 30° north or south of the equator.
3. Human populations are exploding in the desert areas of the U.S.
   a) Sunny, warm, dry weather is attractive.
   b) Human populations are putting stress on water supplies.

E. Tundra
1. Tundra is found close to Earth's poles.
2. Temperatures are coldest in tundra regions.
3. Soils are underlain by permafrost, or permanently frozen soil.
   a) Permafrost prevents water drainage.
   b) Soils above permafrost are often saturated.
4. Tundra is lightly settled by humans but not immune to human impacts.
   a) Many oil reserves are located in tundra areas.
   b) The infrastructure associated with obtaining oil interferes with tundra ecosystems.
   c) Global warming due to the burning of fossil fuels worldwide threatens permafrost.
      (1) Former tundra areas can now support shrubs and trees.
      (2) These areas are changing into boreal forest.

II. Aquatic Biomes

A. Freshwater
1. Lakes and Ponds

a) Lakes and ponds are bodies of freshwater—areas that typically have less than 0.1% of salts per total volume.
b) Fertilizers can run off and cause eutrophication.
(1) An increase in algae and associated bacterial populations decreases oxygen levels.
(2) Fish die from low oxygen.

2. Rivers
a) Rivers, streams, brooks, and creeks are flowing water moving in one direction.
b) Rivers may be threatened by chemical runoff.
c) Rivers may also be threatened by the development of dams and channels.

3. Wetlands
a) Wetlands are areas of standing water that support above-water aquatic plants.
b) Wetlands support many plant species.
(1) High productivity is due to high nutrient levels at the interface between aquatic and terrestrial environments.
(2) The large plant populations present in wetlands help to slow the flow of water running through the wetland.
(a) Wetlands help prevent flooding.
(b) Wetlands prevent sediments and pollutants from running into lakes and rivers.
c) Over 50% of the wetlands in the U.S. have been degraded or destroyed since European settlement.
d) Legislation has been enacted to slow the rate of wetland loss.

B. Saltwater
1. Overview
a) About 75% of Earth's surface is covered with saltwater, or marine, biomes.
b) Saltwater forms through evaporation, which leaves dissolved salts behind.
c) Marine biomes can be grouped into oceans, coral reefs, and estuaries.

2. Oceans
a) The open ocean covers about two-thirds of Earth's surface.
b) About 50% of the oxygen in Earth's atmosphere is generated by oceanic photosynthetic plankton.
c) Oceans also generate most of Earth's freshwater through evaporation of ocean water and precipitation over land.
d) Fish in the ocean are heavily exploited by human fishing fleets.
(1) An estimated one-third of all species of ocean fish are severely endangered.
(2) Global fish catch and biodiversity are declining rapidly.

3. Coral Reefs
a) Coral reefs are composed of the limestone skeletons of reef-building coral.
b) Coral reefs are found in warm, well-lit waters in tropical oceans.
c) The biodiversity in coral reefs is comparable to that of tropical rain forests.

4. Estuaries
a) Estuaries are the zones where freshwater rivers drain into saltwater oceans.
b) Water level fluctuations and mixed fresh and salty water make these ecosystems highly productive.
c) Estuaries provide habitat for up to 75% of commercial fish populations and 80–90% of recreational fish populations.
d) Estuaries are a rich source of shellfish.
e) Vegetation surrounding estuaries provides a buffer zone that reduces shore erosion.
f) Estuaries are threatened by fertilizer pollution and resort or housing developments.

III. Human Habitats
A. Overview
1. Humans have modified 50% of Earth's land surface for their own use.
2. Most of this modification is a result of farming and logging.
3. Around 2–3% has been modified for human settlements.

B. Energy and Natural Resources
1. Comparing Systems
a) A forest is a local system that recycles its wastes into resources using energy from the sun.
b) A city imports energy and materials and produces wastes that are mainly disposed of elsewhere.

2. Energy Use
   a) Most energy used to power cities in developed countries is derived from fossil fuels.
   b) The environmental impacts of acquiring and transporting fossil fuels can be substantial.
      (1) Oil spills degrade oceans and estuaries.
      (2) Whole mountains are destroyed during coal mining.
   c) Most energy used to power settlements in nonindustrial countries is derived from resources in the surrounding environment.
      (1) Wood and plant-based materials are commonly used for heating and cooking.
      (2) This can lead to destruction of surrounding forests.
3. Natural Resources
   a) Human settlements use many materials.
      (1) Clean water is used for human and industrial consumption.
      (2) Food is used for nutrition.
      (3) Metals and wood are used for shelter and products.
      (4) Wood is used for paper and packaging.
      (5) Petroleum is used for energy, asphalt, and plastics.
   b) An ecological footprint describes the amount of land needed to support the human activity there.
      (1) The ecological footprint for the city of London in 2000 was equal to twice the entire land surface of the United Kingdom.
      (2) Most resources that supported London came from other places.
      (3) An ecological footprint can be reduced if people make sensitive consumer choices.

C. Waste Production
1. Wastewater
   a) In developed countries, sewage treatment plants handle wastewater from homes and industrial plants.
      (1) These plants remove semisolid wastes.
      (2) They typically use chemicals to kill disease-causing microbes.
      (3) They eventually discharge treated water into lakes, streams, or the ocean.
   b) Older treatment plants can be overwhelmed by storm water.
      (1) Untreated water is discharged into water bodies.
      (2) Untreated water contains nutrients that can cause increased growth of algae and bacteria.
   c) Semisolid wastes are often processed and applied to land as fertilizer.
      (1) If properly treated, human waste can be a valuable fertilizer.
      (2) Industrial chemicals in this processed waste can introduce toxins to the land.
   d) In less-developed countries, untreated human waste may flow in gutters.
   e) Intestinal diseases resulting from microbes in untreated human waste cause the deaths of more than 2 million children under five years old each year.
2. Garbage and Recycling
   a) Garbage, or solid waste, is often placed in sanitary landfills.
      (1) Landfills are pits lined with resistant material.
      (2) Landfills have systems for collecting liquid that drains through waste.
      (3) Landfills have pipes that vent dangerous gases produced during decomposition.
   b) Many communities have recycling programs.
      (1) Paper, glass, metal, and plastics are collected and sold to recycling companies.
      (2) Food and yard waste may also be collected and composted.
   c) Less-developed countries may have open dumps instead of sanitary landfills.
3. Air Pollution
   a) Fossil fuels produce large amounts of gaseous waste.
      (1) Emissions include carbon dioxide, a chief contributor to global warming.
      (2) Other emissions include nitrogen and sulfur oxides, small airborne particles, and contaminants such as mercury.
   b) When some of these emissions are exposed to high temperatures and sunlight, they react with oxygen to produce ground-level ozone, or smog.
   c) Ground-level ozone can cause severe illness or death in people with asthma, heart disease, or reduced lung function.

d) Air emissions can travel great distances in the upper atmosphere.
   (1) Air emissions from coal-fired power plants in the Midwest have caused acid rain in the northeastern United States.
   (2) Airborne toxins such as benzene and PCBs have been found in high levels in animals around the North Pole.

D. Climate Change
1. Carbon dioxide is produced by burning fossil fuels.
2. Carbon dioxide has been rapidly accumulating in the atmosphere for the past 150 years.
3. A large and rapid change in carbon dioxide levels influences weather patterns on Earth.
   a) Carbon dioxide contributes to the greenhouse effect, which keeps Earth relatively warm by trapping heat from the sun.
   b) An increase in carbon dioxide can cause an increase in global temperatures.
4. Long-term records have shown that carbon dioxide levels and global temperatures have always fluctuated.
5. The concentration of carbon dioxide in the atmosphere is much higher now than it has been in the last 400,000 years.
6. Computer models predict that average global temperatures will increase between 1.5°C and 4.5°C by 2075.
   a) Warming is not predicted to be uniform.
      (1) Ocean currents may change.
      (2) Typical patterns of snow and cloud cover will be modified.
      (3) The poles are predicted to warm faster than other areas.
      (4) Some regions are likely to cool slightly.
   b) Global climate change will alter the patterns of rain and snowfall.
   c) Melting of glaciers and ice caps will cause the oceans to rise.
      (1) Coastal areas will be flooded.
      (2) Some oceanic islands will disappear.
   d) An increase in heat energy in the atmosphere will increase the power and frequency of storms.
   e) Warmer weather and longer summers will allow some insect species to thrive.
      (1) Plant-eating insects may cause damage to ecosystems.
      (2) Disease-carrying insects may increase and carry disease to new areas.
7. Evidence currently exists that the planet is already warming.
   a) Alpine glaciers have retreated.
   b) Average yearly temperatures are rising.
   c) Arctic and Antarctic ice is melting.
8. Global warming already threatens biodiversity.
   a) Scientists have already pinpointed species and ecosystems that have been affected by climate change.
   b) Many affected species are temperature sensitive and must move toward the poles to find a livable climate.
9. Climate change has affected certain patterns of life.
   a) Changes in leafing and blooming times for flowering plants have been observed.
   b) Earlier migration dates for birds and insects have been observed.
   c) Earlier mating seasons for amphibians have been observed.
10. Global warming and human-caused habitat destruction may work together to cause species extinctions.
   a) Sensitive species may try to disperse to try to find a livable climate.
   b) The species may not be able to cross human-modified landscapes.
   c) The species may become extinct if its own habitat becomes unlivable.

E. The Future of Our Shared Environment
1. Thoughtful planning and the use of improved technology can mitigate the negative effects of human settlements on natural biomes.
2. Laws passed in the U.S., such as the Clean Air Act and Clean Water Act, have greatly reduced air and water pollution.
3. Cities throughout the developed world are supporting projects aimed at creating sustainable communities.

Practice Questions

Matching

1. smog
2. marine
3. land biome
4. precipitation
5. ecological footprint
6. tundra
7. steppe
8. wetland
9. chaparral
10. desert
11. equator
12. decomposition
13. wastewater
14. estuary
15. lake

a. breakdown of waste and dead organisms
b. area of standing water that supports above-water plants
c. body of water surrounded by land
d. ground-level ozone
e. rain or snowfall
f. circle equidistant from the poles
g. amount of land needed to support human activity in an area
h. aquatic biome where saltwater and freshwater mix
i. saltwater
j. temperate grassland characterized by short grasses
k. biome that receives less than 50 centimeters of rain per year
l. region characterized by distinct climate and vegetation
m. water generated by residential or industrial use
n. biome characterized by flammable shrubs
o. frigid biome containing permafrost

Fill-in-the-Blank

16. The term ____________ refers to current temperatures, cloud cover, and precipitation, whereas the term ____________ refers to long-term trends in temperature and precipitation.
17. Valleys often have ____________ temperatures than adjacent hilltops because cold air ____________.
18. As ____________ on a mountain increases, temperature ____________.
19. Thunder Bay, Ontario, Canada, is on the northwest shore of Lake Superior, and Sault Ste. Marie, Michigan, is on the southeastern shore. Sault Ste. Marie is on the shore that is opposite the prevailing winds. Therefore, it likely receives ____________ annual precipitation than does Thunder Bay.
20. Cairo, Egypt, is located at 30°N latitude. It has ____________ air and a ____________ biome.
21. The soil in a ____________ forest in Massachusetts is likely to have more nutrients than the soil in a ____________ forest in Brazil.
22. Coniferous trees characterize the ____________ biome, and evergreen shrubs characterize the ____________ biome.
23. Tallgrass ____________ and shortgrass ____________ are temperate grassland biomes.
24. In terms of rainfall, the tundra ____________ is most similar to ____________.
25. Aquatic biomes where saltwater and ____________ mix are called ____________.

Labeling

Use the words below to label Figure 24.1. Words that appear more than once should be used more than once.

30°N
30°S
Boreal forest
Desert

Equator
Grassland
Mountain
Temperate forest
Tropical forest
Tundra

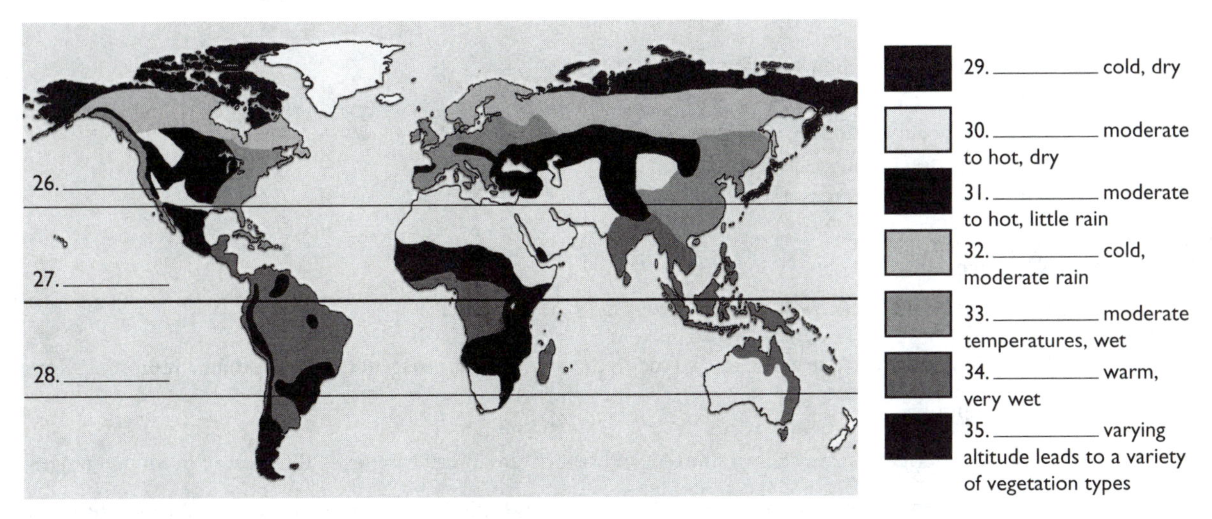

Figure 24.1

Roots to Remember

Use your knowledge of the root words presented in this chapter to answer the following questions.

36. Where on Earth are boreal forests found?
37. Young children have deciduous teeth. What happens to those teeth?
38. How is a terrarium different from an aquarium?

Crossword Puzzle

Across

39. effect that traps solar energy to warm the Earth
40. examples are gas, oil, and coal
41. represents the amount of land needed to support activities
42. melting due to climate change
43. quality of storms as more heat is added to the atmosphere

Down

44. one way to reduce solid waste
45. material that can be recycled
46. areas that reduce flooding and filter water
47. may be treated and used as fertilizer
48. can cause asthma attacks at ground level

Word Choice

Circle the word or phrase that correctly completes each sentence.

49. A region of tallgrass prairie is likely to receive (more/less) annual precipitation than a region of short-grass steppe.
50. In North America, you are likely to find (desert/temperate forest) biomes to the west of mountain ranges.
51. As Earth warms, boreal forests are likely to replace tundra, and (temperate forests/temperate grasslands) are likely to replace boreal forests.
52. Ocean levels are predicted to (rise/fall) as a result of global warming.
53. Commercial fishing would be most harmed by the destruction of (coral reefs/estuaries).

Table Completion

Complete the following table on land biomes.

Biome	Climate	Location on Earth	Threats
54.	Very dry, hot	55.	Human population growth, water depletion
56.	57.	Between approximately 45°N and the Arctic Circle	Deforestation
58.	Very wet, warm	59.	Deforestation
Savanna	60.	61.	Overgrazing
62.	Dry, warm	Near the Mediterranean Sea and in California, South Africa, and southwestern Australia	63.

Critical Thinking

Oregon is a Pacific coastal state. The Cascade Mountain Range runs from north to south, near the west coast of this state. Eugene and Bend are two of Oregon's larger cities. Eugene averages around 50 inches of precipitation per year. Bend averages less than 12 inches of precipitation per year.

64. What are the likely geographical factors that influence the climate in each of these cities? Explain your answer.
65. Which land biome would you expect to find in each city described above?

Practice Test

1. Which aquatic biome would be *least* likely to be affected by fertilizer runoff?
   A. wetland
   B. estuary
   C. coral reef
   D. open ocean

2. Many of the world's biodiversity hotspots are located on islands and around coastal areas. Which of the following is likely to have the most devastating effect on the species in these hotspots?
   A. global warming
   B. untreated wastewater
   C. smog
   D. deforestation

3. Which of the following is true about wetlands?
   A. Wetlands prevent flooding.
   B. Wetlands trap pollutants.
   C. Wetlands trap sediments.
   D. All of the above are true.

4. Which of the following biomes is likely to have the most fertile soil?
   A. desert
   B. tundra
   C. prairie
   D. savanna

5. Water is unavailable to plants for part of the year in all biomes *except*
   A. tundra.
   B. temperate forests.
   C. tropical forests.
   D. temperate grasslands.

6. Which of the following cities is likely to experience the highest amount of rainfall?
   A. a city on the side of a mountain facing prevailing winds
   B. a city at 30°N or 30°S latitude
   C. a city on the same side of a large body of water as the prevailing winds
   D. a city in the savanna biome

7. Smog occurs when
   A. small airborne particulates fall to the ground.
   B. by-products of combustion react with oxygen in the presence of sunlight and high temperatures.
   C. ozone is released by combustion.
   D. nitrogen oxide combines with carbon dioxide during combustion to form ground-level ozone.

8. Cities are often heat islands in the middle of a cooler landscape. What can be done to reduce the heat-island effect?
   A. Increase the use of air conditioning.
   B. Reduce the use of buses and other mass-transit vehicles.
   C. Plant trees and increase green space.
   D. Replace light roofing materials with darker materials.

9. An estuary can be damaged by sediments that flow into the biome from
   A. wetlands.
   B. rivers.
   C. ocean waves.
   D. ponds.

10. Energy for heating and cooking in nonindustrial countries is usually obtained by burning
   A. coal.
   B. oil.
   C. wood.
   D. natural gas.

11. Which land biome is dominated by coniferous trees?
   A. chaparral
   B. boreal forest
   C. temperate forest
   D. tropical forest

12. Which of the following is found in a lake that has undergone eutrophication?
   A. low oxygen levels
   B. high nutrient levels
   C. large algae populations
   D. all of the above

Answer Key

Matching

1. d; **2.** i; **3.** l; **4.** e; **5.** g; **6.** o; **7.** j; **8.** b; **9.** n; **10.** k; **11.** f; **12.** a; **13.** m; **14.** h; **15.** c

Fill-in-the-Blank

16. weather; climate; **17.** cooler; sinks; **18.** altitude; decreases; **19.** more; **20.** dry; desert; **21.** temperate; tropical; **22.** boreal forest; chaparral; **23.** prairie; steppe; **24.** biome; desert; **25.** fresh water; estuaries

Labeling

26. 30°N; **27.** Equator; **28.** 30°S; **29.** Tundra; **30.** Desert; **31.** Grassland; **32.** Boreal forest; **33.** Temperate forest; **34.** Tropical forest; **35.** Mountain

Roots to Remember

36. in northern latitudes, south of tundra but north of temperate zones; **37.** they fall out; **38.** A terrarium would contain soil and terrestrial plants, and an aquarium would contain water, aquatic plants, and fish.

Crossword Puzzle

39. greenhouse; **40.** fossil fuels; **41.** footprint; **42.** glaciers; **43.** extreme; **44.** recycle; **45.** glass; **46.** wetlands; **47.** sludge; **48.** ozone

Word Choice

49. more; **50.** temperate forest; **51.** temperate forests; **52.** rise; **53.** estuaries

Table Completion

54. Desert; **55.** 30°N or 30°S; **56.** Boreal forest; **57.** Cold, moderate rain; **58.** Tropical forest; **59.** Around the equator; **60.** Hot, dry; **61.** Around the equator; **62.** Chaparral; **63.** Fire suppression, urbanization, and human population growth

Critical Thinking

64. Eugene receives a fairly large amount of rain each year, so it is likely to be influenced by the Pacific Ocean. Bend receives very little rain, so it is likely to be on the sheltered side of the Cascade Mountain Range.

65. You would expect to find temperate forest in Eugene and desert in Bend.

Practice Test

1. D; **2.** A; **3.** D; **4.** C; **5.** C; **6.** A; **7.** B; **8.** C; **9.** B; **10.** C; **11.** B; **12.** D

NOTES

NOTES

NOTES